The Politics of Animal Experimentation

The Politics of Animal Experimentation

Dan Lyons
Centre for Animals and Social Justice and University of Sheffield, UK

palgrave
macmillan

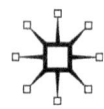

© Dan Lyons 2013
Foreword © Wyn Grant 2013

All rights reserved. No reproduction, copy or transmission of this publication may be made without written permission.

No portion of this publication may be reproduced, copied or transmitted save with written permission or in accordance with the provisions of the Copyright, Designs and Patents Act 1988, or under the terms of any licence permitting limited copying issued by the Copyright Licensing Agency, Saffron House, 6–10 Kirby Street, London EC1N 8TS.

Any person who does any unauthorized act in relation to this publication may be liable to criminal prosecution and civil claims for damages.

The author has asserted his right to be identified as the author of this work in accordance with the Copyright, Designs and Patents Act 1988.

First published 2013 by
PALGRAVE MACMILLAN

Palgrave Macmillan in the UK is an imprint of Macmillan Publishers Limited, registered in England, company number 785998, of Houndmills, Basingstoke, Hampshire RG21 6XS.

Palgrave Macmillan in the US is a division of St Martin's Press LLC, 175 Fifth Avenue, New York, NY 10010.

Palgrave Macmillan is the global academic imprint of the above companies and has companies and representatives throughout the world.

Palgrave® and Macmillan® are registered trademarks in the United States, the United Kingdom, Europe and other countries

ISBN: 978–0–230–35511–8

This book is printed on paper suitable for recycling and made from fully managed and sustained forest sources. Logging, pulping and manufacturing processes are expected to conform to the environmental regulations of the country of origin.

A catalogue record for this book is available from the British Library.

A catalog record for this book is available from the Library of Congress.

Contents

List of Illustrations	xii
Foreword Wyn Grant	xiii
Acknowledgements	xvi

1 Introduction 1
   Animal protection, animal research and political science 1
   Advancing knowledge of British animal research policy 2
   Improving public debate and democratic accountability 4
   Outline of chapters 5

2 Towards a Dynamic Model of British Policy Networks 9
   Introduction 9
   The nature of power in British politics 11
      The Westminster model 11
      The differentiated polity model 12
      The asymmetric power model 13
   Policy network analysis and the Marsh and Rhodes typology 15
      Policy communities 16
      Issue networks 18
      Policy network membership, definition and boundaries 22
      Summary 23
   Policy network dynamics: the dialectical interaction between exogenous factors and endogenous factors 23
      Categorising degrees of policy change 25
      Exogenous factors 27
         Economic factors 28
         The ideological and cultural context 29
         Developments in knowledge and technology 32
         The political and institutional context 32
      Endogenous network factors 36
         Endogenous dynamics through policy communities and issue networks 38
         Learning dynamics in policy networks 43
         Bureaucratic and implementation structures 46
         The role of agency 48

	Conclusion	52
3	**The 'Animal Research Issue Network' Thesis: A Critique**	**55**
	Introduction	55
	The Garner analysis	55
	Garner's 'Animal Research Issue Network' Thesis	57
	The network's origins and the 1876 Cruelty to Animals Act: regulating or facilitating animal experiments?	57
	The evolution of the network and the implementation of the 1876 Act until 1950: persistent issue network or transformation to policy community?	58
	Post-war politicisation and legislative change: dynamic conservatism or genuine response to public concern?	60
	Implementation of Animals (Scientific Procedures) Act 1986: balancing of interests or symbolic reassurance?	76
	The Animal Procedures Committee	77
	The Three Rs	78
	Group-state interactions	80
	The operation of the cost-benefit assessment	81
	Network structure, network interactions, and exogenous factors	81
	Conclusion	87
4	**Theory and Method in the Study of Animal Research Policy**	**89**
	Introduction	89
	Towards a critical realist methodology	89
	Applying new institutionalism	89
	Historical institutionalism and path dependency	90
	Structure and agency: the need for reconciliation, not vacillation	93
	New institutionalism and critical realist epistemology	95
	Methodological implications of critical realism	96
	The role of case studies	98
	Data collection	100
	Constraints on collecting data	100
	Data relating to the historical evolution of animal research policy	101
	Obtaining primary data	102
	The emergence of primary documents	102

	The Imutran xenotransplantation research case study	104
	The limits of the methodology	108
	Conclusion	111
5	**The 1876 Cruelty to Animals Act: Protection for Animals or Animal Researchers?**	**113**
	Introduction	113
	The emergence of animal experimentation on to the political agenda	114
	The philosophical, religious and cultural roots of vivisectionist and anti-vivisectionist thought	114
	The birth of vivisection and anti-vivisection as British social movements	117
	The politicisation of the vivisection debate	118
	The interaction between vivisection and animal protection	118
	The interaction between public opinion and vivisectionists: self-regulation as attempted depoliticisation	119
	Professional groups: strategic actions stemming from perceptions of interests and structural constraints	121
	Animal protection and anti-vivisection: strategic action in the midst of structural opportunities and constraints	122
	Vivisection enters the parliamentary arena: private members' bills	125
	Characterising the embryonic policy network	127
	Vivisection as a policy issue	128
	The Royal Commission on vivisection: 1875–6	128
	The passage of the Cruelty to Animals Act 1876	131
	Carnarvon's original bill	131
	Group reaction to the bill	132
	The impact of the medical profession	133
	The provisions of the Cruelty to Animals Act 1876	136
	Characterising the animal research policy network in 1876	138
	The purpose of the 1876 Act and government-vivisectionist relations	139
	Animal research policy network dimensions at the assent of the 1876 Act	140
	Conclusion	147

viii *Contents*

6	**The Evolution of the Animal Research Policy Network: 1876–1950**	**149**
	Introduction	149
	What happened to the 1876 issue network? The early administration of the Cruelty to Animals Act	150
	Early Home Office implementation	150
	Applications	150
	Infringements	152
	Pro-vivisection strategy and action	154
	Anti-vivisection politics: 1876–82	155
	The animal research policy network: 1876–82	156
	Pro-vivisection strategic action: an issue network under pressure?	159
	The International Medical Congress: the initiation of concerted pro-vivisection activity	159
	The impact of the Association for the Advancement of Medicine by Research	161
	Understanding the dialectical process of network evolution	166
	Implications for animal research policy literature and policy network theory	168
	New insights into animal research policy	168
	Implications for the policy network analytical framework	169
	The animal research policy network: 1900–50	171
	The second Royal Commission, 1906–12	172
	Policy community success breeds external opposition	172
	The second Royal Commission	173
	The interaction between the animal research policy community and exogenous pressures	177
	Animal research policy, 1913–50	181
	Conclusion	183
7	**The Animals (Scientific Procedures) Act 1986: Emergence of an Issue Network or Policy Community Dynamic Conservatism?**	**186**
	Introduction	186
	The Littlewood Report	186
	The origins of the Littlewood Committee: the dialectical relationship between outcomes, network, context and agency	187

The policy network in action: the implementation of the Cruelty to Animals Act 1876 prior to the Littlewood Enquiry	187
Responses to policy community outcomes	192
Policy community mediation of exogenous pressure and new information	193
Ongoing politicisation forces policy review	194
The Advisory Committee: the reaction of a policy community to external pressure	196
The Littlewood Enquiry: scope and participation	200
Scope of enquiry	200
Group and public evidence to the Littlewood Enquiry	202
The Littlewood Report: findings and recommendations	206
Pain	206
'Unnecessary Wastage'	207
Licensing, control and inspection	211
The Advisory Committee ('AC')	213
Summary of recommendations	214
Insights into the evolution of the animal research policy network	214
Implications for future network evolution	216
The government's response to the Littlewood Report	218
The path to the reform the 1876 Act: the Animals (Scientific Procedures) Act 1986	219
Specifying the group and network ideologies	220
Structure and agency: the dialectic between perceived policy outcomes, exogenous pressure and strategic action	221
1975: the intensification of animal research politicisation	222
The evolution of proposals for new legislation	224
New legislation on the political agenda	224
The development of legislative proposals: the 1983 and 1985 white papers	226
The passage of the Animals (Scientific Procedures) Bill	230
The policy network implications of the assent of the Animals (Scientific Procedures) Act 1986	232
The implications of the new cost-benefit assessment	232
The implications of the new Animal Procedures Committee (APC)	235
Conclusion: the Animals (Scientific Procedures) Act 1986 and network change	236

8 Imutran Xenotransplantation Research Case Study — 240
Introduction — 240
The Animals (Scientific Procedures) Act 1986 and
  policy network analysis — 241
  The issue network scenario — 241
  The policy community scenario — 243
  The regulatory framework for animal experimentation
    after 1986 — 244
    Factors in the cost-benefit assessment of
      project licence applications — 245
    The cost-benefit assessment process — 247
The case study: Imutran's primate
  xenotransplantation research — 249
  A summary of Imutran's primate xenotransplantation
    research and related data — 249
  Determining the 'cost': severity assessments — 251
    Severity classifications — 251
    Severity limits and bands — 252
    The severity assessment of Imutran's protocols and
      projects — 254
  The benefit assessment — 257
    The formal requirements for benefit assessments — 257
    The initial benefit assessment of Imutran's research — 259
  Regulation and policy outcomes in the Imutran case — 261
    The starting point for Imutran's primate research — 261
    Heterotopic abdominal cardiac xenografting in
      cynomolgus monkeys — 262
    Technical failure rate — 262
    Immunosuppressive drug toxicity — 263
    Animals 'Found Dead' — 264
    Imutran-Home Office interactions: 'Ensuring
      Smooth and Rapid Passage' of applications — 266
    Cardiac xenotransplantation in baboons: the
      final step to clinical trials? — 268
    Further immunosuppression studies in baboons — 272
    A 'Cavalier Attitude' and 'Violation of Trust' — 273
    Heart xenografting abandoned — 275
    Kidney xenografting — 277
    Adverse effects: moderate or substantial? — 278
    APC scrutiny: a 'rubber-stamping' exercise? — 279
    Implications for Home Office–Imutran relationships — 281

	An overview of the cost-benefit assessment of	
	Imutran's primate xenotransplantation research	281
	The adequacy of the cost assessment	281
	The adequacy of the benefit assessment	284
	The Home Office response	286
	The form of the Home Office's response	286
	The content of the Home Office's response	289
	An unresolved controversy: questions of	
	accountability	294
	Conclusion	297
9	**Conclusion: The Power Distribution in British Animal Research Politics**	**303**
	Introduction	303
	Understanding change and continuity in animal	
	research policy	304
	Policy network analysis	304
	Developing an analytical framework to re-assess	
	animal research policy-making	305
	Serving the purposes of animal researchers: the Cruelty to	
	Animals Act 1876	306
	The unstable issue network transforms into a	
	policy community	309
	The gestation and assent of the Animals	
	(Scientific Procedures) Act 1986	313
	Assessing the implementation of the Animals	
	(Scientific Procedures) Act 1986	319
	Animal research policy since 2000	320
	Additional contributions, limitations and	
	further research paths	325
	Pluralistic or asymmetric power relations?	325
	Limitations and further research	329
Notes		333
Bibliography		345
Index		357

List of Illustrations

Tables

2.1	The Marsh/Rhodes policy network typology	21
2.2	Variable policy network dynamics	53
3.1	Key group participants engaged in British animal research politics in the mid-1990s	82
5.1	Politically-active groups 1874–6	144
6.1	Comparing the policy network before and after the advent of the AAMR	167
7.1	Littlewood Enquiry: groups and their policy goals	205
8.1	Imutran xenotransplantation protocols: severity assessments and actual severity	282

Figure

2.1	Policy networks and policy outcomes: a dialectical approach	24

Foreword

Issues relating to animal protection can prove highly politically controversial as the arguments surrounding the pilot badger culls show. Animal experimentation is undoubtedly the most emotive and controversial area of the animal protection debate. Hence, the need for an authoritative, well-written and well-structured treatment of the kind that Dan Lyons has provided.

The book is written from a particular perspective, but the arguments are made in a logical fashion and consider alternative explanations. Use is made of as much evidence as possible to support the arguments advanced, although as is pointed out there are limitations in terms of both the availability of archival documents and opportunities for élite interviewing. However, a major source of originality is found in the utilisation of historically unprecedented confidential primary data relating to a major animal research programme: pig-to-primate organ transplantation. This information became available following the settlement of legal proceedings between the author and the research company. It forms the core of a critical case study that facilitates a new approach to the research area.

The book fills a gap in the literature as there is only one major earlier study of animal experimentation from a political science perspective. Unfortunately, this area of public policy, and animal protection more generally, have been relatively neglected by political scientists, in terms of either research or the coverage of standard texts on public policy. This is unfortunate, given that it helps to shine light on the actual practice of regulation and how it is influenced by changes in knowledge, technology and public opinion. It has implications for the study of public policy more generally. An examination of animal protection issues can help us to better understand the nature of the contemporary 'regulatory state', including its European dimension, and its limitations as a means of achieving stated policy goals. In the particular case, the adverse effects suffered by animals were found to exceed the level posited by regulatory assessment.

The primary focus of the study is the impact of the Animals (Scientific Procedures) Act 1986, which shifted the framing of policy from an 'animal use' discourse to one of 'animal welfare'. This does not proceed from the assumption that the interests of animals may be sacrificed for those of humans, but requires a cost-benefit analysis of animal research proposals involving the weighing of adverse effects likely to be experienced in animals used in procedures against the likely benefits for humans, animals and the environment. The central hypothesis examined in the book is that the interests of animals have been given relatively little consideration in a policy process that is characterised by the predominance of research interests and the exclusion of animal protection groups.

However, the book does not just provide an analysis of the particular policy area but also makes a significant contribution to a continuing debate about relevant analytical frameworks in political science. Very effective use is made of the policy community/policy network model, and the literature review in Chapter 2 is one of the most comprehensive and authoritative I have seen. In particular, I thought that important and novel insights were offered on the roles of peripheral insiders.

This literature has evolved over time, and this has led to the emergence of a more dynamic conception of policy networks. However, a core insight remains: Relatively closed policy communities tend to produce policy outcomes that favour network members at the expense of excluded groups. Weaker groups that challenge insider interests find themselves excluded or limited to a token role. Policy communities are characterised by a dynamic conservatism in which the co-option of new actors legitimises existing power structures rather than changing them. Policy networks with a broad membership and relatively low entry barriers tend to produce outcomes that do not consistently favour one set of interests.

The book is based on a clearly articulated critical realist epistemology and methodology. This is consistent with a dialectic approach to policy networks that seeks to reflect and explore the relationship between structure and agency. This involves a mixture of quantitative and qualitative methodology, in line with the current consensus in favour of a mixed methods approach to political science that does not privilege particular research techniques. It recognises the variety of institutional forms that shape power relationships and that reflexive actors' interpretations of structures affect their behaviour and hence outcomes, interpretations that are influenced by social constructions of reality. The nature of UK animal research policy-making clearly reflects a persistent policy community rather than an issue network.

There is considerable scope for further research both in terms of cross-national comparison and the impact of the 2010 EU directive in different member states, a directive that offers an interesting example of Europeanisation in terms of uploading the UK approach to the EU level. However, this book represents a major contribution to the literature in terms of theory, methodology and empirical evidence. It significantly enhances the literature on animal protection issues but also deserves a wider audience among those interested in issues of public policy-making, interest representation and regulation.

Wyn Grant
University of Warwick

Acknowledgements

I have a number of debts of gratitude to acknowledge to those who made this book possible. First, David Thomas, a solicitor who represents animals and other vulnerable groups, has been a rock, and his help was particularly crucial during the legal proceedings I defended in order to be able to publish the primary data that forms the central case study discussed in Chapter 8. Thanks also to Martin Smith and Razi Mireskandari at Simons, Muirhead and Burton solicitors and David Bean QC (now The Hon Mr Justice David Bean) for their *pro-bono* representation in the first few months of the legal battle, and to Tamsin Allen and Stephen Grosz at Bindmans for persevering with the Legal Aid application and helping to secure our ultimate victory.

I would also like to express my appreciation of my supervisors at the Politics Department at the University of Sheffield, Dr James Meadowcroft (now at Carleton University in Canada) and Professor David Richards (now at Manchester) – without their encouragement and the occasional kick up the backside, I suspect this research would never have been completed. Professor Rob Garner's previous research in this area provided an inspiration and launch pad for this study, and his comments and suggestions have been very helpful – and gracious given my critique of his analysis. It has also been a real honour for this study to be graced by a Foreword from the eminent Professor Wyn Grant.

I am deeply appreciative of the support provided by Richard Johnston and Petra Pipkin MBE to help fund the legal scrutiny of this manuscript. Thank you also to my mum, Joan Lyons, for her support, and for helping to point me in an educational 'direction' from a young age. Finally, and most importantly, I would like to thank my partner Angela Roberts, who founded animal advocacy group Uncaged, fought and suffered with me through the proceedings and supported me as I took a sabbatical from Uncaged to finish the research upon which this book is based. Without her love, companionship, determination and sense of justice, none of this would have been possible.

1
Introduction

Animal protection, animal research and political science

Since the mid-1970s, increasing public concern about the treatment of animals and a growing animal protection movement have contributed to the evolution of public policy regulating aspects of human/nonhuman animal interactions. However, the politics of animal protection has been largely overlooked by political science, particularly in the field of public policy research. Robert Garner's 1998 work *Political Animals: Animal Protection Politics in Britain and the United States* remains the sole study of animal protection public policy. This book aims to help remedy this neglect.

Of all the animal protection issues, animal experimentation is arguably the most emotive and contentious. On the one hand, Garner (1993: 118) notes that laboratory experimentation on animals 'provides some of the most severe examples of animal suffering'. Furthermore, some critics argue that research on animals is scientifically flawed and hence detrimental to human well-being: 'In medical research animal experiments are generally bad science because they tell us about animals, usually under artificial conditions, when we really need to know about people' (Sharpe, 1989: 111). Deeply-help moral positions on the necessity of animal rights and the perception of a policy process dominated by groups with a vested interest in animal research have contributed to 'direct action' outside the policy process – some of it illegal and aggressive – by some sections of the animal protection movement (Garner, 1998: 4–5).

On the other hand, proponents of animal research insist that the practice has been and continues to be essential to the achievement of major public health benefits, particularly in terms of the development

1

of medical therapies (Paton, 1993: 4). Moreover, it is claimed that, in response to public concern for the welfare of animals, animal experiments are subject to a strict regulatory regime which ensures that the perceived benefits of such research outweigh any animal suffering, which is minimised (Matfield, 1992: 335).

This apparently acute conflict between the interests of human and non-human animals, and the fact that anti-vivisection pressure group activity also challenges the legitimacy of powerful economic and professional interests, makes for controversial politics with the potential to affect core public policy areas such as the economy, health, science and technology, consumer and environmental protection, and law and order.

From a political science perspective, UK animal research policy is especially significant because of the introduction in 1986 of an innovative legal framework that appeared to represent a fundamental change in the way that animals' interests are considered. Previously, under the regime established by the Cruelty to Animals Act 1876, licenses for animal research were granted without any regulatory scrutiny of the potential value of the proposals or the potential pain likely to be caused to animals (Garner, 1998: 187). However, the putative regulatory system introduced by the Animals (Scientific Procedures) Act 1986 is based on a cost-benefit assessment involving the weighing of adverse effects likely to be experienced by animals used in procedures against the likely benefit to accrue to 'man, animals and the environment' (Hampson, 1989: 240–1; APC, 1998: 43). This cost-benefit assessment is supposed to be the core determinant of whether proposals to conduct animal research projects should be legally permitted and, if so, the level of officially-sanctioned animal suffering. In January 2013, a new EU Directive (2010/63/EU) was transposed into UK law[1] which retains this fundamental decision-making framework, meaning that, other things being equal, case studies of this policy area under the 1986 legislation remain relevant today.

Advancing knowledge of British animal research policy

The single study of animal research public policy to date (Garner, 1998) concludes that animals' interests are given significant consideration as a result of the assent of the 1986 legislation. On this reading, the animal research policy process is 'relatively open and pluralistic'; both animal researchers and animal protection groups have access to and influence over policy-making, overseen by Home Office actors who adopt the role of neutral arbiters between the conflicting groups (Garner, 1998: 231).

However, until now, research into this policy field has been severely constrained by a lack of available empirical evidence regarding the operation of the 1986 Act and policy outcomes, as well as being influenced by a method involving comparison with US animal experimentation regulation. So, the description 'relatively open and pluralistic' may not illuminate the true state of the British situation. Moreover, concerns have been raised regarding whether the implementation of the 1986 statute has given animals' interests the level of consideration indicated by the formal legislative and administrative framework (FRAME Trustees, 1996). These considerations indicate the need to explore an alternative hypothesis regarding UK animal research policy: **the interests of animals are given scant consideration in an elitist policy process characterised by research interests' domination and the effective exclusion of animal protection groups.**

One of the major sources of this study's originality is its utilisation of historically unprecedented confidential primary data relating to interactions and policy outcomes concerning a recent major animal research programme: pig-to-primate organ transplantation conducted between 1995 and 2000.[2] This information has been legitimately disclosed following the settlement of legal proceedings involving the author (in his capacity as an animal protection lobbyist) and the research company (Townsend, 2003). Normally, such sources are prevented from entering the public domain due to commercial confidentiality and Section 24 of the Animals (Scientific Procedures) Act 1986 (and as amended in 2012), which prohibits the disclosure of information related to the regulation of animal research. This therefore forms the core of a critical case study that enables the previous research constraints to be partially overcome, thus facilitating a novel re-examination of this policy area.

The case study, and in particular, the primary data, provides new insights into the way that the potential costs and benefits of animal research are measured, weighed against each other and controlled in policy-making. Furthermore, this research programme was subject to relative close regulatory scrutiny, and so it would be most likely to support the pluralist description rather than the hypothesis proposed here. This means that if the case study supports the hypothesis, its generalisability is enhanced. For these reasons, this data is more reliable and relevant than any hitherto available in relation to the hypothesis addressed in this book.

The analysis of this data reveals that the 'costs' – i.e. adverse effects – suffered by animals significantly exceeded the level posited by the regulatory assessment. The severity assessments required that the vast majority of the primate recipients of pig organs had to be euthanased before they suffered systemic illnesses or significant discomfort. But in reality, many were left to deteriorate until they were found dead or in a collapsed state. On the other hand, the benefits that accrued fell considerably short of the scientific and medical advances that were predicted and formed the justification for the research. The project was permitted on the basis that Imutran were likely to achieve progress that would allow human trials of pig organ transplants. In reality, Imutran and Home Office regulators had overlooked the precipitous immunological barriers to cross-species transplants. Consequently, the research failed to make significant headway in four and a half years of experimentation. The implication of this case is that, when the Home Office assesses licences to conduct animal experiments, animals' interests in not being subjected to pain and suffering – and the related concerns of sympathetic members of the public and cause groups – are afforded little effective weight relative to researchers' interests and their claims for potential medical and scientific benefits. Furthermore, the practical operation of the cost-benefit assessment is revealed to be inconsistent with formal policy requirements and official statements on the implementation of the legislation. These outcomes reflect a policy process monopolised by pro-animal research interest groups to the exclusion of animal protection actors, which remains the case to the present day.

Improving public debate and democratic accountability

However, advancing knowledge of animal research policy-making not only brings academic benefits, but may also facilitate positive social impacts in terms of animal protection, biomedical research and democratic accountability. Public debates about animal experimentation tend to be conducted in rather ideal, absolute terms: is the practice justified, or should it be abolished? While framing the debate in this way does, indeed, reflect vital ethical arguments about the moral status of human and other animals and the current utility of animal experimentation, it has actually had merely marginal, indirect effects on policy outcomes. As this study will demonstrate, since 1876 the prospects of achieving the abolition of animal experimentation within a definable timeframe have diminished from slim to negligible. That is not to denigrate the ethical argument for the cessation of such practices insofar as they represent the knowing infliction of pain, suffering and harm on sentient individuals. Rather, it reflects the inescapable reality of the huge, historically entrenched power advantages enjoyed by animal research interests.

Therefore, if the only option proposed for short-term practical change is abolition, then change is rendered unfeasible. This should be a matter of concern not only to anti-vivisectionists, but to the majority of the public for whom cruelty to animals involves, at least, significant ethical costs. Moreover, animal researchers' statements of commitment to the 'Three Rs' (House of Lords, 2002: 37) – the reduction, refinement and replacement of animal experiments in order to minimise animal suffering – give the impression that the practice is deemed a source of moral regret even by those engaged in it.

Consequently, a public discourse dominated by absolute and highly generalised policy positions is likely to obscure the questions which are of most practical relevance. Instead, it is through attention to questions surrounding the severity of animal suffering and the manner in which such 'costs' inflicted on animals are compared with predicted benefits – and, crucially, who makes those decisions – that animal research regulation can become publicly accountable and implemented in a way that honours the apparent consensus in favour of the reduction and eventual elimination of harm to animals.

Outline of chapters

In order to establish a theoretical framework for this study, the next chapter reviews the policy network approach to political science. The policy network approach focusses on the relationships between interest groups and the state in particular policy areas in order to understand the policy process and policy outcomes. In addition to its widespread use in public policy research (Marsh and Smith, 2000: 4), another reason why this tool is adopted is that it is the approach that guides Garner's analysis of animal research policy. Therefore, by gaining an in-depth understanding of policy networks, a critical review of the existing analytical framework underpinning the current understanding of animal research policy will be facilitated.

Chapter 2's examination of the policy network approach focuses on the dominant Marsh/Rhodes typology (Marsh and Rhodes, 1992b). The analytical utility of the Marsh/Rhodes schema – and of policy networks in general – is based on the idea that variations in the dimensions of policy networks affect policy outcomes. Thus, policy networks with a broad membership, fluctuating access for different groups, distant state-group relationships and high levels of conflict – 'issue networks' in the Marsh/Rhodes terminology – will tend to produce outcomes that fluctuate and do not consistently favour one set of interests. On the other hand, policy networks characterised by exclusive membership, close integration between state actors and certain group members, and consensus – known as 'policy communities' – will tend to produce outcomes that consistently favour network members at the expense of excluded groups. The explication of the policy network framework means that a key question can be posed based on the hypothesis addressed in this study: Is animal research policy made in a policy community environment?

The review of the policy network literature also identifies a shift from a static approach to a more dynamic conception of policy networks (see Marsh and Smith, 2000; Hay and Richards, 2000). In particular, it is necessary to understand the process of change and continuity in networks and outcomes, which involves iterative – or 'dialectical' – interactions between political actors and structures, and between the network and both the actors within it and its structural context. In other words, networks are said to be constrained by broader patterns of power distribution and other exogenous factors such as public opinion and changes in knowledge and technology. However, although networks and outcomes are constrained, they are not determined because of the ineluctable role of agency. Chapter 2 concludes by trying to overcome a perceived hiatus in the network literature: it presents a table which postulates variability in the way that issue networks and policy communities mediate different exogenous dynamics and facilitate agency. For example, whereas issue networks are susceptible to changes in public opinion or the governing party, policy communities are thought to be relatively resistant to such perturbations. This table is then used as a heuristic device in subsequent chapters to assist understanding of the evolution of the animal research policy network.

Chapter 3 applies these insights to review Garner's case for an animal research policy network with issue network characteristics: the 'issue network' thesis. Garner's inference is based significantly on his perception of the circumstances surrounding the formation of the animal research policy network in 1876, and indicates his implicit adoption of a historical institutionalist approach and the concept of path dependency. This emphasises the need to reconstruct the evolution of this policy process in order to understand its present operation, which gives rise to four chronologically-ordered research questions:

1. Which group(s) interests were served by the assent of the Cruelty to Animals Act 1876?
2. Did the policy network that emerged during the passage of the 1876 Act evolve into a policy community in the subsequent years?
3. Did the passage of the Animals (Scientific Procedures) Act 1986 signify a core change in policy?
4. Does the implementation of the Animals (Scientific Procedures) Act 1986 reflect an issue network or a policy community model of policy-making?

These questions point to a further research question related to the core hypothesis – that the nature of UK animal research policy-making reflects a persistent policy community rather than an issue network – which can be comprehensively addressed through these four questions. They also provide the framework for the empirical parts of the book, which are covered from Chapter 5 through 8.

The review of the animal research policy literature raises important methodological questions. Therefore, Chapter 4 begins with an examination of Garner's meta-theoretical assumptions which underpin his use of historical institutionalism, path dependency and policy networks. Through this analysis, a critical realist epistemology and methodology is outlined that provides the foundation for this work. This is consistent with the dialectical approach to policy networks that seeks to reflect the relationship between structure and agency. The critical realist epistemology also implies that a suitable methodology for this study involves a mixture of qualitative and quantitative methods, and the use of a case study. Moreover, the constraints on, and opportunities for, obtaining relevant data indicates that it is appropriate to utilise secondary and tertiary data to address the first three research questions relating to the historical background to animal research policy, while the primary data provides a suitable basis for a detailed case study of recent practices that are relevant to the fourth research question.

Thus, in order to address the first three research questions, Chapters 5 through 7 use policy network analysis within a critical realist epistemology to reconstruct a chronological narrative of animal research policy. However, the most salient empirical part of the book is found in Chapter 8, which presents the case study. By analysing unique primary sources, it will be possible to offer new empirical data concerning network interactions and how the cost-benefit assessment introduced by the Animals (Scientific Procedures) Act 1986 (and retained in the amended

2012 law) operates in practice. As a result, it will be possible to assess the nature of power in the network and revise the existing understanding of the nature of the network drawn from the Marsh/Rhodes typology.

Finally, Chapter 9 summarises the findings of this study and discusses its contribution to the understanding of animal research policy, power distributions in the British political system and policy network analysis. In addition, the limitations of the book's conclusions are set out as well as beneficial future research paths.

2
Towards a Dynamic Model of British Policy Networks

Introduction

This chapter explores the policy network approach as an organising framework for this study of the UK animal research policy process. The policy network approach is adopted as the principal analytical tool for the following broad reasons. Firstly, policy network analysis can be applied to the entire policy process (Smith, 1997: 15), thus corresponding to the scope of this study, which examines agenda-setting, policy formulation and implementation in animal research policy. Secondly, as explained below, it has come to occupy a prominent position in the public policy methodological canon. Thirdly, it is the approach utilised by the only significant contemporary analyst of British animal protection policy, Robert Garner (1998).

The emergence of policy network analysis as a favoured analytical tool in British public policy research (Marsh and Smith, 2000: 4) since the late 1970s (Richardson, 2000: 1006) is said to derive partly from dissatisfaction with the validity of traditional theoretical frameworks. For example, the 'Westminster model' (see below) assumed that Parliament and the Cabinet were the primary influences on the policy process (Rhodes, 1997: 5–7). In contrast, policy network analysis reflects the perception of an increasingly marginal role for Parliament and, concomitantly, an ability on the part of policy networks to resist democratic steering (Jordan and Richardson, 1987: 56).

Policy network analysis can also be seen as a reaction to established theories of the state, such as corporatism and, particularly, pluralism which, like the policy network approach, have focussed on government-interest group relations but have tended to work at a general, macro-level (Rhodes, 1997: 29–31). Instead, policy network analysis builds

10 *The Politics of Animal Experimentation*

on the observation that '...in different policy arenas a range of group/ government relationships exist' (Smith, 1993b: 76). Thus, it appears to represents a more realistic model that corresponds to the complex interactions that take place in diverse and disaggregated policy-making arenas (Parsons, 1995: 185). Hence, policy network analysis is generally conceived as seeking to understand and/or explain[1] policy outcomes by primarily focussing on the interactions between interest groups and the state in policy sectors centred on a distinct government institution or set of institutions. Thus, Rhodes (1997: 29) articulates policy network analysis as a *meso-level* approach that links and contextualises the *micro-level* of analysis, which focuses on the actions of, and relations between, individual actors and organisations as they interact over particular policy decisions, and mediates the impact of *macro-level* phenomena such as broader patterns of power distribution in society or national political institutions.

This means that policy decisions on animal research cannot be adequately explained without an understanding of the broader power structure in which they are made. Therefore, this chapter begins with a survey of competing models of power in British politics. The second section reviews the development of the policy network approach and introduces the influential typology of British policy networks developed by Marsh and Rhodes (1992b). This typology consists of a continuum upon which policy networks can be located according to the patterns of state/group relationships within them, which, in turn, can be analysed according to four interrelated variables relating to a network's membership, integration, resource distribution and power balance. The characteristics of each variable and the corresponding policy outcome implications are explored for the ideal-type networks – policy communities and issue networks – that are predicated to represent opposite poles of the Marsh/Rhodes continuum.

The third section of the chapter examines the literature that attempts to move beyond typology to model the dynamics of policy networks. It commences with a discussion of the different categories of change, particularly the distinction between minor secondary changes in instrumental aspects of policy and networks, and major changes representing systemic shifts in core beliefs and values. This important analytical distinction facilitates an examination of whether different degrees of policy change tend to be caused by certain combinations of network type and political developments. For example, does the type of network affect whether a major change in public opinion tends to lead to secondary or core policy change? Analysis is then undertaken of how networks and policies are seen to be affected by exogenous factors and endogenous network features that include both network structures and the strategic actors within them. Thus, the ultimate aim of this chapter is to develop a model of network and policy evolution that can be applied to an analysis of UK animal research policy.

The nature of power in British politics

The broader, macro-political environment, incorporating government institutions and socio-economic structures, inevitably constrains and enables meso-level networks in a variegated manner (Daugbjerg and Marsh, 1998: 54, 61). Therefore, in order to explain policy outcomes in Britain, it is necessary to consider three different models of British governance that have various implications for power distribution across the British political system:

- the Westminster model
- the differentiated polity model (see Rhodes, 1997)
- the asymmetric power model (see Richards and Smith, 2002; Marsh et al., 2003).

The Westminster model

The Westminster model can be seen as the traditional, institution-based framework for understanding British politics, and is said to embody, among other characteristics, the ideas of the British state as a representative government where Parliament plays a central role in policy-making (Rhodes, 1997: 22; Bevir and Rhodes, 1999: 217; Judge, 2004: 687), and the neutrality or 'constitutional propriety' of civil servants (Richards and Smith, 2004: 777). Thus, the Westminster model encapsulates an optimistic and occasionally teleological view of British government as effective, legitimate and progressive (Bevir and Rhodes, 1999: 217–18; Judge, 2004: 684).

However, Marsh et al. (2003: 306) argue that the Westminster model has been an implicit organising framework underpinning an ideal of British politics, rather than an explicit, well-theorised model of the British state. In particular, it is seen as an outdated, legitimising mythology that continues to manifest itself in the discourse of political élites (Rhodes, 1997: 22), despite sustained criticism from political scientists (Richards and Smith, 2002: 48–9). In fact, while Rhodes (1997: 24) suggests a real shift in the nature of British governance away from the Westminster model since the late 1970s, Judge (2004: 697) notes that its empirical accuracy has been consistently questioned for the last hundred years. For example, Moran (2003) develops Marquand's notion of 'club government' as a broad description of the regulatory style of the British state since the 19th century. Far from excluding groups from the policy process, 'club government' developed through a powerful ideology of self-regulation advanced by professional groups who played a key role in 19th-century economic life. Rather than all-powerful state actors adopting a neutral, 'public interest' stance untainted by the demands of special interests, small-scale Inspectorates lacked resources compared to 'regulated' professional and related economic groups. As a result, Inspectorates practiced co-operative regulation with regulatees rather than enforcing a literal interpretation of the law. Regulation was thus determined by the dominant values and interests of these elite groups, and therefore became a merely symbolic matter. Consequently, 'Not only are legally specified standards breached, the breaches are institutionalised: non-compliance with standards is thus organizationally sanctioned' (Moran, 2003: 35).

The differentiated polity model

One of the most detailed alternative organising perspectives to the Westminster model has been developed by Rhodes (1997). He argues that particularly since 1979, the predicates of this traditional perspective have been replaced by what he terms 'the differentiated polity', signified by: 'interdependence, a segmented executive, policy networks, governance and hollowing out' (1997: 7). Thus, the unified, top-down power structure envisaged by the Westminster model is said to have fragmented as the government has devolved service delivery to a 'maze' of public and private bodies, as well as the voluntary sector. In the differentiated polity, interdependent 'governance', comprising the use of markets, hierarchical bureaucracies and networks as governing structures, has replaced centralised government (Rhodes, 1997: 8, 47). Power relations are also conceived differently. Instead of the Westminster model's zero-sum, centralised notion of power, the differentiated polity exhibits power-dependence relationships where actors exchange resources in positive-sum games (Rhodes, 1997: 9). Thus, attempts by central government in the 1980s to assert executive power in the face of entrenched policy communities are said to have had the unintended effect of transferring power to new, complex, self-organizing networks involving a wider range of actors. This is a macro-level development that Rhodes (1997: 45) terms 'pluralization'.

Rhodes' (1997: 195) notion of 'institutional pluralization' may have a *prima facie* affinity with the traditional concept of pluralism, but he appears to acknowledge that plurality is not a sufficient condition for pluralism:

> The differentiation scenario of an ever-more fragmented, complex and unaccountable system looms large. It will act as a check on executive interventions, but it does not herald a pluralist heaven. Differentiation provides checks and balances but without both constitutional guarantees or democratic accountability. (1997: 135)

Indeed, Rhodes (1997: 197) concludes that the 'fetishization of economy, efficiency and management' that has accompanied the emergence of the putative differentiated polity has led to the exclusion of 'new' ideological groups, which would include animal protection. For example, the values underpinning this change stem less from a concern with taking account of animal welfare considerations in animal research policy, than from the alleged 'profligacy' of 'experimental rats bred at £30 each when available commercially at £2' (Rhodes, 1997: 93). Furthermore, it could be argued that in regulatory policy domains such as animal research, a weak state vis-à-vis the regulated industry may undermine pluralism insofar as the state lacks the resources that would allow it (if it so wished) to reflect any public interests that might conflict with business and professional interests.

The asymmetric power model

Rhodes does not explicitly develop his differentiated polity model to account for the broader questions of the distribution of power in the British political system that are raised by policy areas such as animal research. In order to address this issue, Marsh et al. (2003: 307) propose an 'asymmetric power' model, which they describe as 'an adaptation of the Rhodes model in which we have used our own research to account for what we regard as the structural inequalities that still exist in British politics'.

This model postulates a number of asymmetries in power distribution (Richards and Smith, 2002: 282–3), but most relevant to this study is the assertion that persistent patterns of structural inequality in society mean that many groups continue to be denied access to policy-making. It is argued that economic and professional groups possess resources deemed essential by the government, such as knowledge and expertise, which facilitate close and exclusive exchange relationships with the government. These resources give these groups unique influence over the government, representing a significant socio-economic constraint on government action, especially if their goals coincide with the neo-liberal policies implied by the discourse of globalization. However, it is claimed that the government continues to enjoy asymmetric relations with other groups and has the means to dominate networks if desired.

British politics' asymmetric power structure is simultaneously obscured and perpetuated as a result of the mythical discourse of the Westminster model that legitimises elite rule on the basis of a false idea of accountability (Richards and Smith, 2002: 283). Key aspects of elite rule include an unrepresentative electoral system based on the perceived desirability of strong majority government, and obsessive secrecy in policy-making that serves to obstruct outside scrutiny and public accountability (Marsh et al., 2003: 312).

Thus, the asymmetric power model appears to envisage dialectical relationships involving ideas, institutions and structural inequality that dynamically reconstitute a closed and élitist British political system (Marsh, 2008). As a result, the government can exclude from policy networks those weaker groups who challenge insider groups' interests while '[those] powerful economic and professional groups that have the greatest resources to exchange with government...are evident in policy communities' (Marsh et al., 2003: 318). This centralised view of power also implies that policy subsectors are constrained from above by policy sectors.

One important reason for articulating these various models of the British polity is that policy network analysts often work with implicit models that affect their interpretation of policy-making (Marsh et al., 2003: 306). Elucidating such implicit models therefore enables a deeper analysis of previous network studies. For example, interpretations of policy networks and their outcomes in pluralistic terms may be linked with the implicit adoption of 'Westminster model' concepts such as bureaucratic neutrality and electoral accountability. Another reason for considering these models is that they 'offer an explanation of the pattern of inclusion and exclusion within the network and a hypothesis about whose interests are served by the outputs from the network' (Daugbjerg and Marsh, 1998: 54). However, although macro-level power structures provide an ineluctable context to policy-making, they constrain and enable policy networks and actors rather than determine them.[2] Furthermore, as Marsh et al. (2003: 317) observe, 'There are varying relationships between departments and interest groups in each department, both across policy areas and across time'. It is therefore necessary to disaggregate analysis down to the meso-level of policy networks.

Policy network analysis and the Marsh and Rhodes typology

Within British political science, three different policy network approaches have emerged (Hu, 1995: 47). Two of these models, developed by Richardson and Jordan (1979) and Wilks and Wright (1987), have failed to become influential, mainly due to their failure to distinguish different types of policy network (Smith, 1993b: 77; Rhodes, 1990: 309; Hu, 1995: 57). Instead, it is the model developed by Marsh and Rhodes (1992b) that has come to dominate. Thus, when introducing an edited collection of policy network case studies, Marsh (1998a: 7) notes: '...it is the work of Marsh and Rhodes which has probably been the most significant development.... [T]hey developed a typology of networks which has been influential and is used in this book'. This model is additionally relevant to this study of UK animal research policy because it is adopted by Garner (1998: 7) as the theoretical framework for his analysis of this policy area, which forms the starting point for this study. These reasons form part of the rationale for adopting the Marsh/Rhodes model as the present analytical framework.

At the heart of the Marsh/Rhodes model is the conception of 'networks as structures of resource dependency' (Marsh, 1998a: 11). Smith (1997: 38) identifies the following types of resource that are relevant to policy network relationships: legal/authority (both formal and discretionary), economic/financial, political legitimacy (access to policy-makers, public opinion), information (especially control over its generation and distribution) and organisational (resources that enable a group to engage in direct policy-related action). Compston (2009: 32) proposes another resource often possessed by businesses and corporations – patronage – where businesses can exchange directorships with public actors in return for policy amendments. As Smith (1997: 38–9) explains:

> In summary, at the heart of the policy networks concept is the notion that resource interdependent policy actors deploy, withhold and exchange resources in order to influence decisions during the policy process. Policy network analysis examines the relations between policy actors, uncovers the dominant appreciative system, seeks the rules of the game and strategies employed, and considers the resource interdependencies which structure the interaction (or exclusion) arising in a policy process or sector.

One fundamentally important feature of the Marsh/Rhodes model is that network structures, as manifest in the structure of resource interaction and hence power distribution, influence policy outcomes (Marsh and Rhodes, 1992b: 252–4; Marsh, 1998b: 186–7).[3] In order to analyse the effects of variations in network structures, Marsh and Rhodes (1992b: 249) suggest that policy networks can be located on a continuum, with 'policy communities' and 'issue networks' posited as ideal types that can be found at either end, each representing contrasting patterns of government-interest group relations. Therefore, while the policy network approach has been associated with a specifically élitist power structure (Blanco, Lowndes and Pratchett, 2011: 304), in fact the dominant Marsh/Rhodes model adopts a neutral starting point on this question. The position of a network on this continuum – and thus its power structure – must be determined by empirical analysis of eight variables of interrelations in policy networks, grouped under four headings (Marsh and Rhodes, 1992b: 251):

1. Membership (number of participants; type of interest)
2. Integration (frequency of interaction; continuity; consensus)
3. Resource distributions (within the network; within the participating organisations)
4. Balance of power.

The characteristics of policy communities and issue networks are summarised in these terms in Table 2.1.

Policy communities

The *membership* of a 'policy community' comprises a small number of economic and/or professional groups and state actors, to the deliberate exclusion of other non-state interests (Marsh and Rhodes, 1992b: 251). In respect of non-state members, a single organisation with monopoly membership will represent each of one or two interest groups rather than competing pressure groups being present (Smith, 1993b: 79). In terms of state participation, according to Smith (1993b: 79), normally only one government body will be involved. But if two or more government institutions are members of a policy network, for it to be a policy community one of the institutions must acquiesce to be subordinate to the other. Furthermore, Bomberg (1998: 173) asserts that policy communities 'tend to exclude parliamentary influence and scrutiny'.

Policy communities exhibit high *integration*, with frequent and exhaustive policy communications among a stable membership that shares an ideological consensus and agreement on general policy preferences, thereby promoting policy continuity (Marsh, 1998a: 14). Citing Hall

(1993), Jordan and Greenaway (1998: 671-2) characterise the dominant set of ideas in a policy community as a 'policy paradigm', defined as:

...'a framework of ideas and standards that specifies not only the goals of policy and the kind of instruments that can be used to attain them, but also the very nature of the problems they are meant to be addressing'....They dominate a policy community structurally by framing its internal assumptions, policy agenda and internal discourse.

Ideological consensus is identified by many analysts as a contributory factor towards the ability of policy communities to withstand external pressure for change (Smith, 1993a: 98; Daugbjerg, 1998: 79-80). Smith (1997: 36) argues that this consensus, or dominant 'appreciative system', defines both the policy problem and, consequentially, the preferred solution or goal of the policy network. This, in turn, creates a bias in favour of those groups with resources that are deemed instrumental to achieving the network's perceived goals. Thus, the *resource distribution* dimension of policy networks is closely related to the *balance of power* dimension: '...networks are rooted in resource exchange. So, the distribution of resources among actors in a specific network remains central to any explanation of the distribution of power in that network' (Rhodes, 1997: 37). Resource distribution is particularly salient in the Marsh/ Rhodes typology, to the extent that it is said to be the most significant influence on network interactions and policy outcomes (Smith, 1997: 30-1). In a policy community, the members exchange their resources in a positive-sum power game in order to achieve their shared goals (Marsh, 1998a: 14).

Therefore, in practice the non-governmental members of policy communities are said to be those groups who have essential economic and professional resources that state actors perceive are necessary in order to resolve policy issues and implement policy (Richards and Smith, 2002: 282). Smith's analysis (1993a) of policy networks in UK agriculture and health provides evidence of how groups with key resources – such as farmers and doctors – were invited into policy communities that were set up by state actors who decided to intervene in these policy sectors.

Expert knowledge is one of the key resources identified by Marsh and Rhodes (1992b: 265) as facilitating access to policy-making, particularly in highly technical areas (Smith, 1993b: 81). For example, Smith (1997: 208) describes how, from the mid-19th century until the early 1990s, the government required technical information in order to set the pollution limits that were conditions of licenses issued to the chemical industry. As a result:

> ...industrial air pollution policy was the domain of a policy community. The membership of Inspectors and operators regulated emissions in an exclusive partnership. The organisationally-constrained Inspectorate had the authority to ensure operators pursued the vaguely defined BPM[4] principle. Operators possessed the information needed to elaborate this principle and the wherewithal to apply it. This resource interdependency bound the policy community together. (Smith, 1997: 208)

If vital resources and a shared ideology with state actors are important criteria for membership of a policy community, then a group's exclusion is conversely dependent on a perceived lack of resources and a failure to adhere to the ideology and the 'rules of the game' – e.g. no overt criticism of policy and constitutional behaviour – that structure the policy community (Smith, 1993b: 80–1). Inclusion and exclusion occurs through state officials' decisions regarding which groups are admitted to the network's formal and informal institutions, such as advisory committees, ad hoc committees and meetings (Smith, 1993b: 83–4).[5]

By excluding opposing ideologies from the network, policy community members can work together to promote their shared interests (Smith, 1993b: 82). For example, Cavanagh's (1998: 105–6) case study of health and safety policy in the North Sea oil and gas industry identifies a policy community-type of network incorporating the industry and UK government health and safety regulators, to the effective exclusion of labour representatives.[6] This had a considerable effect on policy outcomes: 'The exclusive relationships which have been evident in the British case have had the effect of giving a disproportionate influence to production and exploration matters' (Cavanagh, 1998: 108). Similarly, in the industrial air pollution policy community, the exclusive membership of industry and government inspectors developed a shared ideology and rules of the game which meant that 'Members exchanged their resources in the pursuit of mutually beneficial outcomes' (Smith, 1997: 78).

Issue networks

If policy communities appear to encapsulate a generally oligopolistic and conservative policy process (Smith, 1993a: 75; Marsh and Rhodes, 1992b: 260), then, at the other end of the policy network continuum, 'issue networks' represent a more open and unstable environment. The pluralistic character of issue networks in the Marsh/Rhodes typology reflects the origin of the term in the work of Heclo (1978; cited by Rhodes, 1990). His discussion, which focussed on micro-level, interpersonal relationships, argued that policy is made in the midst of a broad, communicative network of influential actors, thereby diffusing power (Rhodes, 1990: 296). Heclo posited 'issue networks' as a general pattern of policy-making across different policy sectors. In contrast, 'issue networks' in the Marsh/Rhodes model are meso-level, ideal types of policy networks whose existence in individual policy sectors or subsectors is a matter for empirical evaluation.

The *membership* of 'issue networks' is in a state of constant flux, with a large number of actors from a wide range of interests, potentially including numerous government institutions (Smith, 1993b: 81). Indeed, it is argued that if there is a lack of consensus between the participating government actors, then 'This conflict between government agencies is often a key reason why an issue network develops' (Smith, 1993b: 82).

With respect to the *integration* dimension in issue networks, the relationships both between groups and the state and also between the various groups themselves tend to be looser than in the case of policy communities. Network interactions will fluctuate in frequency and intensity (Marsh and Rhodes, 1992b: 251). In particular, groups' access to the state will be generally consultative (Rhodes, 1997: 45), in contrast to the more integrated policy role, or 'insider status', of a privileged set of interested groups that is characteristic of policy communities. In this vein, Smith (1993a: 10) observes that issue networks can emerge 'in new issue areas where interests have not had the time to establish institutionalised relationships'. Similarly, Hay and Richards (2000: 7) suggest that when networks form they tend to resemble issue networks rather than policy communities.

The large membership and breadth of interests in an issue network will therefore make consensus difficult to achieve. Thus, according to Garner (1998: 7), in issue networks state decision-makers will 'take on the role of a neutral arbiter seeking to balance the interests of the groups involved'. An absence of clear state policy preferences, combined with fluctuating membership and access, are said to lead to 'little continuity' in policy outcomes compared to policy communities (Marsh, 1998a: 14), or, as Garner (1998: 7–8) puts it, '...there may well be policy stalemate as competing groups cancel each other out, accompanied perhaps by unpredictable and violent policy shifts'.

Policy outcomes in issue networks may still be influenced by their uneven patterns of *resource distribution* (Bomberg, 1998: 174). However, it could be argued that, in the absence of the tight ideological structures found in policy communities, an issue network will exhibit a broader range of valued resources, and hence membership and power distribution. In contrast, challenges to producer interests would more likely be completely excluded from a policy community, and the insider/outsider distinction (see below) is relatively absolute. Meanwhile, Smith (1993b: 83) argues that in general:

> In an issue network, although some actors have resources, they are likely to be limited. Most of the interest groups are likely to have little information to exchange and little control over the implementation of policy. Consequently, they are forced into overt lobbying activities.

In terms of the structure of resource-based interactions, instead of the close resource interdependencies of policy communities, issue networks involve 'more informal, less standardised sharing of resources among a wide variety of members' (Bomberg, 1998: 174). In deploying their resources, issue network members engage in conflictual, zero-sum *power* games in contrast to the positive-sum, exchange interactions that characterise policy communities.

According to Bomberg (1998: 172), an issue network is to be found in EU environmental policy and is indicated by 'a varied and fluid membership'. This includes industrial interests, scientific experts, several Directorates (EU executive departments), environmental NGOs, the European Parliament and different member states with conflicting goals. The power of producer interests to influence the packaging waste policy subsector is said to demonstrate the uneven resource distribution typical of issue networks. However, environmental NGOs and parliamentarians were able to enter the network and exert some influence, varying from case to case, as a result of resources such as technical information and democratic legitimacy (Bomberg, 1998: 175). Indeed, the wide array of policy network membership – a defining aspect of an issue network model – is said to be facilitated by the particularly open nature of the institution nominally responsible for environmental policy, DG XI (Bomberg, 1998: 173). This proposition suggests that the institutional structure established by the state or international authority in a policy sector may have a significant influence over the type of policy network found therein. This institutional structure may, in turn, be constrained by the political culture and nature of power in that authority.

Table 2.1 The Marsh/Rhodes policy network typology

Dimension	Policy community	Issue network
Membership		
No. of participants	Very limited number, some groups *and Parliamentarians* excluded. *One government institution leads policy by consensus.*	Large, *including Parliamentarians. More than one department involved.*
Type of interest	Economic and/or professional interests dominate.	Encompasses range of affected interests.
Integration		
Frequency of interaction	Frequent, high-quality, interaction of all groups on all matters related to policy issue, *including implementation.*	Contacts fluctuate in frequency and intensity, *often limited to consultation.*
Continuity	Membership, values and outcomes persistent over time, *though core network and policy change can occur through combination of major exogenous pressures.*	Access fluctuates significantly, *and core policy change occurs through both endogenous policy learning and entry of new actors.*
Consensus	All participants share basic values and accept the legitimacy of the outcome. *Élite consensus.*	A measure of agreement exists, but conflict is ever present. *Élite dissensus, including between government departments.*
Resources		
Distribution of resources within network	All participants have resources *perceived as valuable by government;* basic relationship is an exchange relationship.	Some participants may have resources, but they are *perceived by government to be of* limited *value,* and basic relationship is consultative.
Distribution of resources within participating organization	Hierarchical; leaders can deliver members.	Varied and variable distribution and capacity to regulate members.
Power	There is a balance of power among members. Although one group may dominate, it must be a positive-sum game if community is to persist.	Unequal powers, reflecting unequal resources and unequal access. It is a zero-sum game.

Source: Adapted from Marsh and Rhodes (1992b: 251) and Marsh (1998a: 16). Italicised sections are additions to the original Marsh/Rhodes schema that incorporate insights in subsequent analysis.

Policy network membership, definition and boundaries

There is, however, a degree of confusion in the literature regarding the identification of the boundaries of policy networks, and thus whether policy communities and issue networks are mutually exclusive. For example, Bulkeley (2000: 729–30) interprets Marsh and Rhodes as claiming that issue networks can *co-exist* with policy communities in the same policy network. In this scenario, a tiered policy network is conceived with a policy community at the core and an issue network at the periphery. Smith (1997: 44) employs a similar notion: 'Often issue networks ring a policy community core ...'. However, Bulkeley seems to misread Marsh and Rhodes, for later they dismiss the core/periphery distinction as:

> ...primarily serving to obscure network boundaries. Pross...refers to the 'attentive publics' of policy networks, a more apposite phrase because it draws attention to the range of possible actors but does not treat them as members of the network. (1992b: 256–7)

It is argued here that if the Marsh/Rhodes schema is to have any utility as a diagnostic tool for policy networks that allows them to be used as a variable to explain policy outcomes, policy communities and issue networks must be conceived of as mutually exclusive. The type of core/periphery distinctions adopted by Bulkeley and Smith 'do not add a great deal to the analysis of networks' (Marsh and Rhodes, 1992b: 256).

This discussion highlights the need to determine which interest groups active on a policy issue are members of a policy network, in order to place that network on the Marsh/Rhodes schema. Another typology that may assist in this task has been developed by the 'Aberdeen Group' (Maloney, Jordan and McLaughlin, 1994: 25; cited by Grant, 2000: 23), which distinguishes among core insiders, peripheral insiders and outsider groups. Their 'outsider' category suggests that policy networks tend towards some degree of exclusion. But the most interesting distinction is between core and peripheral insiders. This is because while the form of both types of groups' participation appears to be similar, peripheral insiders lack influence relative to core insiders, who are said to interact with state actors in exchange relationships that are characteristic of those predicated by the policy community model. Peripheral insiders have an illusory type of insider status, often as members of advisory committees to keep them and their public constituency satisfied and to discourage any potential public criticism of current policy, while keeping them at arm's length from concrete policy-making (Grant, 2000: 22–4; Garner,

1993: 194–5). This raises two important questions. Firstly, how can influence be measured in order to place a group in either category? Secondly, can peripheral insiders be regarded as members of a policy network?

In relation to the first question, Grant (2000: 24) points out that survey data is unreliable because groups may overestimate their influence and/ or mistakenly attribute perceived policy changes to their actions, when the causes may lie elsewhere. Therefore, the only reliable distinction between core and peripheral insiders is that the latter *consistently* lose the battle to realise their goals *in terms of policy outcomes*. So, does it make sense to classify peripheral insiders as members of a policy network? This question can be addressed by examining how the notion of peripheral insiders relates to the policy community-issue network continuum. In the case of an issue network, although such a group may appear to be a member of such a network by virtue of its position in formal policy-making institutions, the fact that it *consistently* loses suggests that this model is inapplicable. So instead, the lack of influence of the peripheral insider group(s) compared to the core groups suggests that, in terms of the Marsh/Rhodes typology, the most coherent way of conceiving such a policy arena is to posit peripheral insiders as excluded from a policy community of state actors and core insiders.

Summary

The Marsh/Rhodes schema has come to dominate the policy network approach, with case studies confirming its utility for categorising networks and helping to explain different patterns of outcomes (Marsh, 1998b: 186–7). However, towards the end of the 1990s there seems to have been a growing awareness on the part of network analysts that the approach tended to reify the structures of networks by focussing on understanding networks at a particular point in time rather than adequately exploring their dynamic evolution (Marsh, 1998b: 192; Hay and Richards, 2000: 4). Thus, a more realistic approach requires a greater emphasis on how networks originate and evolve, which is the subject of the next section.

Policy network dynamics: the dialectical interaction between exogenous and endogenous factors

The need to develop a more dynamic approach in policy network analysis was, in fact, alluded to by Marsh and Rhodes (1992b: 260) when they concluded that 'focusing on policy networks will never provide an adequate account of policy change, because such networks are but one

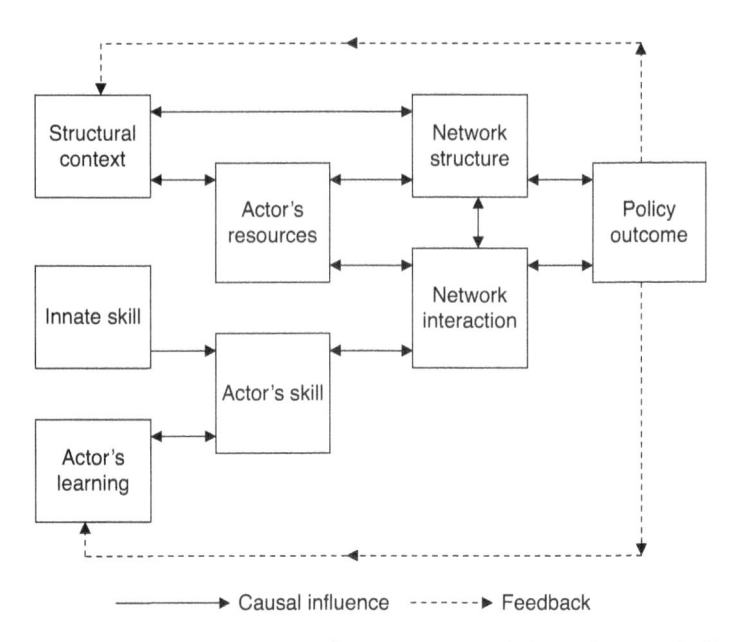

Figure 2.1 Policy networks and policy outcomes: a dialectical approach (from Marsh and Smith, 2000: 10). Reproduced with permission, © Political Studies Association

component of any explanation'. Thus, they argued that the first step in developing a dynamic conception of policy networks was to analyse the interrelations among the different levels of analysis – (macro-), (meso-) and (micro-) – in order to fully understand public policy (1992b: 268).

The concept of the relationships among these three levels of analysis has been developed by Hay and Richards (2000: 14) in terms of a: 'dialectical interplay of structure and agency'. Thus, Hay and Richards (2000: 5) argue that in order to develop a better understanding of the evolution of policy networks and hence outcomes, it is necessary to analyse networks' dynamic, dialectical relationships with exogenous structural influences on the one hand (for example, the electoral cycle) and, on the other, strategic network actors who apply their resources and skills to their political activity within their perceived structural context. That structural context is conceptualised as 'strategically selective' (Hay and Richards, 2000: 15), which means that certain courses of action and, thus, certain actors are privileged over others (McAnulla, 2002: 280). This 'strategic-relational' approach does not merely seek to account for the role of agency, but rather aims to transcend the structure-agency dualism by recognising that an absolute distinction is artificial and that each pole has to be examined in terms of its dynamic interaction with the other (Hay and Richards, 2000: 14).[7]

A related (as acknowledged by Hay, 1998: 35) dialectical approach, which takes account of the structure/agency interaction, has been developed by Marsh (1998b) and then Marsh and Smith (2000). Marsh and Smith (2000: 5) define a 'dialectical relationship' as 'an interactive relationship between two variables in which each affects the other in a continuing iterative process'. These authors unpack the dialectical interaction between the three levels of analysis to identify a specific set of two-way relationships relevant to a dialectical model of policy networks: network and context; network structure and network actors; and network and outcomes (see Figure 2.1 above).

This section discusses these crucial relationships between the three levels of analysis and their interactions with policy outcomes. The first part focuses on broader macro-level, or exogenous structures and pressures. The second part examines the impact of endogenous policy network factors. This includes both the meso-level of policy network structures, which have been analysed in the previous section using the Marsh/Rhodes typology, as well as the micro-level of agents' actions and interactions. Bearing in mind the point made above – that such distinctions run the risk of obscuring relatedness – each discussion will explore how those phenomena are conceived as impacting on other levels of analysis. For example, where the direction of the causal relationship is from exogenous context to either network structure or agents, this will be explored under exogenous sources of change. However, it is initially necessary to analyse what is meant by 'change' in order to help understand how different types of networks, exogenous perturbations and actor behaviour may interact to produce different degrees of change in policy outcomes.

Categorising degrees of policy change

It is therefore initially necessary to provide a 'definition of, or criteria for measuring, the degree of change.... When is a change a radical change?' (Marsh and Rhodes, 1992b: 260–1). Firstly, it is necessary to note that although changes in policy outcomes are related to changes in policy networks (Marsh, 1998b: 187), the two types of policy change are, to some extent, conceptually distinct (Smith, 1997: 43), and so it is reasonable to consider them separately at this point. In respect of network transformations, Hay and Richards (2000: 21) distinguish between secondary and core changes. Secondary changes are associated with 'tinkering – the minor reconfiguration of the network and the rethinking of the strategies likely to advance the long-term (collective) strategic goals of its members'. On the other hand, core changes are said to involve: '...the shedding of partners, the development of a new strategic agenda, and the wholesale modification of existing network hierarchies, practices and modes of conduct, i.e. the establishment of the parameters of a new network regime' (Hay and Richards, 2000: 21).

Because the network provides the immediate strategically selective context for the policy-making decisions of its members, it is therefore unsurprising that Hay and Richards' secondary/core distinction of network change mirrors typologies of outcome change. Furthermore, certain policy outcomes may relate directly to changes in policy network dimensions, such as decisions to alter the membership of policy-making bodies or patterns of consultation, which may themselves influence future policy outcomes (Marsh, 1998b: 197; Hill, 1997: 22–3). Jordan and Greenaway (1998: 672) address the issue of degrees of policy change when they cite Hall's (1993) three-tiered model of policy learning:

- The precise settings or calibrations of policy instruments (first order)
- The particular techniques or policy tools employed to provide policy solutions (second order)
- The overarching goals which guide policy-making (third order).

> Shifts in the first two levels occur regularly and incrementally and are associated with 'normal' policy making.... A paradigm shift of seismic proportions is required to knock them [policy-makers] from well-trodden paths, altering the underlying goals of a policy area.

Jordan and Greenaway (1998: 673) remark on the similarities between Hall's model and the belief system structure in Sabatier's 'advocacy coalition framework'. One of the significant premises of the advocacy coalition framework is that policies:

> ...can be conceptualized in much the same way as belief systems [in that] [t]hey involve value priorities, perceptions of important causal relationships, perceptions of the state of the world (including the magnitude of the problem), perceptions of the efficacy of policy instruments, etc. (Jenkins-Smith and Sabatier, 1994: 179–80)

Borrowing from Putnam's work on elite belief systems,[8] Sabatier (1993: 288) differentiates between *core* and *secondary* facets of advocacy coalition belief systems to try to account for the observation that: 'some aspects of public policy clearly change far more than others'. Core beliefs are subdivided once again into 'deep' core and 'near' core. In this formulation, belief systems are seen as hierarchical, with the 'higher/broader levels' constraining the more specific positions below. Thus, the deep core: '...includes basic ontological and normative beliefs, such as the perceived nature of humans or the relative valuation of individual freedom or social equality, which operates across virtually all policy domains' (Jenkins-Smith and Sabatier, 1994: 180).

The next level down, the 'near' or 'policy' core, refers to 'a coalition's basic normative commitments and causal perceptions across an entire policy domain or subsystem' (Jenkins-Smith and Sabatier, 1994: 180). These represent: 'basic strategies for achieving normative axioms of deep core [beliefs]' (Sabatier, 1993: 290). Examples of near/policy core beliefs would include positions on the balance between environmental protection 'versus' economic development, and preferences between coercion, inducement or persuasion as policy instruments. The lowest stratum of the belief system deals with 'secondary aspects' of policies that consist of specific, practical decisions or recommendations designed to realise policy core beliefs. These will include budgetary allocations, institutional structures, determinations of cases, public appointments, and assessments of agency or departmental performance. As in Hall's model: 'The hierarchy is arranged in order of decreasing resistance to change, with secondary elements being the most fluid' (Jordan and Greenaway, 1998: 673). Both models are also said by O'Riordan and Jordan (1996: 83) to have similarities 'with Lindblom's "Grand majority" and secondary issues'.

Having established important distinctions in the degrees of change, the dialectical interactions that may lead to policy network and outcome changes can now be analysed in detail.

Exogenous factors

Initially, it is important to acknowledge that an absolute distinction between exogenous and endogenous structural influences on policy networks is problematic because of network agents' interpretative mediation of exogenous factors in the context of their network structure (Marsh, 1998a: 12). Furthermore, network interactions partially reflect exogenous factors through the latter's effect on agents' resources, skills, interests and actions (Marsh, 1998b: 193; Toke and Marsh, 2003: 232) as they 'interpret and negotiate constraints or opportunities' (Marsh and Smith, 2000: 6). Similarly, Hay and Richards (2000: 20) contend that actors' perception of the evolving exogenous environment is one of the mechanisms of network transformation.

Nevertheless, Marsh (1998b: 193) asserts that elucidating exogenous influences on policy networks is particularly important in order to help explain both network structure and policy outcomes. One type of exogenous influence has been discussed above and comprises models of the general system of power in British politics. This could be conceived as a relatively stable type of exogenous parameter (Sabatier, 1998: 102–3).[9] However, this section discusses a range of additional exogenous factors that tend to be more dynamic. Thus, this discussion follows Marsh and Rhodes (1992b: 257), who identify four general types of exogenous events – economic, ideological, knowledge-based and institutional.

Economic factors

The economic context, as interpreted by network members and mediated by the ideology of governing parties, can act as a structuring force on the resources in policy networks. For example, Richardson (2000: 1020) argues that policy-making changed across many sectors in the late 1970s as a result of 'the declining competitiveness of Western Europe in the face of perceived (and possibly exaggerated) globalization'. In order to improve competitiveness, Thatcher's Governments are said to have broken up entrenched, consensual policy communities so that they could impose optimal decisions instead of the sub-optimal compromises that supposedly characterised the previous, post-war consensual British policy style (Richardson, 2000: 1010). However, it should be noted that other commentators argue that changes in policy outcomes were actually much more modest than the legislative changes in this period, due to an impositional policy style that neglected the power of policy networks and the importance of interest groups' resources for the effective implementation of policies that affected them (Marsh and Rhodes, 1992a: 185–7).

Another salient exogenous economic development involves changes in the global economy accompanied by the rise of multinational corporations: 'Their importance in terms of employment, investment, and overall economic growth is such that nation-states are dependent on them rather than the other way round' (Richards and Smith, 2002: 127). Therefore, multinational corporate actors appear to have gained additional resources in policy networks relative to the state and, *ceteris paribus* (other things being equal) other actors. In general, if networks represent a 'new mode of governance' that increases the role of markets in policy-making (Marsh, 1998: 190), then the impact of economic developments and actors on policy networks is likely to be accentuated.

For example, the UK government's perception of potential economic benefits, in terms of increased employment and export earnings, to be gained from plant biotechnology has consistently endowed biotechnology business groups with significant structural resources in the GM crop policy network (Toke and Marsh, 2003: 241, 244). Meanwhile, farming interests have also perceived economic self-interest in the commercialisation of GM crops. Relatedly, both biotechnology and farming interests have close institutionalised relations with state actors as manifest in the support and sponsorship they received from the then Ministry for Agriculture, Fisheries and Food (MAFF),[10] resulting in privileged access to the policy process. Furthermore, both groups are said to benefit from structural economic inequalities that mean they possess substantial economic resources to be employed in pursuit of their interests in this policy network. On the other hand, groups such as Friends of the Earth and the Soil Association have been excluded from the network, partly because they 'do not represent entrenched economic interests.... [I]t is clear that it is economic, and to a lesser extent professional, interests that dominate the networks' (Toke and Marsh, 2003: 244). Prior to 1998, this policy network strongly resembled a policy community that was dominated by farming and biotechnology interests who believed that they stood to gain commercially from the growing of GM crops. Environmental groups and their concerns were excluded from the policy network, affecting how problems were defined, and the character of policy outcomes.

Changes in the policy network came about through a range of exogenous factors, possibly including alterations in the economic context through the refusal of supermarkets to stock GM products: 'If there is no market for GM food, then commercial growing of GM crops will effectively have been prohibited whatever the regulations may say' (Toke and Marsh, 2003: 250). These changes may have contributed to the admission of powerful 'insider' wildlife protection bodies such as English Nature and the Royal Society for the Protection of Birds (RSPB) (Toke and Marsh, 2003: 249). However, the power of the GM crop lobby's economic resources endured, forcing the government to exclude from the policy network environmental groups who opposed GM crops in principle, and helping to limit the criteria to be applied to trials designed to test the environmental impact of GM crop trials.

The ideological and cultural context

In terms of assessing exogenous ideological pressures on policy networks, one of the most important mechanisms is through changes in governing party ideology, because 'party is the blade for prizing apart the mollusc's shell of Whitehall and the policy networks' (Marsh and Rhodes, 1992b: 257).

One related element of the context of policy networks is the broader 'battle of ideas' (Jordan and Greenaway, 1998: 671). Jordan and Greenaway (1998: 671) suggest that changes in ideology are the most likely source of core changes in networks and outcomes. It is not surprising, therefore, that networks' ideological context tends to be a significant focus of the activity of 'new social movements' (Richards and Smith, 2002: 183–6). However, many such groups specifically avoid interaction with the state because of a sharp conflict between the state's and the groups' ideologies and value systems, and thus they are happy to be excluded from the policy networks. Nevertheless, Richards and Smith (2002: 184) observe that policy communities can be affected by new social movements who try to disturb the 'climate of ideas, [which] can have an important impact on established policy networks by politicising the closed and settled agenda and introducing new ideas'. The potential impact of those ideas depends significantly on the coverage that related actions receive in the media, and their effect on public opinion. Consequently, if a policy community's appreciative system becomes sufficiently politicised and discredited, then new actors with ideas that reflect those broader public concerns may gain entry to a policy community (Smith, 1997: 210). This may occur when the public legitimacy resources conferred by hitherto-excluded actors come to be perceived as valuable in a network that had previously been isolated from broader political scrutiny and criticism. However, analysts must take care to distinguish between peripheral insiders and core insiders under such circumstances.

Richards and Smith (2002) hold up GM crops as an example of the impact of new ideas on policy networks. Campaigns by environmental groups that emphasised the risks of GM food gained relatively sympathetic coverage in the media, exemplified by the *Daily Mail's* 'Frankenstein Foods' campaign (Richards and Smith, 2002: 186; Toke and Marsh, 2003: 245). The cumulative effect was that this exogenous activity was 'successful at shaping the debate on GMOs away from the notion that this was a good development for humankind to one in which GMOs were seen as a tremendous risk' (Richards and Smith, 2002: 186). Toke and Marsh (2003: 246) identify this perception as possibly reflecting and intensifying a 'risk society'[11] consciousness, or 'cultural context', where new technologies are perceived as hazardous, and industry and government scientists viewed with suspicion. Consequently, the government's position changed away from an unequivocally supportive stance towards GM crops, to a position whereby a moratorium was introduced, pending evaluation of certain potential environmental risks.

It is, however, important to bear in mind that the actual impact of this putative cultural context on the policy network is mediated by the nature of the policy network and the resource distributions within it. In the case of GM crops, powerful 'insider' environmental organisations, such as English Nature and the RSPB, who shared some of the concerns of, and networked with, the more radical 'outsiders' who were engaged in direct action and activity in other arenas such as the consumer arena, now had the additional resources that permitted entry to the policy network because government policy on GM crops would have lacked legitimacy without their participation. This meant that the pattern of the network's 'appreciative system' changed to one that was more sensitive to the precautionary principle that 'absence of evidence of risk should not be mistaken for absence of risk' (Toke and Marsh, 2003: 246). The altered network 'ideology', in turn, affected the way in which the 'risk society' cultural context interacted with the (realigning) policy network to produce policy changes in the form of increased regulation. This case appears to reinforce Marsh and Rhodes' (1992b: 260) conclusion, which emphasises the need to understand the interrelations between structure and agency, and/or among network, context and actors: 'The environment [of the policy network] is not given; it is both constituted and constitutive, and the analysis of the appreciative system of actors in the policy networks is central to understanding this interactive process'.

There are, of course, constraints on the power of ideas or narratives. For example, initially, Richardson (2000: 1017–8) appears to emphasise the power of ideas:

> ...exogenous changes in policy fashion, ideas, or policy frames presents a very serious challenge to existing policy communities and networks. New ideas have a virus-like quality and have an ability to disrupt existing policy systems, power relationships and policies.

However, adopting Kingdon's metaphor of ideas floating around in a 'policy soup', Richardson (2000: 1018) acknowledges that the ideas that stand a realistic chance of being taken up as policy must meet criteria, including '...fit with dominant values and current national mood,...political support/opposition'. This suggests that new ideas (just like viruses, to extend Richardson's analogy) themselves face contextual constraints and opportunities: some environments suit their replication better than others. Nevertheless, Jordan and Greenaway (1998: 672–3) suggest that ideology is a particularly important type of exogenous pressure because major, or 'third order', policy change involving alterations in the underlying goals of policy 'is only brought about by evolving societal debate and reflection – social learning – regarding the overall direction of policy'.

Developments in knowledge and technology

The third external destabilising factor identified by Marsh and Rhodes is changes in information or knowledge about an issue, including the development of new technologies that pose new problems and opportunities. However, new information is mediated by policy networks, which can either adapt or interpret that information in such a way that it is consistent with the network's ideological structure, or the network actors can learn from that information and initiate policy changes. These endogenous factors are discussed in more detail below.

The political and institutional context

The fourth category of external pressure on policy networks is 'institutional' or 'political', which encompasses a number of interrelated dimensions, such as the role of political authority and state interests, public opinion, party and parliamentary support, relations with other networks (particularly the effect of sectoral on subsectoral networks), and the international context.

Political authority and state interests: Marsh and Smith (2000: 8) assert: 'Political authority is perhaps the most important external constraint. If a minister, or particularly the Prime Minister, is prepared to bear the costs of breaking up a policy community, he or she has the resources and the authority …'. A dominant role for government interests in general is perceived by a number of the contributors to Marsh's collection of case studies (1998: 189), and this feature is emphasised by the asymmetric power model of the British political system discussed above (Marsh et al., 2003). Alterations in the national government context can lead to changes through the actions of new agents such as government ministers.

The possible effects of governmental interests and authority are cited by Hay and Richards (2000: 24) when they speculate that some networks may have faced termination following New Labour's 1997 victory due to 'the creation of a series of alternative power bases across a whole range of policy fields each of which have received the patronage of the Prime Minister'. These new power bases comprised task forces and advisory groups, and were accompanied by a tendency towards centralised coordination through the Cabinet Office. As Hay and Richards (2000: 25) note, entrenched insider groups can shift their networking activity to the new networks. This point could be taken further to propose that apparently radical change in policy networks could, in fact, promote continuity in the pattern of policy outcomes. In other words, the transformation of policy-making architecture could, in some instances, be a strategic move to constrain evolutionary changes in the policy direction of networks that are promoted by other exogenous dynamics, such as shifting public opinion.

Generally, network literature tends to emphasise perceived advantages to the government from stable policy communities. In this vein, Richards and Smith (2002: 173) comment that government 'is unlikely to want to have contact with a wide range of groups with competing demands, because this will introduce greater complexity and conflict into the policy process'. Conversely, issue networks are said to be more likely to be found in policy areas that are characterised by a high degree of ideological conflict and/or are considered to be relatively inconsequential by the government (Marsh and Rhodes, 1992b: 254; Smith, 1993a: 10). However, unrestrained conflict among government departments regarding overlapping policy areas may also lead to issue network-type policy-making arenas.

Public opinion: Changes in government and its impact on existing policy networks is, of course, partly related to public opinion, which comprises another aspect of the political context of policy networks. Thus, Toke and Marsh (2003: 244) identify public opinion as a causal factor in changes in the GM crop policy network: 'There is an association between network change and the changing parameters of public opinion which, in the post-BSE period, have involved major concerns about food safety in general and GM food in particular'.

As discussed above, public opinion is influenced by other contextual factors, such as the interaction between changes in information and ideologies that seek to interpret and give meaning to that information. Moreover, the impact of public opinion on policy networks will be mediated by the resources and skills of actors excluded from the network. However, once again, the structure of networks and members' interpretations of their strategically selective context (Hay and Richards, 2000: 14) mediate the effects of public opinion and the salience of 'public legitimacy' as a resource within the policy network. This mediation is discussed in more detail below through an analysis of endogenous policy impacts.

Parties and Parliament: Daugbjerg and Marsh (1998: 63) identify party and parliamentary support as a key 'macro-feature' or 'state characteristic' that influences networks: 'The structure of party loyalties has an impact upon the formation of meso-level policy networks. Political parties tend to favour some groups' interests by giving them access to policy networks and by excluding others'. The impact of the electoral cycle on policy networks is modelled by Hay and Richards (2000) in the context of the 1997 change in governing party although they note that the relationship between this form of macro-political change and policy networks will vary from network to network, and requires empirical investigation. Indeed, one of the tasks of this study will be to follow their recommendation for a 'disaggregated approach' to assessing the impact of this exogenous shock, by looking at its effects on the animal research policy network.

Other policy networks: Marsh and Smith's review of the preceding policy network literature perceives a limitation in the original Marsh/Rhodes model's account of the exogenous constraining role of 'other networks', particularly the effects of sectoral networks in providing:

> ...a crucial aspect to the context within which subsectoral networks operate. Overall, it is evident that exogenous changes can affect the resources, interests and relationships of the actors within the networks. Changes in these factors can produce tensions and conflicts which lead to either a breakdown in the network or the development of new policies. (Marsh and Smith, 2000: 8)

There does, however, appear to be a tension in the policy network literature between, on the one hand, the notions of disaggregation and policy sectorisation, and, on the other hand, the recognition of interconnections between policy sectors/networks. While some policy network analysts, such as Garner (1998: 229–30), consider the concept of policy sectorisation as both credible and necessary to the utility of the approach, the dialectical models advanced by Marsh and Smith (2000) and Hay and Richards (2000) attempt to understand the policy process and outcomes in terms of the relationships between the network, the actors within it, and the external environment. Disaggregation tends to be associated with a pluralist model of state power where power is widely dispersed, whereas in the élitist model, sectoral policy networks are said to constrain subsectoral policy networks (Daugbjerg and Marsh, 1998: 57–8).

International political pressures: Britain's accession to the European Community in 1973 represented a major perturbation to the structural context of national UK policy-making. Richardson (2000: 1013-6) argues that the institutional structure of the EU has moved and altered network structures. Firstly, policy sectors are becoming Europeanised, albeit with differences in timing (see also Moran, 2003: 170). Secondly, the openness and diversity of EU institutions such as the Commission and Parliament means that a large number of actors are interacting in many different policy venues. This complex policy system is said to make it impossible to constitute policy communities across the different venues and actors 'except where highly specialised and detailed technical issues are being resolved' (Richardson, 2000: 1015).[12] Moran (2003: 166) concurs with the description of the EU policy process as intensely complex. However, he postulates that 'technical' issues are more prominent in EU policy-making than Richardson appears to acknowledge. Hence, the fragmentation of the process means that specialised expertise and significant resources are required to monitor policy closely and thus intervene effectively. Furthermore, as the Commission lacks resources, it 'relies heavily...on business and the professional expertise which business has the money to buy'. These resource demands are exacerbated by:

> ...the juridified character of the process created by the prominent role of the European Court of Justice as an important source of policy creation and adjudication: monitoring, exploiting, and, where, necessary, challenging the Court's jurisprudence is no job for amateurs or part-timers. (Moran, 2003: 167)

Consequently, one of the broad effects of the EU policy process is that it is said to mobilise distinctive biases and empower distinctive oligarchies: '...in favour of business, especially big business' (Moran, 2003: 167). This represents another means of insulating elites from democratic and accountability pressures, though this time through high resource requirements rather than 'customary integration' (Moran, 2003: 167).

However, it is important to bear in mind that the European Union's activity has varied considerably across different policy domains. Furthermore, its impacts are inevitably mediated by the policy network structure: '...the structure and the culture of individual departments do have an important effect on the patterns and pathways of Europeanization...'. (Jordan, 2002: 210) Thus, the influence of the EU varies both from network to network and across time. A further, crucial source of variation occurs within the policy network, particularly between formulation and implementation phases of policy-making. In environmental regulation, for example, while policy formulation reflects EU-stimulated attempts to move towards more open, formal and rule-based approaches to regulation, implementation processes continue to benefit business interests through informal and highly cohesive relations between regulators and business that are indicative of policy communities (Moran, 2003: 171; Smith, 1997). Furthermore, Moran argues, research into the development of EU monetary policy demonstrates that policy community-type relations can also exist in the realms of EU policy formulation.

This multi-dimensional variation makes it essential to disaggregate any analysis of the impact of the EU, not only by focussing on individual networks but also by examining the different policy-making processes within that network. It is also important to consider the dialectical relationships between national and European policy-making arenas. As Jordan (2002: 209) comments, national policy networks, and the actors within them, have the potential to 'upload' policies to the European level to try to modulate perceptibly discordant EU policies to 'fit' national circumstances more closely. In summary, the 'Europeanization' of policy is highly variable and occurs through a dynamic, complex process involving institutions and actors at various levels of governance – such as European, national and sub-national policy formulation, and policy implementation – and this must be taken into account by any attempt to describe and explain the impact of the EU on UK policy networks and outcomes.

Endogenous network factors

One of the subsidiary themes to emerge from the preceding discussion has been the role of policy networks in mediating the impact of exogenous pressures on policy outcomes. Marsh (1998b: 193–5) postulates two interrelated features of policy networks to be relevant in this respect: *structures* and *interactions*. Network *structures* and their associated resource dependencies broadly correspond with the dimensions of the Marsh/Rhodes typology and are said to be constrained by exogenous structural phenomena. Meanwhile, network *interactions* occur between strategically calculating network members as they exchange resources, and are the product of the application of those actors' resources, innate skills and policy learning. Thus, the concept of network interactions helps to capture the 'agency' element of the policy process, thus introducing essential dynamism to the network approach (Daugbjerg and Marsh, 1998: 53; Hay and Richards, 2000: 3).

But agency is not discrete from structure, and, therefore, network interactions are held to be dialectically related to the network structure.

Thus, on the one hand, network agents are variably constrained and enabled by network and exogenous structures, and the pattern of resource dependencies between network actors is partly derived from the positions they occupy in their structural context (Hindmoor, 2009). Marsh and Smith (2000: 6) illuminate the structural aspect of policy networks through the notion of 'institutionalisation':

> Networks involve the institutionalization of beliefs, values, cultures and particular forms of behaviour. They are organizations which shape attitudes and behaviour. ... In doing so they are not neutral, but, like other political institutions and processes, they both reflect past power distributions and conflicts and shape present political outcomes. Thus, when a decision is made within a particular network, it is not simply the result of a rational assessment of available options,...but rather reflects past conflicts and the culture and values of decision-makers.

On the other hand, the strategic decisions of network actors are not *determined* by this network structure and can have intended and unintended effects on the network structure. Furthermore, 'the way in which [resource dependencies] are discursively constructed by the participants affects their behaviour and policy outcomes' (Marsh, 1998b: 195).

From this discussion, three types of endogenous network influence on the pattern of policy outcomes over time can be identified:

1. Policy networks' mediation of the effects of exogenous change on policy outcomes
2. Network structures' constraining and facilitating impact on the behaviour of network actors and, hence, policy outcomes (see also Bulkeley, 2000: 729).
3. The interpretive and strategic action of network members.

In order to give a complete account of the dynamics that underpin policy-making, it is also necessary to acknowledge that the dialectical relationships between network and exogenous context involves causal effects from network-to-context (Marsh and Smith, 2000: 7–9), in addition to the more familiar reverse relationship discussed above.

This section begins by focussing on the effects of the structural aspect of networks, initially with a review of the variable ways in which policy communities and issue networks are said to mediate exogenous forces and facilitate change or stability endogenously. In addition to this body of work, there are two other approaches that either implicitly or explicitly attempt to elucidate endogenous impacts on policy change. Firstly, there is a literature that focuses on the way in which networks interpret and generate new knowledge, arguments, ideas and policy learning. Secondly, there is analysis of the way in which policy-making bureaucracies are structured, particularly at the implementation level, and hence how they deal with policy problems and the demands of competing groups.

Endogenous dynamics through policy communities and issue networks

Policy communities are conceived as strongly institutionalised types of policy process, compared to issue networks. Thus, Smith (1997: 44) argues:

> Endogenously driven radical change is rare owing to the routinised relationships which constitute policy networks (especially policy communities) .... . Policy output is unlikely to change considerably if it is the domain of a policy community. Issue networks are more likely to generate substantial shifts owing to the lack of any strong consensus, fluctuating interaction and the fluidity and diversity of membership.

The tightly-integrated structure of a policy community is causally linked to policy continuity by Smith (1993a: 98) when he proposes that a strong consensus in a policy community and a high degree of group control over implementation maximise the policy network's ability to resist change. For example, one of the principal institutional structures that formed the basis of the policy community and promoted continuity in agriculture policy was the Annual Review, which decided commodity prices for the subsequent year: 'It gave the farmers a statutory right to consultation and excluded other groups (e.g. consumer or environmental groups) from the process' (Smith, 1992: 29). In addition, the remit of the Annual Review was narrow, reflecting the tight ideological structure of the policy community: its goal was merely to recommend price increases, so alternative policy goals could not be developed. As Smith explains, in the late 1960s, MAFF made no effort to consult consumer associations. The tight structure of the agricultural policy network enabled it to resist exogenous pressures for many years, and the close relationships between MAFF and the National Farmers' Union encouraged the state actors to ignore other groups with interests in agricultural policy.

Thus, in the case of agriculture, 'It was through the ideological and institutional structures of the community that change was prevented in agricultural policy' (Smith, 1993a: 134). Marsh and Smith (2000: 14–15) also argue that, once it was entrenched, the ideological structure of the agriculture policy community also affected the network's context, particularly the perception of the Treasury, which continued to accept and fund a policy of maximum food production and farm subsidies despite the disappearance of the original conditions that gave rise to that policy. Once again, this highlights the dialectical nature of the network/context relationship.

Smith postulates a different type of network-to-context dynamic (1997: 43) when he argues that it was the closed, secretive nature of the industrial pollution policy community that first provoked suspicion and public concern about the policy in the 1970s, indicating that attempts by policy communities to monopolise power may incur legitimacy costs, which can have destabilising repercussions. This is another example of how a network can affect its context, and is proposed as a general phenomena by Richardson (2000: 1008): '[T]he very success of policy community politics might be the cause of its erosion over time .... [I]nterest group activity begets yet more interest group activity thus increasing, not decreasing uncertainty as the number of stakeholders increases'.

Together, these observations evoke the idea of policy-making as an inherently dynamic process, with policy communities conceived as entities or open systems that tend towards the maintenance of an equilibrium, or 'homeostasis', in the midst of an intrinsically uncertain and constantly changing world. The ultimate potency of network-to-context dynamics are probably contingent, with their eventual effect on policy outcomes depending on the ongoing dialectical relationship between the network and its context as the network continually (re)mediates the evolving environment. Key factors affecting policy community homeostasis would appear to include the strength of any exogenous perturbations in relation to the resource distributions in the policy community. For example, when the air pollution policy community came under particular stress in the early 1990s as a result of pressure from environmental groups and from the European Union, state actors at Her Majesty's Inspectorate of Pollution (HMIP) tried to impose a change in the rules of the game in the policy network, away from consensual and cooperative 'regulation' and towards 'a more strict and arms-length approach' (Smith, 1997: 49). However, the chemical industry's persistent monopoly on technical knowledge, which was partly a result of a historic unwillingness by government to invest sufficient resources in independent regulation and monitoring, put the industry in a powerful position, to the extent that it was the dominant partner in the policy community (Smith, 1997: 158–9). Consequently, the attempt by HMIP to impose change failed, because continuity in the structure of resource interdependency meant that the industry retained powerful influence over implementation, and thus was able to manage exogenous perturbations. The new policy 'was born of top-down disruption, but has grown under a conservative, bottom-up influence' (Smith, 1997: 50).

Therefore, not only did the policy community's structure mediate the exogenous perturbations, but the network structure affected the behaviour and perceptions of the participants. In the latter case, the information asymmetry between HMIP and industry facilitated industry's attempts to modify the way pollution limits were set, from prescription to general guidance. This meant that: 'Standard-setting was deferred to the site-level instead ...' (Smith, 1997: 211). Industry's influence had the effect of 'frustrating HMIP's arms' length approach, and regaining deep industry participation in standard setting: a reversal in HMIP's intended rules of the game' (Smith, 1997: 127–8). On the other hand, state actors, in particular HMIP, were constrained by the policy network structure: 'Without industry's information HMIP was unable to exercise its authority. HMIP's organisational constraints and lack of political support made this information dependency more acute' (Smith, 1997: 211).

Furthermore, HMIP is said to have responded to industry conservatism by coming:

> ... to realise that the policy community arrangements that developed were of mutual benefit, even if they were not what it had initially desired. The policy community has enabled HMIP to keep to its implementation timetable and avoid criticism from political masters, whilst industry has been able to win influential access to standard setting. (Smith, 1997: 212)

The notion that policy communities promote continuity should not, however, be taken to mean that such networks and their policy outcomes are inert entities. As mentioned above, policy networks, as meso-level phenomena, mediate wider environmental or 'exogenous' factors. This means that in the face of external shocks, policy communities may, in certain circumstances, have reform forced upon them. However, it is worth noting Marsh's ('1998b': 188–9) observation that in 'most cases closed policy communities attempt to ignore public opinion'.

For example, policy communities may be in a position to manage and hence mitigate the effects of such changes to the relative benefit of their members, by appearing to allow groups representing public concerns into the network, but as 'peripheral' insiders:

> Limited accommodation of some actors into a policy community is more likely to maintain the level of influence enjoyed by original members than is a complete breakdown into pluralistic arrangements. ... So changes in a network's environment can be internalised in two ways, either through building new resources or through including new (interdependent) members possessing those resources. The core policy community practices a 'dynamic conservatism'. (Smith, 1997: 45)

Industrial pollution policy provides an example of another policy community strategy in the face of growing exogenous pressure for change. Smith (1997: 28) cites evidence that industry actors in this policy community took 'pre-emptive' action to promote minor reforms which would maintain their close, privileged involvement in the network and protect their interests in the face of changes in the political context of this policy sector:

> British industry began lobbying for pollution control reform in the mid-1980s. This was a defensive measure against the imposition of 'irrational' standards from the European Community. Industry wished to maintain the British tradition of accessible (to its interests) standard setting procedures which ensured their participation.

In contrast to policy communities' ability to resist or manage external perturbations, issue networks are more permeable and so susceptible to change due to exogenous pressures. Bomberg (1998: 175) expresses this in terms of relations with other networks: 'outcomes from issue networks ... cannot be fully understood without examining the network's wider interaction with other networks'. In terms of other networks' exogenous effects on the EU environment issue network, Bomberg (1998: 182) reports: 'In the packaging waste case, "threatened" economic and producer groups entered the environment network at any number of stages of the policy-making process and had a significant impact on policy outcome'.

In relation to the converse network-to-context effect, the EU environment issue network lacked the resources to implement explicit constitutional requirements to integrate the principle of environmental sustainability into other policy areas, such as agriculture, economic and internal market, that have a greater influence on environmental policy.[13] Bomberg (1998: 179) explains:

> ...an issue network's diversity and open character mean that its outputs or 'signals' to other actors may be ambivalent, incoherent or confusing. In short, its issue network characteristics disadvantage it in relation to other, better-established policy networks. Consequently, the relationship between networks is characterized by subordination rather than integration.

However, although Bomberg's analysis suggests that the lack of coherence and resources in issue networks may weaken their impact on other networks, evidence from the GM crops case (Toke and Marsh, 2003) suggests that the same open, loose structure may increase the flow of information out of the network into the public domain, thereby affecting the broader political agenda, albeit in a chaotic manner. For, as the coherence and consensus of the GM crop network lessened, and the hegemony of economic and producer interests weakened, the network started to have new types of external impact in the shape of increasing levels of politicisation and thus public concern. In this case, new GM-sceptical network members such as English Nature made novel (for the policy network) policy demands, such as the call for a moratorium on commercial planting pending trials. It could be argued that English Nature's new status as a member of the policy network gave its demands greater credibility (in terms of their potential influence on policy outcomes), and contributed to the decision by the then leader of the Opposition (William Hague MP in 1999) to openly support their position. This pronouncement, in turn, coincided with the beginning of an increase in media coverage for the issue while the government temporarily resisted the moratorium call (Toke and Marsh, 2003: 247).

Thus, in general, the loose structure of resource interdependency and ongoing conflicts in issue networks contribute to the variable outcomes through time, across particular policy issues within the issue network, and even within the same policy initiative, as reflected in EU environmental policy:

> The contradictory and confusing nature of environment legislation is widely documented...the diversity and complexity of an issue network can render its outputs unpredictable, complex and sometimes contradictory. Whereas policy-making in the auto emission

case resulted in a gradual tightening of standards, the opposite is true in the case of packaging waste. This case illustrates the extent to which key characteristics of an issue network – its shifting, uneven balance of resources and permeability – can shape policy outcomes'. (Bomberg, 1998: 175)

Learning dynamics in policy networks

Although the Marsh/Rhodes model tends to emphasise the stabilising role of policy networks, they acknowledge that: 'Change in policy networks can also be endogenous. Consensus within networks is the product not of one-off negotiations but of a continuing process of re-negotiation which can be characterised as *coalition building*' (1992a: 260 – emphasis added). Noting the significance of this comment, Bulkeley (2000: 732) suggests that 'learning models' may potentially be helpful in understanding these processes of coalition building:

> These approaches 'stress the critical importance of ideological factors, discourse, rational argument and belief systems in bringing about policy change...whilst at the same time taking cognisance of the interplay and importance of particular political forces and bureaucratic interests'. (2000: 731–2; Citing Jordan and Greenaway, 1998: 670)

Jordan and Greenaway (1998: 670) assert that theories of policy-oriented learning have tended to be neglected by political science. In particular: 'What is currently missing in network accounts of change is an adequate account of the dialectic between external events and the network's understanding of them' (Jordan and Greenaway, 1998: 671). Utilising such 'learning models' may help to overcome a perceived analytical gap by providing a more detailed focus on the destabilising effects of ideological factors on network structures and interaction, especially the ideological and concomitant institutional structures that sustain policy communities.

Interestingly, the roles of beliefs, argument and coalitions in network and policy dynamics are significantly pre-figured in earlier policy network analysis, though as a force for stability rather than change. For example, Smith (1993b: 82) argues:

> In fact a policy community often has more than a consensus; it actually has an ideology which determines the community's 'world-view'. ... Ideology defines not only what policy options are available but

what problems exist. ... The ideology thus privileges certain ideas within the policy process. In doing so it ensures that the interests of the dominant actors within the policy community are served and it acts as a further means of exclusion.

However, the power of ideas is considered to be something of a double-edged sword for policy communities: 'ideologies are simultaneously the most powerful mechanism of exclusion in policy communities and also the most vulnerable to attack' (Jordan and Greenaway, 1998: 671). New knowledge tends to be interpreted by decision-makers in such a way as to fit pre-existing ideologies, or policy paradigms (Jordan and Greenaway, 1998: 672–3). But if new information cannot be adapted to the needs and paradigm of the existing policy community, then it may pose a threat to that network (Richardson, 2000: 1018). Fundamental network and policy changes are said to occur through the accumulation of anomalous information that cannot be explained by the existing paradigm, leading to the politicisation of policy problems and the participation of new actors in the debate as policy actors seek alternative sources of information and solutions (Jordan and Greenaway, 1998: 672–3). This learning process potentially results in the breakdown of the pre-existing policy community and its guiding paradigm.

However, the impact of new knowledge is mediated by competing coalitions of actors who make truth claims based on what they present as 'valid scientific and technical knowledge' (Richardson, 2000: 1020). This suggests that those actors with technical expertise will have the advantage of being in a position to mediate the political impact of new knowledge. Moreover, according to Jordan and Greenaway (1998: 671): 'Those that can demonstrate technical competence stand a much greater chance of being admitted into the institutional realms – the policy networks – where policy is determined'.

Therefore, to the extent that experts already dominate policy communities, this implies that the potential for knowledge to act as a major perturbation on networks may be attenuated by the pre-existing resource distributions in policy communities. Thus, Marsh and Rhodes assert that endogenous network change occurs through '*élite* dissensus' (Marsh and Rhodes, 1992a: 260; emphasis added). Lindblom and Woodhouse (1993: 122–3) argue that élite disagreements tend to involve secondary policy issues; they are 'not about challenges to basic aspects of the political and economic systems'. This implies that ideas-driven endogenous network change tends to be a relatively exclusive process, depending on where the network can be found on the policy community/issue network spectrum.

As Bulkeley comments (2000: 732), 'Policy learning occurs within the institutionalised contexts of policy networks, which constrain and also enable new understandings of the policy problem'. Furthermore, it also appears plausible that closed and secretive policy communities dealing with technical and scientific areas of policy can also affect the network-to-context learning dynamic through their control over policy outcome data and how it is interpreted and perceived by external actors.

One potential source of endogenous, knowledge-driven policy change is through the process of 'lesson-drawing' (Rose, 1993). Dolowitz and Marsh (1996) discuss lesson-drawing in the course of a review of the policy transfer literature, where policy transfer is defined as 'the process by which actors borrow policies developed in one setting to develop programmes and policies within another' (Dolowitz and Marsh, 1996: 357). Lesson-drawing refers to a particular type of policy transfer – voluntary adoption of policies from elsewhere as a result of dissatisfaction with existing policies.

However, lesson-drawing tends to be limited to secondary policy change within an existing paradigm, rather than a paradigm shift itself (Rose, 1993: 25–6). It also takes place within the constraints of pre-existing policies, and wider structural constraints which can be institutional, ideological, economic or stemming from social inequality (Dolowitz and Marsh, 1996: 355–6). Dolowitz and Marsh argue that the policy transfer and lesson-drawing literature tends to assume a pluralistic macro-perspective, overlooking potential structural constraints on lesson-drawing. For example, insofar as policy-making takes place in exclusive policy communities across different nations, particularly across international networks of experts or 'epistemic communities', then the lessons drawn are likely to involve a narrow range of ideas and actors, thereby tending to reinforce the status quo rather than open up policy-making. In this vein, pressure groups are said to be able to influence lesson-drawing if they have some of the resource attributes of insider groups: technical knowledge useful to policy-making and implementation, and significant political resources (Rose, 1993: 56).

Finally, lesson-drawing presupposes that the 'new' policies, or ideas, under consideration exist somewhere else in time and space. In the face of genuinely innovative sources of dissatisfaction, or perceptions of dissatisfaction from attentive publics that have no pre-existing access to policy networks in any polity, then there is no experience upon which lessons can be drawn. To the extent that policy-makers prefer not to take the risk of being the first to trial a new policy (Rose, 1993: 24–5), this may pose an additional constraint on innovative policy ideas. In summary, there are many potential constraints on the breadth of lesson-drawing.

This discussion of learning models has focussed upon their effects at the meso-level of network structures and interactions. However, it is individual agents who interpret ideas and information, albeit within a more or less institutionalised structure. Therefore, in order to understand cognitive impacts on policy, it is necessary to consider the micro-level of analysis that focuses on individual agency within a structured context. This is discussed below. However, understanding the evolution of policy networks and outcomes requires consideration of administrative institutions as well as ideas (Jordan and Greenaway, 1998: 670; O'Riordan and Jordan, 1996: 81). Therefore, it is time for a discussion of the policy implications of bureaucratic and implementation structures.

Bureaucratic and implementation structures

It was noted above that a policy network's structure influences how it mediates exogenous factors and initiates policy change. Implementation structures can be an important component of the network structure: differences in the membership and interrelations between policy formulation and policy implementation may insulate implementation structures from formulation networks, leading to a failure to implement intended policy changes: an implementation gap (Marsh, 1998b: 192; Marsh and Rhodes, 1992a). Smith (1997: 21) notes that attention to implementation structures is particularly important 'where regulation involves considerable...discretion to set standards within a legislative framework carrying vague statutory principles'. This implies that implementation may play a significant role in animal research policy. In such circumstances, 'the implementation process becomes a continuation of the formulation process, involving negotiation and bargaining between multiple, resource interdependent actors' (Smith, 1997: 28). The dialectical interaction between exogenous factors and the implementation network, which can affect network actors' resources, may have a significant impact on the pattern of outcomes.

For example, Hill (1997: 144) suggests that where the resource interdependency between regulator and regulatee favours the latter, they can evade the intentions of policy formulators, thus promoting regulatory 'co-production' between implementers and regulatees in order to achieve 'voluntary compliance'. One instance can be found in pollution policy: 'policy is no more than the terms that the regulator is able to reach with the regulatee'. As Smith's (1997) study of pollution policy has shown, if an insider group dominates policy implementation through its monopolisation of technical knowledge, then it is in a strong position to neutralise or mitigate perturbations inimical to their interests.

Social and ideological factors are also important. Hill (1997: 165–77) proposes that the social and professional relationships between implementing officials and regulatees will inform officials' activities through shared membership of professional groups or career paths. The cohesion of relationships with state policy-makers is a key resource for such group actors (Daugbjerg, 1998: 88–9). The closer the relationships, the more likely it is that implementation will reflect the interests and ideologies of regulatees (Hill, 1997: 191). Exogenous structures, such as a postulated tendency towards capital accumulation, may also affect network relationships and hence implementation (Hill, 1997: 209).

These network relationships, network structures and exogenous structures are likely to affect the exercise of discretion by implementers, potentially leading to profound effects on the translation of new laws and rules into policy outcomes that may indicate 'regulatory capture' by regulatees (Hill, 1997: 218). Complex, or 'polycentric' policy areas involving 'co-produced' expert judgments on a range of interacting factors pose particularly acute problems for the control of discretion. Additional considerations that complicate rule enforcement include the capacities of grievance procedures and the existing constitutional and legal system to prevent the arbitrary exercise of power by government bureaucrats and politicians (Hill, 1997: 178).

Hill suggests that one useful way of categorising implementation in relation to wider social concerns or the intentions of policy formulators is to adopt the 'model of justice' typology. In particular, policy formulation may envisage a 'moral judgment' model where the exercise of discretion aims for conflict resolution between competing social values, and prioritises fairness and independence as rules of the game between implementers and affected interests (Hill, 1997: 184–5; 211). The 'moral judgment' model therefore corresponds to a pluralistic, issue network type of policy-making structure.

However, an implementation gap may appear if the model applied at the frontline resembles the 'professional treatment' model, which, it is said, 'calls for the application of specialist skills in complex situations' (Hill, 1997: 185). Because of the political power of professions, stemming from their claims to exclusive expertise, self-governance and hence successful demands for occupational autonomy (Hill, 1997: 207), it is reasonable to hypothesise the evolution of a 'professional treatment' model in areas of policy involving scientific and technical expertise, such as animal research. This model appears to correspond to an élitist, policy community-type of network. Where unified professional worldviews and implementation network resource distributions are divergent from the spectrum of social values envisaged in the 'moral judgment' model, the exercise of discretion in the interpretation of ambiguous rules may promote implementation gaps. This may be exacerbated in situations where the conflict of interests between the source and 'victim' of regulated activities is acute, and the 'victims' are unable to complain effectively. Secrecy reinforces the difficulty faced in holding implementers to account for the adequacy of their discretionary activity.

So far, this analysis of endogenous policy network dynamics has tended to emphasise the constraining and facilitating role of network structures. But, as the introduction to this section argued, it is essential to recognise the role of network actors in interpreting their structural context and undertaking strategic action in pursuit of their goals. Thus, the next sub-section elaborates on the role of agents in policy network approaches.

The role of agency

Marsh (1998a: 12) argues that it is essential to incorporate the role of agents into explanations of network and policy change, and to recognise that their relationship with structures (both network/endogenous and macro/exogenous) is 'dialectical':

> Clearly, the context within which a network is located affects the shape of the network and the behaviour of the agents in the network. However, it is the agents who have to interpret that context and their behaviour is not determined by that context. In addition, the behaviour of the actors affects both the structure of the networks and the broader context within which the network operates.

Thus, as the policy networks literature has developed, it has attempted to redress the perceived over-emphasis on structures in previous policy network literature, by recognising a role for agents and their room for manoeuvre in interpreting and altering their structural contexts (e.g. Bevir and Richards, 2009). Reflexivity regarding the relationship between structure and agency is also important to guard against the converse problem of intentionalist approaches, which stress the role of agency as the fundamental explanation of political phenomena (McAnulla, 2002: 276–8). This flaw is said to be associated with certain pluralist approaches that neglect the constraining and enabling role of structures (McAnulla, 2002: 278), thereby eliding the possibility of related resource and power inequalities between actors. Instead, the dialectical approach argues that an actor's resources are 'a reflection of the structural position occupied by the group which the actor represents' (Marsh, 1998b: 193).

Nevertheless, in the dialectical network account, agents' political behaviour occurs through their interpretation (within cognitive limits) of exogenous or structural conditions '...in the context of the structures, rules/norms and interpersonal relationships within the network' (Marsh, 1998b: 197). Later, Hay and Richards (2000: 8) take this theme forward and introduce a more dynamic element when they propose that network evolution occurs: '...through strategic learning on the part of network participants as they revise their goals in the light of changing perceptions of what is feasible or desirable'. For example, as the GM crop policy community came under exogenous ideological and political pressure in the late 1990s, government actors in the network decided it would be preferable to reconcile biotechnology interests with ecological concerns. This required changes in network membership, which affected the strategic intentions and actions of the network actors, including those previously excluded (Toke and Marsh, 2003). Thus, the biotechnology industry decided to agree to a moratorium on commercial planting pending limited trials of the crops, while newly-admitted GM-sceptical groups, such as the RSPB – by virtue of their membership of the scientific advisory committee – came to explicitly accept the conduct of field trials of GM crops, despite fears that such a position might cause the charity to lose significant numbers of members opposed to any open-air planting of GM crops.

The extent to which each actor or coalition modified its goals will have depended at least partly[14] on their strategic calculations regarding the potential impact of their skills and resources in the context of the constraints and opportunities they faced. However, it could therefore be argued that, within any network, the extent of an actor's autonomy to modify their goals and pursue them successfully will depend not only on their skills in interpersonal interactions, but also on their network and exogenous structural context (though mediated by their interpretation of those related structures) and the extent to which these enable or constrain the realisation of their core policy beliefs. This proposition would appear to have some affinity with Hay's (1995: 200) conception of the relation between structure and agency, particularly this aspect: 'Action settings can be conceived of in terms of a nested hierarchy of levels of structure that interact in complex ways to condition and set the context within which agency is displayed'. These structures are said to be 'strategically selective' in that they favour some strategies over others and '[t]hey provide resources and opportunities for the powerful, while simultaneously constraining the powerless and the subordinated' (Hay, 1995: 206).[15] In the case of GM food policy, environmental groups whose core position was for a ban on the cultivation of GM crops found their policy network-related action severely constrained by the structural context, which could be characterised as involving a capitalist economic structure and a government-perceived imperative to promote business interests (in this case, of biotechnology companies) and thus the economic health of the nation. Anti-GM groups would have increased scope to act within a lower level of structure by accepting the higher structural constraints as given. On the other hand Hay's point about the interaction between the different structural levels may manifest itself in the way such groups might be regarded and positioned as outsiders by other actors, despite any willingness to argue within the network's 'rules of the game', and potentially problematic relations with supporters committed to an absolute ban while the group pursues incremental change.

Thus, within a policy community, it appears likely that insider groups enjoy relatively high autonomy, while any peripheral insiders lack autonomy because of the higher level of constraint imposed by the network's ideological structure. Indeed, peripheral insiders' most powerful strategic action to influence policy outcomes could be the threat to withdraw themselves, and hence their legitimacy resources, from the network. A good example of this is reported by Toke and Marsh (2003: 240) in relation to the successful threat of the RSPB to resign from a committee unless a biotechnology company were to cancel plans to trial GM crops near an organic research facility.

However, as apparent members of the network, it is also possible that a peripheral insider's perception of the range of 'politically-realistic' outcomes as constrained by the network's appreciative system, and their sense of inclusion, may combine to privilege their discursive construction of events that legitimates the network and their membership (McAnulla, 2002: 283; Hay and Richards, 2000: 12–13). Alternatively, in issue networks, autonomy is more evenly spread among members, reflecting the less structured nature of the network. Actors' skills may have greater impact than comparative resource constraints in such networks, a dynamic associated with pluralistic views of power. It is therefore arguable that there is a generally greater scope for some degree of policy transformation through agency in issue networks.

The lack of outsider groups' autonomy in a policy community raises the question of how they can act to affect policy. Following Baumgartner and Jones, Richardson (2000) suggests that groups excluded from policy community-type networks can nevertheless try to affect policy by utilising two external pressure mechanisms, namely public opinion or ideology, and alternative venues such as Parliament and the courts. For example, he argues (2000: 1012) that outsider environmentalists successfully disrupted road-building policy, which had previously been dominated by a pro-road policy community centred on the Department of Transport, by mounting direct protest and exploiting the Public Inquiry process for proposed new roads to directly interfere with road-building, thus politicising the issue through public debate of the ideological justification for road expansion. It could also be argued that by altering the exogenous context, such actors may induce change in the policy network, which inevitably is dialectically related to its context. This raises the further question of whether certain dimensions of the exogenous context tend to have a greater or lesser impact on the policy network.

Thus, there may be various constraints and opportunities to the impact of agents' use of alternative venues, depending on the relationship between their resources and skills and the hierarchical action settings. As Grant (2000: 138–9) notes in the case of road policy, actors engaged in direct protest may have influenced the media and hence exogenous ideological agenda in certain respects, but they appear to have had little lasting effect on the policy community and the perceptions of its members:

> ...the media is one part of an overall strategy of exerting influence. The media may be particularly important in getting an issue established on the public agenda. However, at a later stage of the decision-making process, different strategies and tactics may be necessary as the group encounters the forces which produce inertia and continuity in political decision-making.

Richardson (2000: 1012–13) also cites the activities of the Countryside Alliance as an example of the use of direct action and alternative venues by traditional elites. It could be argued, though, that this indicates how agency can be employed to try to promote policy continuity as well as change. Indeed, access to the media, public consciousness and the judicial system is significantly dependent on actors' resources, suggesting that well-resourced, traditional insider groups, such as landowners and industrial interests, may well be able to utilise these alternative venues more effectively than outsiders.

Conclusion

The policy network approach has evolved from a relatively static model of politics to one that tries to account for the interactions among exogenous factors, network structures and agents, thereby postulating a dynamic, dialectical policy network approach (Marsh and Smith, 2000: 10). However, although the Marsh/Rhodes typology was initially a crucial tool for understanding the variations in networks and their effects on policy outcomes, there appears to have been little attempt in the later policy network literature to systematically analyse the variability in the way that policy communities and issue networks affect the dialectical interactions that drive the policy process through time.

Nevertheless, it appears that the raw ingredients of such a variable model do exist, and have been brought together in this chapter. Therefore, Table 2.2 below contains a number of postulated tendencies regarding the manner in which policy communities and issue networks interact with exogenous factors or affect the policy processes in a more endogenous fashion. It must be emphasised that these postulations are highly contingent because of the inherent complexity and variability of political phenomena, which is linked with the role of agency and the fact that networks and 'higher' structures are open systems. This therefore represents a heuristic model that maps tendencies rather than formal, predictive relationships, and thus its utility depends on complimentary empirical analysis

The aim of this chapter has been to establish an organising framework to help guide the subsequent examination of the evolution of animal research policy. In the next chapter, these proposed causal influences are compared with the extant studies of UK animal research policy in order to generate a series of research questions to be answered by the historical and case study data presented in Chapters 5–8. Addressing these questions will allow this study's hypothesis concerning the distribution of power in the policy network to be tested.

Table 2.2 Variable policy network dynamics

	Policy community policy dynamics	Issue network policy dynamics
Exogenous dialectics	Dynamic conservatism: can resist or mitigate exogenous change up to a point, depending on network versus external resources.	Relatively open to exogenous forces that cause changes in structure of resource interdependency in the policy network.
Public opinion	• Tries to ignore. • Pre-emptive public relations to legitimise network. • Introduce peripheral insiders. • Professional communities may set agenda, control outward flow of information and its public interpretation. • Secrecy and/or success at pursuing narrow rather than public interests may erode legitimacy, leading to unstable environment.	• Reflects shifts in public mood. • Open, chaotic information flow from network to public domain.
Ideological/cultural	• Potentially more destabilizing. • Depends on resources of excluded ideological critics, such as technical competence. • Power over interpretation of information helps protect existing paradigms.	• New actors and ideologies may move into network relatively freely.
New information and technology	• Technical expertise privileges network truth claims, mediating new information. • Lesson-drawing from similar experts in international epistemic community. • Anomalous information *may* accumulate to change paradigm and hence network and outcomes.	• May lead to broader lesson-drawing and shift in balance of power between competing coalitions.
Economic	• Perceived contribution to national economy proportionate to insiders' autonomy and government dependency, thus promoting policy community stability. • Disappearance of market for members' products may weaken resource structure and destabilise network.	• Actors with increased economic resources may gain some more influence, but this offset by other groups' legitimacy resources, for example.

Continued

Table 2.2 Continued

	Policy community policy dynamics	Issue network policy dynamics
Other policy networks	• May reflect structure of broader, related policy communities. • Resistant to influence from issue networks. • Can influence narrower networks and issue networks.	• Susceptible to influence from other networks, particularly policy communities. • Lack of influence over other networks.
Change in government	• Can resist attempts at network realignment, depending on saliency of policy community's resources for policy-making and implementation.	• Normally relatively vulnerable to transformation.
Endogenous and agency-related dynamics	• Third order change rare; homeostasis normally maintained through adaptive, secondary change.	• Variable, may include radical, third order change. Loose integration between groups and state actors means the former are likely to act outside network.
Ideological	• Consensus structures network and constrains actor preferences. • Policy learning on secondary issues. • Tendency to control outward flow of information, thereby encouraging a stable environment through depoliticisation.	• Dissensus and broad policy learning leads to fluctuating coalitions and inconsistent outcomes on both secondary and core policy issues. • Chaotic outward flow of information potentially destabilising the environment through politicisation.
Institutional	• Enables exclusion of groups with different goals, leading to policy stability. • Formulation institutions may change incrementally while implementation structures remains constant to stabilise outcomes. • Resource distributions may affect state actors' ability to change rules of the game. • Cohesion breeds stronger relations between state actors and insiders, promoting stability.	• Open, ease of access contributes to diverse and sometimes contradictory outcomes. • In young networks, powerful interests may institutionalise their relationships with state actors in order to block threats to their interests.

3
The 'Animal Research Issue Network' Thesis: A Critique

Introduction

This chapter applies the insights presented on policy network analysis to the current state of knowledge regarding animal research policy, in order to identify outstanding research questions to be subsequently addressed through analysis of secondary data and primary case study data.

The contribution of political science, particularly public policy research, to the study of animal experimentation policy has been minimal (Garner, 2002: 395). The only major extant animal research policy study has been undertaken by Robert Garner in *Political Animals: Animal Protection Politics in Britain and the United States* (1998) which, therefore, provides the principal starting point for a re-evaluation of this policy area.

The Garner analysis

Garner's 1998 study utilises policy network analysis, particularly the Marsh/Rhodes typology (Garner, 1998: 7), to understand and compare animal farming and animal research policy-making in the US and UK. One of Garner's primary goals is to establish the extent to which these animal protection policy networks can be characterised as pluralistic issue networks. If, as Garner concludes, UK animal research policy displays some significant issue network characteristics, then this implies that the key government institution, the Home Office, would 'take on the role of a neutral arbiter seeking to balance the interests of the groups involved' (1998: 7).

Garner (1998: 7) argues that the type of policy network is primarily determined by the relative quality of access, both informal and formal, that groups have to the state. In particular, he focuses (1998: 13) on understanding personal interactions between group and state actors. In order to obtain empirical data concerning these interactions, Garner (1998: 239) places considerable emphasis on questionnaire data gained from political participants (see Table 3.1 below), which were designed to identify:

- groups that were politically active in the UK animal research policy network (among others)
- issues they lobbied on and how they lobbied
- targets/extent of their activity and influence.

The groups perceived to be most active and influential were then interviewed to explore their questionnaire responses in greater depth. This data is augmented with information from official government and group publications.

As Garner (1998: 13–14) acknowledges, his study is constrained by unavoidable methodological limitations. Firstly, the qualitative data gained from participating actors, whether through interviews, or reviews of government statements and pressure groups literature, may not be the most reliable guides to actors' beliefs, motivations, degree of influence or policy outcomes. Moreover, a lack of documentary evidence makes the empirical task of assessing policy outcomes and, hence, the balance of power in a policy network, 'horrendously problematic'. Indeed, this difficulty appears to be particularly acute in the case of the UK animal research policy network throughout its lifetime, dating back to its origins at the time of the passage of the 1876 Cruelty to Animals Act, as French (1975: 179n8) notes in his study of anti-vivisection agitation in the Victorian era:

> My account of the administration of the [1876] Act is largely based upon Home Office ~156 letterbooks. It is a measure of the sensitivity of the vivisection issue that these documents remain under one hundred year restriction and I am most grateful to the Home Office for permitting me to examine the nineteenth-century letterbooks for purposes of this study.

More recently, Section 24 of the Animals (Scientific Procedures) Act 1986,[1] which succeeded the 1876 Act, makes it a criminal offence for an individual involved in the licensing or conduct of licensed animal research to disclose information gained through their position. In addition, rights of commercial confidentiality are a major constraint on openness, as this author has discovered.

Working within these constraints, Garner (1998: 230, 176) concludes that there is a 'relatively open' or pluralistic policy-making process – 'closer to the issue network end of the [Marsh//Rhodes] continuum' – which has resulted in a legislative framework that tries to take account of the interests of animals and the demands of cause groups who aim to protect them. Maybe because placing networks on the Marsh-Rhodes spectrum involves complex, multi-factorial, qualitative judgements, Garner's 'issue network thesis' should be understood as (at least partly) a relative concept, compared to the types of network found in British farm animal welfare policy and the corresponding two policy areas in the US. That said, as described in this chapter, Garner's analysis also positively attributes some significant 'issue network' and pluralistic traits to the UK animal research policy network. Hence, for these and brevity reasons, Garner's position in his 1998 work will be referred to as his 'issue network thesis'.

This book can be thought of as – in part – a re-examination of Garner's issue network thesis, mainly through an analysis of a case study providing fresh data regarding the balance of power in the animal research policy network. However, the first task is to review Garner's application of policy network analysis to this field. This will allow a series of research questions to be generated that will enable the core hypothesis – that the actual nature of the policy-making area reflects a persistent 'animal research policy community' rather than displaying issue network features – to be tested.

Garner's 'Animal Research Issue Network' Thesis

The network's origins and the 1876 Cruelty to Animals Act: regulating or facilitating animal experiments?

One of the core reasons for the proposed existence of an issue network in UK animal research policy is said to be found in the evolution of the concomitant institutional framework, which invokes the notion of 'path dependency': 'institutional choices made early in the development of a policy area delimit policy choices thereafter' (Lowndes, 2002: 101). Garner's study (1998: 176–7) places considerable emphasis on the notion that the original regulatory framework – embodied in the 1876 Cruelty to Animals Act – was specifically created for that purpose and administered by the Home Office, which had no prior relationship with animal research interests:

> The significance of the legislative and administrative framework created by the 1876 Cruelty to Animals Act lies in the fact that it was devised specifically for the purpose of protecting animals used for research. The Home Office had no prior interest in this issue... There is a marked contrast here with the farm animal policy arena ... [R]esearch interests did not have from the outset the kind of structural advantages that the British farming community had. As a result, it seems safe to conclude that governments have been more able to respond to rising concern about the use of animals in the laboratory and thereby more able to exercise the balancing of interests required by pluralist theory.

However, Garner (1998: 187–8) acknowledges that the 1876 Act also served the purposes of animal researchers, because 'without some legislative framework permitting painful procedures to be carried out on laboratory animals, researchers would be liable to prosecution under general anti-cruelty statutes'. Thus, the question of the real purpose of the 1876 Act and, relatedly, whose interests it served remains unresolved. To understand the relevance of those initial 'institutional choices', it is necessary to address the question: to what extent was the purpose of the Act to regulate or facilitate animal experiments? Thus, the first research question concerns which interest groups benefited from the 1876 Act. Responding to this question involves reviewing the secondary data in order to analyse:

- the ideologies and resource distributions of the groups with an interest in this policy area
- the temporal sequence of inter-relationships between exogenous factors, the policy network and agency
- the evolving network structure and interactions up until the assent of the 1876 Act, in order to categorise the policy network according to the Marsh/Rhodes typology.

The evolution of the network and the implementation of the 1876 Act until 1950: persistent issue network or transformation to policy community?

Following the assent of the 1876 Act, Garner (1998: 56, 177) argues that the animal research lobby has been 'extremely influential' through, for example, its nurturing of a tightly integrated relationship with the

Home Office. This saw the lobby in the form of the Association for the Advancement of medicine by Research (AAMR) successfully ensconced in 1882 as the covert advisory body on the operation of the 1876 Act. Later, in what appears to be the last event in this first period of a politicised animal research policy arena, a second Royal Commission was established in 1906, which resulted in a number of policy changes. Garner (1998: 177) concludes:

> [W]hile the 1876 Act had its serious shortcomings, it did have some impact on the activities of animal researchers, prevented outright abuses and probably discouraging some animal research which might otherwise have been done.

This is said to indicate (Garner, 1998: 177) that pro-animal research interests have not had significant 'structural' advantages. However, one of the striking features of this analysis is the linear extrapolation of a persistent issue network, which rests significantly on an interpretation of the animal research policy network's formation in 1876. Yet, conversely, as discussed in the previous chapter, issue networks tend to be relatively open to exogenous forces and thus unstable in terms of both policy outcome and network structure.

Furthermore, the assertion of an issue network appears paradoxical, given Garner's description of a close relationship between animal research interests and the Home Office after 1882 and the Littlewood Report's comment: '... that the Home Office was not "concerned to assess the potential value of proposed research or the results of past research" but was only concerned to make sure the right certificates were being applied for' (Garner, 1998: 187).

In a previous study of the normative political theory of animal protection, Garner (1993: 126) further questions the impact of the 1876 Act:

> Whatever the details of the legislation, it quickly became apparent that Home Secretaries would rely on the advice of the scientific community when considering applications and so, in effect (despite the creation of the more formal Advisory Committee on Animal Welfare in 1912), scientists still regulated their own activities.

Thus the question is prompted of whether a policy community arose in the animal research case following the assent of the 1876 Act. This is the second research question to be addressed when analysing the evolution of the UK animal research policy network up to the end of the Second World War.

In order to tackle this research question, it will be necessary once again to re-analyse secondary accounts to elucidate the interactions among the macro-level structure, the policy network, the strategic actors and policy outcomes. Key moments of this process include:

1. patterns of macro-level structural power distribution
2. the interaction between the network and other exogenous factors
3. resource distributions between pro- and anti-animal research lobbies
4. actors' learning in respect of initial policy outcomes
5. evolving network interactions and broader political activity in response to that learning, involving the application of their skills and resources and in the context of perceived structural constraints and opportunities
6. the evolving network structure, (institutional, ideological, and rules of the game)
7. developments in policy outcomes, whether maintaining relative stability, or witnessing change of some order.

Post-war politicisation and legislative change: dynamic conservatism or genuine response to public concern?

Garner (1998: 93, 178) argues that, having disappeared from the political agenda at the outbreak of the First World War, a combination of rising public concern for animal welfare and an exponential increase in the scale of animal experimentation[2] led to a re-emergence of the issue in the 1950s. The majority of his analysis focuses on this post-war period, assisted by the questionnaire and interview data collected in the years immediately preceding his work, and the enhanced access to pressure groups' and official publications in this more recent period. Thus, particularly for the period since the mid-1970s, more detailed characterisations are provided of the various groups and coalitions active in animal research policy, which is a crucial factor in policy network analysis. In this subsection, which examines Garner's description and explanation of the change in this policy area signified by the introduction of the Animals (Scientific Procedures) Act 1986, his portrait of the interest group constellations, relations with government actors, and policy proposals will be outlined for key temporal stages of this political process.

From the animal protection side, the RSPCA is identified by Garner (1998: 178) as the only significant group lobbying for policy change in terms of increased restrictions on animal research until the 1970s.

Elsewhere he (1993: 52) attributes to the RSPCA a: 'traditional welfare ideology'. This implies that the welfare of animals has some moral value, but that it is justifiable to sacrifice animals' welfare to achieve significant human benefits, although there is an inevitable element of subjectivity involved in what counts as 'significant'. The RSPCA were sufficiently concerned about post-war increases in animal research to submit requests to the Home Office for an inquiry into the adequacy of the regulatory system established by the 1876 Act (Garner, 1998: 178). This is said to have eventually led to the establishment by the Home Office, in 1962, of a departmental committee of inquiry chaired by Sir Stanley Littlewood, which reported in 1965.

Opposing the RSPCA at that time was an animal research lobby which supported the existing legislation and opposed reform (Garner, 1998: 180). Garner (1998: 47) notes that the primary ideological claim of the animal research community, be it in academia, the pharmaceutical industry or the chemical industry, is that animal experimentation benefits humans because it is essential to, or at least strongly facilitates, the production of safe and effective medicines and other products such as industrial, agricultural and household chemicals. Animal protection groups challenge these claims to varying degrees. However, it could be argued that these pro-animal research claims are not *necessarily* inconsistent with a belief in more stringent regulation; this point is discussed in more detail below.

The pro-animal research lobby was spearheaded by the Research Defence Society (RDS), which had succeeded the AAMR back in 1908 and represented the interests of both individual animal researchers and the British pharmaceutical industry (Garner, 1998: 52–4). Post-1945, the British pharmaceutical industry expanded exponentially and made a significant contribution to the UK economy. By 1980, it employed 72,800 people, and its exports, the Association of the British Pharmaceutical Industry (ABPI) claimed, generated over £450 million towards the country's trade balance (Garner, 1998: 48–9). The weight of influence of the ABPI, which is said to comprise the manufacturers of almost all of the medicines prescribed by the NHS, is indicated by Garner (1998: 48) when he cites Abraham's (1995: 59) observation that during the 1950s and 1960s, 'the Ministry of Health had come to accept the basic philosophy that the export trade of the pharmaceutical industry was so precious that regulation of its affairs was to be avoided'.

Garner does not detail the relationship between the Ministry of Health (and subsequently the Department of Health and Social Security – 'DHSS') and the Home Office, though he notes (1998: 48–9) that the DHSS was a 'powerful ally' of the pharmaceutical industry. Thus, the DHSS informed a 1980 Lords Select Committee (examining a Private Members' Bill to replace the 1876 Act) that it acted as the industry's 'sponsor ... responsible for monitoring the well-being of that industry and for representing its interests within the Government machine'[3]. The Medical Research Council, as a funder of animal research, also gave evidence in support of the practice at this committee (Garner, 1998: 53, 198). Meanwhile, Garner (1998: 27) implies that government autonomy has been eroded as a result of threats from animal research interests to relocate abroad if regulations are tightened: 'The economic consequences of this, particularly in relation to the pharmaceutical industry, is something no government can ignore'. Thus, Garner (1998: 48) concludes: 'There is no question that this had an impact on animal experimentation policy'. The economic resources of the animal research lobby appear to be fortified by the existence of additional British actors with a financial interest in animal research, such as contract research organisations, for example, Huntingdon Life Sciences, who conduct toxicity tests on animals on behalf of pharmaceutical and chemical companies, and suppliers of animals and related apparatus for laboratory experimentation (Garner, 1998: 49–50).

This group situation provides an important context to the 1965 Littlewood Report and the Government's responses. Garner (1998: 178, 1993: 126) notes that the report and its 83 recommendations had little impact for at least a decade. During that period, successive Home Secretaries utilised its equivocal conclusions when informing Parliament of the Government's lack of a detailed response to the report. Thus, while the 1876 Act was noted to be outdated, the report had also asserted that the Home Office had amended its administrative procedures to take account of the vastly increased number and types of animal experiments. Furthermore, the report is said to have downplayed the existence of any electorally-significant public concern about the existing legislative and administrative arrangements. Garner (1998: 179) speculates that the apparent lack of both public concern and any minister with a personal interest in reform may have contributed to the Government's reluctance to devote scarce time and energy to an issue that would stir controversy.

The process that led to the eventual replacement of the 1876 Act with the Animals (Scientific Procedures) Act 1986 is said to have gained decisive momentum in the mid-1970s. In explaining this development, Garner emphasises the impact on government of increasing public concern, a transition which he implies was dialectically related (in the sense used by Marsh) to a more active anti-vivisection movement and critical press coverage of animal experiments: 'A number of initiatives by the animal protection movement in the 1970s were both causes and effects of this changing political climate' (1998: 179). Another factor in this process was the revitalisation and radicalisation of the anti-vivisection movement via the formation of new national groups and increased grassroots activism informed by the 'new' philosophy of animal rights or liberation, as advanced most prominently by Tom Regan and Peter Singer (1976), respectively.[4] Thus Garner (1998: 77) notes:

> plenty of anecdotal evidence that the emerging animal rights movement in the 1970s was bolstered by the work of academic philosophers, such as Singer and Regan, who intellectualised the 'gut' feeling that something was seriously wrong with the way that animals were being treated.

At the same time, older groups such as the British Union for the Abolition of Vivisection (BUAV, founded in 1898) were: 'revitalized by new activists with an animal rights agenda and a harder campaigning edge' (Garner, 1998: 94). A similar process affected the RSPCA, which was: 'dogged in the 1970s by internal disputes as animal rights activists sought, with some success, to change the society's direction' (Garner, 1998: 94). However, as the RSPCA's experience indicates, the emergence of an animal rights ideology not only had a galvanising effect on the animal protection movement, but also opened up divisions between adherents to the welfare ideology and rights advocates. This lack of unity potentially weakens the resources of the animal protection lobby and thus may affect the policy network and outcomes (Garner, 1998: 85). It should be noted, however, that all the main national animal rights groups, including the aforementioned BUAV, the National Anti-Vivisection Society (NAVS) and Animal Aid, did not pursue the abolition of animal research as a short-term policy goal: they were 'pragmatic absolutists', to adapt Garner's typology of animal protection ideology and strategy (1998: 82–3). The pragmatic absolutists have much more limited financial resources than the RSPCA, and are seen by adversaries, other groups and MPs as considerably less important than the RSPCA (Garner, 1998: 103, 95–8).

Garner's account goes on to imply that key animal protection actors' strategic decisions over collaborative lobbying with other groups significantly affected the movement's resources and, hence, influence over the formulation of new legislation. In this vein, the formation of the

Committee for the Reform of Animal Experimentation (CRAE) in 1977 was a key moment affecting subsequent legislative change. This group, which initially consisted of sympathetic MPs, peers, representatives from the RSPCA and other leading animal protection figures, gained access to the Home Office following the submission in 1977 of a memorandum proposing reform, and a perception by the Home Office of the issue's increasing political significance. According to one of the participants cites Garner (Ryder, 1996; cited by Garner, 1998: 179), the involvement in CRAE of the Labour peer Lord Houghton, who had close connections with ministers, is also said to have affected CRAE's thinking regarding the importance of influencing government, and facilitated their access to the executive. Elsewhere, additional repercussions of this evolving situation included a change in the composition of the Home Office's advisory committee on animal experiments to include lay representatives (some twenty years after it had first been proposed by the Littlewood Report, see Garner, 1993: 126) and a manifesto commitment in 1979 by all three major parties to replace the 1876 Act (Garner, 1998: 180).

However, some aspects of Garner's analysis (1998: 187–8) indicate that CRAE's access may have been more a consequence of their conformity to a dominant pro-animal research ideological consensus, rather than being due to any government willingness to broaden the membership and ideological viewpoints of the network:

> the animal protection movement['s] ... participation was conditional upon working within an ideological consensus which was still basically pro-animal research. Thus, at no time did the government give the impression that it thought animal research in general, *or indeed particular types of animal research*, should cease'. (1998: 187; emphasis added)

This assertion raises the issue of how the 'pro-animal research' consensus can be defined. It appears that this concept is ambiguous insofar as it encompasses the distinct positions of, firstly, the goal of policy stability (as indicated by the italicised clause above) and, alternatively, the goal of incremental policy change; i.e. a desire to achieve greater consideration for animal welfare and fewer animal experiments (which would not be consistent with the position of the italicised passage). It therefore appears that, assuming that CRAE did want certain types of experiments to be outlawed (their final negotiating position is discussed in more detail below), the Government may have widened consultative access within the parameters of the broad 'pro-animal research' ideology to endow its deliberations with greater legitimacy, while at the same time maintaining a conservative position within that ideology. The validity of this alternative interpretation of symbolic, rather than substantive, interaction between the Government and animal protection groups will depend on the outcomes of those interactions in terms of legislation and its implementation. Apart from anything else, it highlights the ambiguities and dilemmas faced by reform groups when considering strategic political interaction with government.

Whatever its degree, CRAE's conformity to the dominant ideological consensus appears to have effectively split the animal protection movement, according to Garner's narrative. The terms of the policy debate are said to have conflicted with the opposition of the RSPCA's animal rights-influenced ruling council to any infliction of pain. This partly accounted for the fact that, although formally part of the CRAE coalition, in practice the RSPCA lobbied the Government separately in the run-up to the Animals (Scientific Procedures) Act 1986[5] (Garner, 1998: 104). In this process, the RSPCA's underlying stance to the broad thrust of the new Bill is said to have been one of passive opposition (Garner, 1998: 85–6). The extensive resources and dominant position of the RSPCA within the animal protection movement lends credence to Garner's proposition that, had a CRAE/RSPCA Alliance been forged, the 1986 Act may have favoured the animal protection movement to a greater degree. Given the RSPCA's apparent difficulties with the policy paradigm informing the formulation of the new Bill, it is not surprising that the animal rights groups such as NAVS, BUAV and Animal Aid also withdrew from CRAE to run a separate oppositionist campaign against the Bill. However, as discussed above, their goals for new legislation were incremental rather than outright abolitionist, which reflected their 'pragmatic absolutist' strategy.

Despite CRAE's access to Government ministers, initial relations with ministers and officials ranged from non-committal to outright hostility.[6] What is interesting, but not mentioned by Garner, is that at this point the Government's position appears to have reflected the research community's opposition to change. Furthermore, subsequently improved access for CRAE appears to have coincided with the research community's realisation that increasing public pressure meant that change was inevitable. Thus, once the Government accepted that it would need to reform or replace the 1876 Act, access for CRAE seemed to improve. The relationship between the research community's position, the Government and changing access for reform groups requires more detailed exploration.

Garner (1998: 180–1) argues that as the reform movement gained improved access to government in the late 1970s and early 1980s, there were simultaneous indications that the influence of the animal research lobby on state actors waned. Thus, the Government is said to have opposed a pre-emptive bill, introduced in 1979 by a past president of the RDS, which was designed to preserve existing policy or weaken it in certain respects. While ministers publicly justified their position on the grounds of lack of Parliamentary time and impending European legislation, Garner avers that the Government was sensitive to the public opinion implications of the universal opposition to the bill from all quarters of the animal protection movement. But it could also be argued that any Bill that was perceived to have originated from one side of the debate, no matter what its content, would not be deemed by government to have sufficient legitimacy to be able to command broad credibility and support. Indeed, another private members bill of 1980, sponsored by the RSPCA and said to propose more radical change than that found in the eventual Animals (Scientific Procedures) Act 1986, also failed due to lack of Government support on the stated grounds that it did not represent the broadest achievable consensus. The significance of the Government's public criticism of reformist measures, which contrasts with its neutral response to industry's conservative proposals, needs to be taken into account when analysing this policy network.

Nevertheless, Garner (1998: 181–2) suggests that it was the reformers in the shape of CRAE, rather than the RDS, who achieved increasingly better access to the Government during the late 1970s and early 1980s. While CRAE are said to have been seen as reasonable and willing to compromise, Garner quotes one MP's observation that the RDS had become aggressive and irrational in its opposition to stricter regulation. However, it could be reasonably argued that the comments of a backbench MP in 1977 do not necessarily demonstrate the perception of government through the ten years preceding the Animals (Scientific Procedures) Act 1986, particularly given the change in administration in 1979. It should also be noted that, in addition to its support for the RDS, in 1975 the ABPI had responded to growing public hostility to animal experimentation by establishing its own body to lobby the government on animal research regulation (Garner, 1998: 56). However, the ABPI's interactions with government, which may have been secretive and affected by structural resources, are another significant policy network factor requiring greater attention.

The delay until 1983 of the publication of the White Paper proposing new legislation appears to be indicative of the Government's cautious approach to reform. Indeed, this White Paper fell with the announcement of the General Election that year, and Garner's interview data (1998: 180) elicits the claim that:

> Even as late as 1985 there were some doubts as to whether the bill was going to be in the Queen's Speech (the government believing the bill to be too contentious and time-consuming) and it took numerous meetings between CRAE and the Home Office before the government agreed to include it.

Nevertheless, the 1983 White Paper attributed an important formative role to legislative proposals put forward by a newly-formed Alliance of CRAE and two other groups, the Fund for the Replacement of Animals in Medical Experiments (FRAME: '...formed in 1969 to promote and finance research into alternatives to the use of animals in laboratories' (Garner, 1993: 52)) and the British Veterinary Association (BVA) (Garner, 1998: 183). Garner assumes the validity of this attribution, but it is possible that such acknowledgements were instead motivated by a need to lend the Government's proposals greater legitimacy.

In terms of resources, both FRAME and the BVA offer scientific expertise, thereby enhancing the CRAE Alliance's resources and legitimacy, and hence access to government. However it is significant that, unlike animal research scientists in the pharmaceutical industry, their expertise is not directly linked to perceived economic imperatives. The ideological stances of FRAME and BVA are another relevant factor in understanding the evolution of this policy process. In his 1993 study, Garner describes FRAME's position within the terms of the traditional animal welfare ideology: 'animals should continue to be used until alternatives are found since the potential benefits to humans should not be put at risk' (1993: 52). Similarly, Garner states that the BVA ultimately represents the interests of its members, and further notes that the animal rights perspective is scarce in the veterinary profession, which has been criticised both by animal rights organisations and from within: '... for sacrificing their concern for animals in return for retaining the business of clients involved in animal use' (1993: 53). However, these observations need to be incorporated into the analysis of the policy network.

It was noted above, however, that the 'animal welfare' ideology is ambiguous because of the inherent element of subjectivity involved in conducting a utilitarian cost-benefit assessment. Earlier discussion also suggested that Garner's use of the phrase 'pro-animal research ideology' is similarly ambiguous and, indeed, there appears to be some overlap between the two positions. Furthermore, the CRAE Alliance's access to Government in negotiations over the new legislation appears to have been dependent on positioning itself in that overlap. Although Garner does not detail the position of the Alliance in his policy network study, in his previous work he has stated:

> The negotiating position of the CRAE/BVA/FRAME Alliance was that the infliction of pain on animals in the laboratory should only be allowed in exceptional circumstances when 'it is judged to be of exceptional importance in meeting the essential needs of man or animals'.[7] Further, that there should be a substantial reduction in the number of animals used, that alternative methods should be developed and used wherever possible and finally that those still using animals should be subject to public scrutiny. (1993: 206–7)

The questions of the level of pain that is allowed to be inflicted on animals, and the stringency of the conditions attached to permission to inflict pain, are central to determining how policy outcomes distribute costs and benefits to actors, and hence the balance of power in the policy network. It was also mentioned above that the degree to which CRAE's goals were realised in the new legislation would be some indication of their true quality of access. It appears to be significant that, in the 1983 White Paper, the Government is said to have rejected the relatively tight conditions proposed by the CRAE Alliance regarding this matter,[8] in favour of merely requiring that animals be destroyed once they had reached the stage of severe pain that cannot be alleviated (Garner, 1993: 207). Therefore, this would appear to undermine the validity of the Government's claim in that White Paper that CRAE provided important input. Furthermore, Garner (1998: 185) also notes that the 1983 White Paper envisaged licensing the re-use of animals who had already been subjected to painful procedures, which CRAE and other anti-vivisection groups strongly opposed. It therefore seems at least plausible that the references to CRAE's input in the White Paper were, indeed, largely motivated by the Government's requirements for legitimacy. The additional CRAE goals of reducing the scale of animal research, promoting alternative non-animal research methods and public accountability are discussed below.

Given the Government's apparent reluctance to legislate on animal experimentation, the enthusiasm of David Mellor, the Minister responsible for guiding the legislation through Parliament, seems crucial to the passage of what became the Animals (Scientific Procedures) Act 1986 (Garner, 1998: 184). A major aspect of Mellor's work involved extensive consultation with animal protection advocates; for example, CRAE's Lord Houghton claimed a close relationship with ministers in the drafting of the Bill (Garner, 1998: 183). Garner (1998: 183) also asserts that the Government went to considerable lengths to seek a middle ground between the opposing lobbies. According to one of his interviewees, Clive Hollands of CRAE, during the formulation and passage of Bill: '...the evidence suggests that David Mellor ...was particularly reliant on the advice of the [CRAE] Alliance' (Garner, 1998: 183).

However, it is not denied that the research lobby had some unspecified level of influence (Garner, 1998: 185). Thus, the 1985 Supplementary White Paper included:

- reference to close Government contacts with the RDS 'and a large number of other bodies in the scientific community'
- a new section on scientific and medical benefits of animal research
- two concessions to the research community – a confidentiality clause and the opportunity to appeal against the refusal of a licence.

These pro-animal research measures are of major importance and further challenge the notion that CRAE enjoyed meaningful access to policy formulation. For example, the confidentiality clause severely conflicted with the CRAE Alliance's aim of ensuring public scrutiny of animal research. Nevertheless, for Garner, 'what is striking about the government's pronouncements on the legislation is their perpetual search for balance and consensus' (1998: 184). However, such pronouncements may have been designed to promote legitimacy rather than being candid statements of intent.

Garner (1998: 184) also states that the Government made policy concessions to keep 'moderates' from the animal protection community on board. However, the example Garner (1998: 185) provides, the issue of re-using animals already subject to procedures, does not fully demonstrate this. While the 1983 White Paper allowed re-use, the proposal was removed from the Supplementary White Paper following strong criticism from CRAE and other anti-vivisectionists. However, the clause was re-introduced at the Committee stage by pro-research peers and MPs. Garner notes one MP's significant comment that the amendment would place the CRAE Alliance in a difficult position because they would then be open to attack by anti-vivisectionists opposed to the Bill as 'dupes' or acting in bad faith. Despite this, Mellor abstained, and the vote went against CRAE, allowing re-use at the discretion of the Secretary of State, though formally dependent on strict conditions.

Garner (1998: 182) also claims that 'the more moderate sections of the animal protection movement played a central role in the formulation and passage of the legislation'. The sign of their 'success' was that they had come to be accepted as 'a valid spokesman for a legitimate set of interests involving consultations, negotiations, formal recognition and inclusion'. However, it may be indicative of the balance struck by the Animals (Scientific Procedures) Act 1986 that, during the passage of the Bill, the main opposition came from MPs who, citing the RDS's assertion that it would not affect their activities, felt the Bill did not go sufficiently far towards protecting animals and scrutinising research proposals (Garner, 1998: 183, 191).

A further insight into the membership of the network is provided by the exclusion of those 'pragmatic absolutist' animal protection groups who proposed to ban particular types of experiment (Garner, 1998: 187). However, as Garner (1993: 207) has observed, these types of experiments: '...were strangely irrelevant in anything but a symbolic sense, since these procedures accounted for a small proportion of the total'. Therefore, the fact of their exclusion, in spite of the relatively modest nature of their reform proposals, suggests that the ideological consensus structuring the network might have been quite narrow and might not have countenanced a change in core policy beliefs. Indeed, Garner (1998: 188) acknowledges 'the importance of this ideological consensus as a form of power benefiting those with a vested interest in continuing to use animals'.

Nonetheless, Garner (1998: 185–7) identifies two changes introduced by the Animals (Scientific Procedures) Act 1986 that are said to indicate significant influence on the part of the animal protection lobby. Firstly, CRAE are said to have successfully lobbied for the introduction of a cost-benefit clause, as signified in the Supplementary White Paper, which included the requirement that researchers use severity bands ('mild', 'moderate' and 'substantial') to predict the 'cost' side of their research in terms of the pain and distress likely to be experienced by animals in a proposed research *project licence* application. According to Garner (1998: 187), the significance of the cost-benefit clause in terms of the policy change it represented, and as a guide to the power distribution in the network, was that:

> This cost-benefit clause represents a compromise, a half-way house between a complete prohibition on the infliction of pain – the

position of the RSPCA and, ultimately, what CRAE would have liked – and no restriction on the suffering that can be inflicted – a position realised by the granting of a certificate under the 1876 legislation.

At first sight, the Animals (Scientific Procedures) Act 1986 may appear to have embodied the animal welfare approach and thus offer a framework that has provided the animal protection movement with the opportunity to pursue a strategy whereby certain categories of animal research can be incrementally prevented or abolished (Garner, 1998: 88). However, the Government and network's exclusion of those groups who wished to abolish peripheral areas of research undermines this conclusion. Furthermore, the policy outcomes to emerge from such a model will depend, firstly, on researchers' own estimates of the pain likely to accrue from their proposed procedures and, secondly, on a subjective, utilitarian judgment made by licensing officials as they measure and compare future and, to some extent, uncertain harms and benefits. This is perhaps why any failures of 'animal welfare' approaches to improve the well-being of animals may be, as Garner (1998: 90) suggests, due to 'the political weight exercised by those with a vested interest in exploiting animals coupled with the absence of sustained social pressure for change'.

In this vein, Garner (1998: 192) states that the RDS initially asked parliamentary supporters to lobby against the cost-benefit clause, but then ceased when they realised that the Government's position was set. However, given the RDS's assertions about the lack of impact of the Act, it is likely that they ceased opposing the cost-benefit test because they were confident that it would not be operated in a way that adversely affected their interests. The likelihood of this scenario appears to be reinforced by evidence presented elsewhere by Garner (1993: 207) that opposition to the bill from the pragmatic absolutist groups included the argument that the discretion involved in the cost-benefit assessment 'would leave [the Home Secretary] susceptible to pressure exerted by scientific and industrial interests'. The implication of this is that, *on its own*, the existence of a cost-benefit clause may not represent a compromise between the aspirations of the animal protection movement and the status quo. Rather, the *implementation* of the cost-benefit assessment is a more valid test of the balance of power in the network.

Secondly, Garner (1998: 201) notes that the new advisory Animal Procedures Committee (APC) was the most significant institutional innovation associated with the Animals (Scientific Procedures) Act 1986 because of the increased accountability it appears to introduce into this policy process. This stems from his observation that, between the 1983 and the Supplementary White Papers, animal protection groups managed to introduce an unspecified legal requirement for the representation of animal welfare interests, and a fifty per cent limit on the number of current or recent animal researchers on the APC (1998: 186). Given that a pro-animal research group, the British Pharmacological Society, recommended in late 1970s that the advisory committee be composed solely of nominations from scientific societies (Garner, 1998: 189), this is a plausible claim, though the Act also stipulates that two-thirds of the members must be scientists. In this vein, it should be noted that the opposition to the Bill mounted by pragmatic absolutist groups was based partly on the perception that the APC would be dominated by animal research interests (Garner, 1993: 207). However, Garner asserts that accountability was introduced by the APC through its publication of reports advising the Home Secretary, and therefore, 'if the Home Secretary decides to reject its advice on a particular matter, he has to explain his decision to Parliament' (1993: 127). The effectiveness of the APC in this respect, which is instrumental to the achievement of CRAE's initial aim of ensuring public scrutiny of animal researchers, needs to be ascertained through analysis of events following the assent of the Animals (Scientific Procedures) Act 1986.

Nevertheless, Garner (1998: 200–1) concludes that the introduction of the cost-benefit assessment and the APC through the Animals (Scientific Procedures) Act 1986 indicates that public opinion had a significant impact on policy-making and that animal protection groups had high-quality access to government decision-makers. This is said to denote a policy process that fits the issue network model more closely than the policy community model.

In order to review Garner's interpretation of the evolution of the animal research policy network between 1950 and 1986, it is necessary to compare his approach with the framework developed in the policy network literature review in the previous chapter. It was noted that an important initial task for policy analysis involves categorising degrees of policy change. In order to carry out this task, it is necessary to describe and compare the ideological positions of policy actors and policy outcomes over time, in order to measure policy change and the evolving balance of power in the network.

It was noted above during the discussion of the post-war politicisation of animal research, that Garner's 'pro-animal research' ideological categorisation actually incorporates a range of different policy positions.

Interestingly, Garner refers elsewhere to Orlans' (1993: 22; cited by Garner, 1998: 90) distinction between the 'animal welfare' ideology, which is said to accept regulatory control of practices harmful to animals, and the 'animal use' ideology, which 'recognizes moral responsibilities but favours self-regulation'.[9] Exploring this distinction may provide a clearer image of the ideological conflict between the various groups trying to influence this policy area. In this vein, it could be argued that the research community's opposition to tighter regulation could be said to indicate an 'animal use' rather than 'animal welfare' stance, and this distinction, once explored in more detail, may assist in understanding the various government-group relations, the degree of change in policy outcomes and the power distribution in the network.

There are two further broad lines on enquiry that deserve attention. Firstly, the criteria used to ascertain the network's position on the Marsh/Rhodes typology and to indicate network dynamics. Secondly, the interpretation of the data in their relation to a robust policy network theoretical framework.

In connection with the first question, one of the most noticeable analytical gaps concerns the network's interactions with its exogenous context, which may cause structural factors to be overlooked in favour of agency. This is important because structural and exogenous factors are essential elements in explaining any policy process (Hay, 1995). However, Garner's case rests largely on qualitative questionnaire and interview data, which are inherently difficult to interpret reliably. But if non-observable or structural forms of potential power and influence are considered to play some role in public policy, then additional questions remain to be addressed. For example, what have been the patterns of resource interdependency between the government and the pharmaceutical industry over time?

One exogenous factor Garner discusses is the policy network's relationship with other networks. However, while he notes the health department's role in sponsoring the pharmaceutical industry, and the economic leverage the industry has over government, he also argues that other government departments have 'tended to be peripheral members of the policy network, being subject to, as opposed to being involved in the development of, policy' (Garner, 1998: 27). Interestingly, the concept of an issue network that dominates other policy networks contradicts Bomberg's (1998: 175) conclusions, related above, regarding issue networks' susceptibility to exogenous pressure from other networks.

Furthermore, it appears that most of the analytical focus falls upon animal protection figures rather than animal researchers. This *may* signify a methodological bias insofar as evidence for the alternative hypothesis for this period of the network – a policy community practicing dynamic conservatism in response to potentially destabilising exogenous forces – may be overlooked because animal research members of policy communities are more covert in their lobbying activities, and thus harder to detect, relative to attentive publics or peripheral insiders. In other words, one has to be very careful in interpreting the overt lobbying activities of animal protection groups as evidence of their membership of an issue network.

A further theoretical concern could be raised regarding Garner's interpretation of government pronouncements: his analysis appears to embrace the Westminster model of British government, which is widely regarded as a legitimating mythology of power (see Marsh et al., 2003; Judge, 2004) rather than a factual account of the real nature of power in the political system. In particular, a core aspect of that model is the impression of a government acting as a neutral arbiter between conflicting interests – a concept of the state that is associated with pluralism. To an extent, such an interpretation is understandable, given the lack of available data describing policy outcomes. But the question of power distribution in the British political system needs to be addressed more fully. This is an issue identified by Marsh et al. (2003: 310):

> Far too much work on British politics focuses exclusively on agents and often appears to assume that the playing field on which they compete is even. In contrast, we would argue that to conceptualise British politics more adequately, one needs to start with an appreciation that it is not an even playing field and that there are enduring slopes and gullies which favour some interests over others.

Indeed, Garner's aforementioned comments regarding the self-regulation of scientists under the 1876 Act and the economic influence of the pharmaceutical industry appear to be consistent with two élitist models of British government that contrast with pluralist concepts and the Westminster model:

- Moran's development of the concept of 'club government', which manifests itself in policy communities comprising regulators and associated economic and professional groups (Moran, 2003).
- the asymmetric power model which indicates a form of network homeostasis that stabilises the position of privileged powerful and economic and professional groups (Marsh et al., 2003).

In answer to the second question regarding Garner's interpretation of his data: it would appear, at this stage of the analysis, that there is some ambiguity in Garner's case for an issue network model of animal research policy rather than a policy community model. With the assistance of the policy network dynamics table (2.2) presented at the conclusion of the previous chapter, these points of ambiguity can be summarised thus:

1. If the policy network resembled an issue network model, then one would expect some degree of policy learning and hence policy change to have occurred following the Littlewood Report. However, the Government's response appears to have been inert and consonant with the position of the animal research community.
2. The growing politicisation of animal research may be partly the result of policy outcomes favouring animal research interests (e.g. an exponential rise in animal experiments), indicating the possibility of a pre-existing policy community.
3. The evidence of consultation with certain animal protection figures is insufficient to determine whether they were either *peripheral* insiders in the network that provided legitimation resources to a policy community, or genuine members of an issue network influencing policy outcomes.
4. Furthermore, the CRAE Alliance may not have been broadly representative of the animal protection movement, but rather a small segment nearest the 'centre ground' of the debate.
5. Animal protection groups proposing unambiguous changes to the network, in the shape of the abolition of peripheral categories of research, were excluded from the formulation network.
6. FRAME and BVA, as 'scientific' members of the CRAE Alliance, offered the potential for lesson-drawing by the pre-existing policy network. But how significant are their technical and political resources, which are required to influence lesson-drawing, and to what extent did their expertise lead to the drawing of different lessons to be applied in this policy network?
7. There is no compelling evidence that CRAE achieved a substantial proportion of its aims. These were: animal pain only inflicted in 'essential' circumstances; a substantial reduction of animals used (NB the Government was at this point unwilling to stop any category of experiments); increased public accountability (animal researchers won on the battle over a confidentiality clause, thus this seems to depend on performance of APC); the requirement to use alternatives

(this depends on future enforcement and investment in development and validation of non-animal tests); and the proposed ban on the re-use of animals in procedures.
8. The grievance procedures seem to bias implementation in favour of animal researchers as they alone have the right to appeal against the refusal of a licence.
9. A network ideological structure existed that excluded groups seeking definite policy change, and there is inconclusive evidence of group conflict or changes in elite actors' strategic intentions that is associated with issue networks.
10. The broad scope of discretion contained in the polycentric cost-benefit assessment makes it impossible to ascertain both whether it really does represent a new compromise between the opposing lobbies, and the extent to which policy change has actually taken place. The institutional structure resembles a 'professional treatment' model of implementation, which is associated with policy community-type networks, but the formal requirements of the legislation appear to call for a pluralistic 'moral judgment' model.
11. Answering these questions with any degree of reliability requires examination of policy implementation structures and subsequent policy outcomes.

Garner's narrative of this period of the evolution of the policy network – which represents the current state of knowledge of this policy area – will be reviewed through analysis of the secondary literature. This critical review will seek to clarify the eleven points of ambiguity related above. But the final point highlights the crucial issue: how has the regime first introduced by the Animals (Scientific Procedures) Act 1986, and in particular the cost-benefit assessment, been operated? This issue will be addressed by the primary case study data presented in Chapter 8. To the extent that this question can be answered with confidence, then the position over time of the animal research policy network on the Marsh/Rhodes typology can be deduced. But first, it is necessary to examine Garner's own analysis of the impact of the Animals (Scientific Procedures) Act 1986: what light does he shed on implementation and outcomes?

Implementation of Animals (Scientific Procedures) Act 1986: balancing of interests or symbolic reassurance?

Having deduced the existence of an issue network-type policy arena up to the point of the assent of the Animals (Scientific Procedures) Act

1986, Garner extrapolates to the issue of the Act's subsequent and future implementation. Thus, he (1998: 200–1) concludes:

> The animal protection movement came from nowhere to play a central role in the formulation, passage and administration of the 1986 Animals (Scientific Procedures) Act. The response of governmental actors to growing public concern about animal experimentation and the privileged position granted to a section of the animal protection movement, tends to reveal that the policy community model is not applicable to this policy arena.

To sustain his case that animal research policy continues to be made in a persistent issue network, and that, relatedly, significant changes in policy outcomes that have benefited animals' welfare have been brought about by the Animals (Scientific Procedures) Act 1986, Garner (1988: 188–201) focuses on:

1. the structure and actions of the new *APC*
2. the benchmarks set out by the 'Three R's' approach: a *reduction* in numbers of animals used; *refinement* of the severity of procedures; and the encouragement of *replacement* methods for animal tests
3. the *interaction* between the Home Office and the opposing lobby groups, including enforcement of compliance with the Act

These factors combine to assist an understanding of both how the Home Office operates the cost-benefit assessment, and a retrospective estimation of the costs and benefits that actually accrued, thus providing clues to the balance of power in the network. Garner also refers to further data relevant to the network structure, network interactions and related exogenous factors: group resources; the role of the EU; the distribution of party support and the change in government of 1997; and ideology. By reviewing this data and arguments, it is therefore possible to articulate Garner's description of this policy process and its outcomes along the lines of a dialectical model of policy network dynamics, as outlined in the previous chapter. Hence, it will be possible to assess the validity of his argument for a continuing issue network.

The Animal Procedures Committee

As discussed above, Garner (1998: 201) places particular emphasis on the introduction of the APC, and its perceived contribution to increased accountability, in reaching the conclusion that the animal research policy network resembles the issue network model. In terms of its composition, however (in 1994), he notes (1998: 191) that only four out of twenty members are from an animal welfare background, with one or two additional 'sympathisers'. Nevertheless, he implies that this, combined with the broad discretion enjoyed by the Home Office over the constraints it may place on animal experimentation, contributes to the animal protection movement's significant leverage over the policy process.

Thus, important changes in regulation are said to have arisen from the APC's recommendations to the Home Secretary. These appear mainly to comprise changes in process. Thus, the APC examined all cosmetic and tobacco product testing project licence applications; it was informed about all new project licences using non-human primates; all procedures involving wild-caught primates were referred to APC; and the Committee was consulted on all procedures of 'substantial'[10] severity (1998: 189). Following recommendations from the Committee, octopuses have come under the remit of the Act (in other words, a licence is now required to conduct potentially painful procedures on octopuses), and special justification has become necessary for permission to experiment on primates, with an even higher standard to be reached – 'exceptional and specific justification' – to obtain a licence to use wild-caught primates (1998: 195). Increased accountability is said to have been achieved through the APC's status as a statutory body whose opinions are a matter of public record, and its influence over decision-making that provides a route for the complaints of anti-vivisection groups (1998: 191).

In terms of the actual impact of the Committee, Garner (1998: 196) states that in the most controversial areas of animal experiments – the testing of tobacco and cosmetic products – most of the very few project licence applications of this type (one or two a year out of about a thousand) that have come before the APC have been approved, albeit after significant modification (presumably decreasing the permitted number of animals or the permitted severity of the procedures). It therefore could be argued that the real policy impact of the APC may have been minimal. It is also interesting to note that two other particularly controversial areas of research identified by Garner – household product testing and military experiments – were not subject to case-by-case APC scrutiny.

The Three Rs

Garner also uses the Three Rs perspective to assess whether the Animals (Scientific Procedures) Act 1986 has brought about policy change in favour of animals' interests. Unfortunately, as Garner (1998: 193) notes, his study is hindered in drawing reliable conclusions because the available evidence lacks sufficient detail. For example, he notes that the rate of **reduction** in the number of animals used following the Act was actually less than before its assent. However, the lack of information about licences, and how the Home Office arrives at decisions to approve or reject applications, means that it is difficult to isolate the role of legislation in affecting numbers relative to, for example, broader technological or economic developments.

Furthermore, raw figures are a blunt instrument for assessing overall animal welfare. But once again, attempts to draw a more finessed description, through testing for the refinement of the severity of the procedures, are hampered by the lack of detail in official statistics. Thus, projects are categorised by severity band, which is arrived at through a complex and subjective formula.[11] For example, of the project licences issued in 1993, 629 were given a 'mild' severity band, 758 'moderate' and 21 'substantial'. However, the definitions of the severity bands are vague, and within an individual project, different types of procedures may involve different levels of severity. Garner (1998: 194) concludes, nonetheless, that 'there is no evidence of a downward trend in the most severe procedures which does not suggest that such research is being more critically examined by the inspectorate'.

A similar picture emerges when replacement is examined. Garner notes that little has been achieved in this area, and quotes from a 1991 letter from the APC to the Home Office:

> Failure to make adequate funds available inevitably casts doubt over the Government's commitment to the fundamental principle of the Act that non-animal alternatives are to be preferred to the use of living animals. (1998: 197)

He goes on to recount the observation by FRAME in 1996, who, as observed above, played a major role in negotiations over the legislation:

> the Act itself has rarely been used as the stimulus for such research [and] we are left with the feeling that the main effect of the Act has often been to permit the continuation of what was done before, albeit with higher standards of experimental work and of animal care. (1998: 197–8)

Notwithstanding the vagueness of the available data concerning the impact of the Three R's, there is little evidence that the Animals

(Scientific Procedures) Act 1986 has led to significant policy changes in favour of animal welfare.

Group-state interactions

It was noted above that Garner uses groups' 'quality of access' to government as an important indicator of group influence over policy. The interactions between the Home Office and animal researchers are particularly closely related to the issue of compliance with the Act. This raises questions about the adequacy of the size of the Inspectorate and whether any bias is present. Garner looks at instances of undercover investigations by anti-vivisection groups to explore this issue. One case in 1990 involved unnecessary suffering caused to cats and rabbits by a Professor Wilhelm Feldberg at the Medical Research Council's (MRC) laboratory in Mill Hill, London (MacDonald, 1994). The MRC's own investigation blamed the Home Office for regulatory failure, but in response the Home Office stated that 'the aim of the inspectorate is not to police the Act but to offer advice and information' (Garner, 1998: 198). Coupled with evidence from further exposés, Garner acknowledges that the revelation of infringements not detected by the Inspectorate demonstrates a 'barely adequate' level of staffing (19 Inspectors for 2.8 million procedures in 1994) and an ideology of self-regulation which fails to prevent regulatory breaches. In this vein, Garner also notes complaints from the APC regarding ignorance and disregard for the Act among licensees.

On the other hand, despite dissatisfaction from animal protection groups about Home Office responses to allegations of regulatory failure, the fact that the APC and the Home Office took any disciplinary action is said to indicate a positive attitude by the Home Office to animal protection groups, and that they 'play a full role in the administration of the legislation' (Garner, 1998: 199–200). Garner (1999: 82, 200) also cites the occurrence of a meeting between the Home Office and the BUAV concerning two undercover investigations as evidence of the group's attempt to establish insider status and of animal protection group influence. However, it is noteworthy that the BUAV were dissatisfied with the Home Office response, and that the perceived cause of damage to the accused company involved was negative media coverage rather than regulatory action. Similarly, further questions could be raised regarding whether *ad hoc* interaction between the Home Office and animal protection groups is really indicative of a substantial role for animal protection in the administration of the Act: an insider strategy from a group is no guarantee that they will gain genuine insider status in the policy network (Grant, 2000: 22).

The nature of infringement action taken by the Home Office also seems likely to impact the adequacy of compliance with the 1986 Act. In this regard, Garner (1998: 200) notes that penalties, such as requiring changes in the management of establishments or occasional revocation of licences, are relatively weak. There had been only one known prosecution under the Act: of an unlicensed rabbit dealer rather than a scientist. Garner notes that this situation has attracted some criticism from the APC. Interestingly, Garner (1998: 200) suggests that 'the failure to take legal action in some cases does give the impression that the Home Office, and, indeed, society itself, does not take animal abuse in the laboratory seriously enough'.

The operation of the cost-benefit assessment

The assumption that the Home Office acts as a cipher for social values seems to inform Garner's evaluation of the Home Office's weighing of costs and benefits of research proposals, and hence the balance of power in the network. Therefore, although Garner acknowledges concerns about the wide scope of discretion afforded the generally pro-animal research Inspectorate, he maintains that the policy process is not dominated by the animal research community (1998: 192). However, Garner (1998: 195–6) also identifies frequent, extreme bias against animals' interests in the Home Office's operation of the crucial cost-benefit assessment. Overall, Garner's characterisation of the network, its interactions and its outcomes does not provide a convincing basis for the existence of an issue network-type policy network.

Network structure, network interactions and exogenous factors

Garner also provides further data relevant to the network structure, network interactions and related exogenous factors which indicate the nature of this policy arena. In relation to actor resources and the structure of resource distribution, Garner (1998: 67) states that pro-animal research groups have developed formidable lobbying and public relations efforts that are paid for by extremely wealthy interest groups. These substantial financial resources are combined with a captive organisational structure, shared purpose, high-status participants (e.g. eminent scientists) and major economic and health benefit claims, to create a set of actors who have the potential to exert some considerable influence. He also observes (1998: 56–7) that from the mid-1980s, the animal research lobby decided to deploy these resources to public relations as a reaction to the growing politicisation of the issue and animal protection campaigning. Garner (1998: 53) describes these groups 'as a reserve army of organisations ready to join the fray if necessary'. Thus, the RDS set up a specific public relations arm, the Biomedical Research Educational Trust. Also in the mid-1980s, the ABPI established its own campaigning division, the Animals in Medicines Research Information Centre (AMRIC). In the early 1990s, the British Association for the Advancement of Science, the British Medical Association, and two new campaigning groups, Seriously Ill for Medical Research and a coalition of the major medical research charities (the Research for Health Charities Group), also became engaged in the public debate to argue for the benefits and necessity of animal experimentation. Garner does not detail any political lobbying activities on the part of these organisations, though it is noticeable that many individuals linked to these groups were members of the APC as of the mid-1990s (Garner, 1998: 190).

In contrast, the animal protection movement appears to be dogged by 'severe organisational, financial and ideological problems' (Garner, 1998: 108). Those animal protection groups striving to keep the issue on the political agenda, such as the BUAV and NAVS, have not been directly represented on the APC and control far fewer financial resources than groups such as the RSPCA, whose staff have been appointed to the Committee.

Turning to the potential exogenous influences on the network, Garner (1998: 27) states that the EU's impact has been marginal, with EU legislation[12] merely establishing relatively weak minimum requirements that all member states must achieve (1993: 124). It therefore appears that,

Table 3.1 Key group participants engaged in British animal research politics in the mid-1990s (compiled from Garner, 1998: 52, 96–7)[a]

Animal research interest groups in Britain	Animal protection/anti-vivisection groups
Research Defence Society[b]	RSPCA
Association of the British Pharmaceutical Industry (ABPI)	British Union for the Abolition of Vivisection
Medical Research Council	National Anti-Vivisection Society
Cosmetic, Toiletry and Fragrance Association	Animal Aid
Research for Health Charities Group[c]	Fund for the Replacement of Animals in Medical Experiments (FRAME)
	Advocates for Animals

[a]This was compiled by Garner firstly, through a reputational survey of animal research groups, animal protection groups and MPs; and secondly, through interview data and study of responses to committees.
[b]Now known as 'Understanding Animal Research'.
[c]Merged into the 'Association of Medical Research Charities' in 1997.

unlike in pollution policy (Smith, 1997: 121–2), the EU has not exerted pressure on Britain to change policy in favour of tighter regulation in animal research. Indeed, the evidence cited by Garner implies the opposite scenario. EU legislation stipulates the use of animals for product testing, and attempts by the UK to act unilaterally while protecting domestic industry are hindered by the single market, which prevents a ban on the importation of products from countries with weaker regulation (1998: 27). Furthermore, the pharmaceutical and biotechnology companies are powerful lobbying forces at the EU level (1998: 53, 65).

Parliament is implicitly considered by Garner (1998: 109–10) to be mainly exogenous to the animal research policy network because it is said to have politicised the issue and affected the political agenda, rather than having had any direct impact on policy-making.[13] Thus, the two Private Members Bills in 1979 and 1980 that sought to replace the 1876 Act had the effects of politicising animal research policy and forcing the Government to respond by offering future government legislation as a preferable alternative.

In order to assess the existence, intensity and distribution of MPs' commitment to animal welfare, Garner (1998: 116–21) applies a scoring scheme, based on whether MPs have tabled Parliamentary Questions (PQs), Early Day Motions (EDMs)[14] or Private Members Bills during the 1985–94 period that seek to enhance animal welfare. Interestingly, although 277 MPs (out of 650, minus ministers and their shadows, who, by convention, do not engage in these actions) were found to have shown commitment (as defined by Garner) to animal protection issues, only 36 are said to have exhibited 'moderate' or 'extensive' commitment, the minimum requirement for which is to table two EDMs and two PQs. Out of the 277 MPs, there is a slight bias towards Labour and the Liberal Democrats.[15] But there is a more acute bias towards Labour among the most active 36 MPs: 24 were Labour, 3 Liberal Democrats, and 9 Conservatives. From these figures, Garner (1998: 121) concludes that 'Labour's resounding victory in the 1997 general election, therefore, is encouraging for animal advocates'. However, in respect of animal research, it should be noted that, barring a small minority, the Parliamentary Labour Party supported the Animals (Scientific Procedures) Act 1986 (Garner, 1993: 207). Whether Garner's projection of enhanced animal welfare consideration after 1997 was borne out in the animal research policy network is a matter for the empirical analysis that is undertaken in Chapter 8. This will, in turn, shed further light both on the type of policy network in this domain and the relationship between shifting distributions of parliamentary support and policy network dynamics.

Garner's observations of the ideological parameters within and beyond the network also form part of his argument for animal research policy-making in an issue network-type context. For example, he argues that the dominance of the pro-vivisection ideology was waning, as indicated by the emergence of the abolitionist welfare themes of banning cosmetic testing or military experiments onto the policy agenda (1998: 188). Furthermore, he perceives greater concern over experiments causing severe suffering, whatever the predicted benefits, and a higher degree of scepticism regarding the efficacy of animal experiments. Thus, Garner (1998: 201) predicted: 'The responsiveness of the policy network means that public pressure may well lead to further restrictions on animal research in the future'.

In relation to Garner's description of the policy network, it was noted above that his key argument – that the accountability introduced by the APC has contributed to an issue network model of this policy network – contained ambiguities because animal welfare sympathisers have been in a minority on the Committee, and the impact on policy outcomes seems to have been marginal. This raises a number of questions that will receive further analysis in the course of the case study:

- Has the APC's composition been representative of public opinion?
- What are its powers, and what is its relationship with group and state actors in this network?
- How has it reacted to allegations by animal protection groups of non-compliance or maladministration?
- What has been its impact on the implementation of the cost-benefit assessment and, hence, policy outcomes?

It was also noted that there were little precise data available to Garner to help measure the impact of the Act in terms of the 'Three R's'. This creates a dilemma, as revealed by the table (2.2) setting out variable policy network dynamics in relation to potential sources of change. In particular, the introduction of new actors into a network in response to shifts in public opinion is common to both policy community and issue network models of policy-making. The difference between the two models is to be found in the variability between the intra-network arenas of such access and in their effect on policy outcomes. Thus, in policy communities under stress from politicisation, new actors may only be peripheral, invited into consultation and formulation exercises to bolster the legitimacy of policy, but excluded from implementation processes which remain dominated by traditional insider groups to prevent core changes in the pattern of policy outcomes. While in issue networks, no group has a persistent institutionalised privilege in terms of influence and access, which is thus reflected in fluctuations in both access and policy outcomes. In other words, in order to discern the difference between issue networks and policy communities, it is necessary to focus on implementation structures and policy outcomes.

Indeed, Garner's own observations suggest that from what little information is available, policy outcomes appear to have remained relatively constant, thereby raising questions about whether the animal protection movement really did manage to enter the network and subsequently participate in the implementation of the Animals (Scientific Procedures) Act 1986. Nonetheless, the case study data, discussed in more detail in the next chapter, fulfil many of the hiatuses identified by Garner in pre-existing evidence relating to project licence applications and their scrutiny, and what the severity bands mean in practice in terms of harms to animals.

The case study will also offer a more detailed insight into the quality of access enjoyed by actors from the opposing lobbies, both during the licensing process and in dealing with grievances submitted by animal protection groups. This data will aim to establish whether apparent attempts at gaining insider access, which is said to be demonstrated by meetings between animal protection groups and the Home Office, have achieved any concrete success in terms of obtaining membership of the network and having an impact on policy outcomes.

Consideration of these empirical issues also facilitates an appraisal of the pluralist assumption that the Home Office is responsive to broader patterns of social values. This assumption is arguably reflected in the following assertions:

- that regulatory officials will act with 'professionalism and genuine intent' (Garner, 1998: 233).
- that because the Home Secretary has the powers, under the Animals (Scientific Procedures) Act 1986, to substantially control and prohibit animal research, that s/he is therefore open to lobbying by the animal protection movement to exercise that control (1998: 192).

A related point made by Garner is that the change in government and composition of Parliament in 1997 was likely to result in changes to the policy network and outcomes. In other words, it is asserted that the government makes and implements policy from the position of a neutral arbiter based on the strength of public support enjoyed by the competing factions. These considerations combine to underpin Garner's conclusion that policy outcomes in animal research reflect a relatively open issue network. Therefore, one indicator of the validity of this thesis will be the impact of the change in government in 1997: evidence of network and policy stability would tend to undermine the postulated issue network model.

However, the literature concerning implementation and bureaucratic structures challenges the assumption that the Home Office reflects public opinion, and thus undermines the suggestion that any neglect of animals' interests in the operation of the cost-benefit assessment must be a direct reflection of the wider spectrum of public opinion. Instead, even if it is accepted that Home Office officials do not act in a consciously biased manner, it is possible that structural constraints may affect implementers' exercise of polycentric discretionary judgments, as discussed above in the review of the policy network literature. These constraints may include exogenous and endogenous network structures, such as a dominant consensus or appreciative system and its position on core normative issues[16] that affect the operation of a cost-benefit assessment, which will be discussed in more detail below. Additional factors may include: macro-level resource distributions between major economic actors such as the pharmaceutical industry and government; any structural and resource inequalities which bias access to policy-making in favour of certain categories of actors; resource distributions and dependencies within the network between implementers and regulatees; established 'rules of the game'; and any shared membership of professional groups by inspectors and inspected. The aim of this work is to use the case study to explore these key themes.

Another crucial task of such an empirical analysis is to address the information gaps identified by Garner that relate to the definitions of severity bands attributed to procedures and what they mean in practice in terms of the actual harms inflicted on animals. The relationship between official descriptions of experimental severity and the true policy outcomes will shed important light on whether the Home Office acts as a cipher in a pluralist or élitist sense and, relatedly, whether the issue network or policy community model is more relevant to the animal research policy network.

If there is congruence between official statements regarding regulation and the policy outcomes, or, to put it another way, the policy outcomes reflect the balance of public opinion rather than the distribution of other types of political resources, it would tend to support Garner's issue network thesis. That would raise interesting theoretical implications not just for policy network analysis but also for other work indicative of an élitist model of the state. Garner's own analysis demonstrates that animal research interests enjoy significant economic and professional resources, while animal protection groups do not represent economic interests. However, the conventional wisdom of both policy network analysis as exemplified by the Marsh and Rhodes typology, and the asymmetric power model as developed by Marsh et al. (2003), is that such a policy network would tend towards the policy community model. This appears to be reinforced by two observations of exogenous pressures. Firstly, unlike environmental policy, where the EU is said to have undermined the cohesiveness of established policy communities, in this case it appears to have had a relatively minor impact on this network. Secondly, Parliamentary participation – a sign of politicisation and issue network-style policy-making – has been sporadic and limited to legislative formulation. Therefore, if Garner's thesis is confirmed, it may be necessary to critically review the policy network and asymmetric power models.

Of course, the requirement to disaggregate must be borne in mind, and therefore any such attributions in animal research policy may be of limited relevance to other policy networks, though the scope of any valid generalisations remains to be addressed. In any case, the primary task involves an empirical evaluation of the implementation of the Act and network structure and interactions (e.g. the role of the APC and group-state interactions), along the lines outlined above in relation to the fourth research question. But, given Garner's assertion of a strong relationship between the circumstances of the inception of the policy network and its latter-day characteristics, the forthcoming analysis of the network must proceed from the beginning, with the events culminating in the assent of the 1876 Cruelty to Animals Act.

Conclusion

This chapter has reviewed the literature dealing with the evolution of British animal research public policy, as represented by Robert Garner's groundbreaking 1998 study. There have been two broad, related aspects to this review. Firstly, Garner's utilisation of a policy network analytical framework has been critically compared to the dialectical model developed in the previous chapter. Secondly, an assessment has been conducted of the data aduced and interpreted by Garner to sustain his 'animal research issue network' thesis. The purpose of this analysis has been to develop a series of research questions that are designed to test the alternative hypothesis proposed in this study; that animal research policy has tended to be made in a policy community-type of network. The research questions thus provide an analytical framework for the subsequent empirical chapters.

In relation to Garner's theoretical approach, this review has noted that his attribution of an animal research issue network through to the present day was based, to a significant extent, on his perception of the network structure and structural context at the time of the passage of the 1876 Cruelty to Animals Act. As a general concept, 'path-dependent' policy-making is plausible. Therefore, the research questions are generated chronologically, corresponding to the key moments in the history of this policy area as identified by Garner.

However, further research is essential to overcome the paucity of data regarding network interactions and policy outcomes, and to investigate the network's relationship with its structural context. Therefore, the four research questions will seek to elucidate these phenomena in greater detail, in order to build a richer narrative regarding the dynamics that have driven the evolution of the policy network up until the present day.

In order to respond to these research questions, subsequent empirical chapters will, first, re-examine the secondary and tertiary literature on the evolution of the network. Furthermore, and most significantly, analysis will be conducted of unique primary data relating to a recent animal research programme, and the network's response to external challenges related to this research. Meanwhile, the next chapter explores methodological issues and explains the relevance and validity of the data in relation to the research questions and, hence, the present 'animal research policy community' hypothesis.

4
Theory and Method in the Study of Animal Research Policy

Introduction

The previous chapter noted that the existing political science on British animal research policy, as represented by Garner (1998), employs policy network analysis to examine the nature of power distribution in this particular policy field, concluding that the policy area displays issue network characteristics. However, the previous chapter raised questions about this conclusion and produced a number of outstanding research questions that need to be addressed in order to provide a richer understanding of this policy process. Before these research questions are examined later in this book, this chapter will examine the epistemology implicitly embraced by Garner. Through this, the justification for a critical realist methodology will emerge, which will underpin this study. There then follows a discussion of the constraints on, and opportunities for, accessing relevant sources. The latter parts of the chapter will outline the validity of the available data and reflect on the limitations of the chosen methodology.

Towards a critical realist methodology

Applying new institutionalism

This preliminary section briefly outlines Garner's general theoretical framework in order to provide the basis for a detailed examination of his underlying epistemology and methodology.

In the broadest sense, Garner's 1998 study implicitly adopts a new institutionalist approach to studying politics. Thus, he sets out to describe the legislation and associated administrative structures that have shaped the nature of animal welfare in research laboratories in Britain[1] as an essential basis for understanding this area of policy-making (Garner, 1998: 16).

There are two additional aspects of Garner's approach that associate it with new institutionalist theory. Firstly, the previous chapter noted that Garner traces a linear path of development starting from his perception of the network at the time of its formation in 1876, which implies an 'emphasis on path dependency resulting from key historical choices made by states' (Burnham et al., 2004: 19). Path dependency is a core tenet of the variant of new institutionalism known as 'historical institutionalism', which is predicated on a view that choices made in the past have shaped the future evolution of a policy. Secondly, policy network analysis, which is Garner's particular analytical tool, is said by Rhodes (1997: 78–9) to represent a 'new' institutionalist approach to the policy process whereby a traditional, descriptive focus on the formal institutions of the state is augmented by an analysis of non-state actors and organizations, and, most importantly, the structured relationships among all these entities. Similarly, Lowndes (2002: 94, 99) notes that policy network studies reflect the response by political science to the fragmentation of the state and the recognition of the role of informal institutions in the policy-making processes.

Indeed, one of the most significant aspects of new institutionalism is its rejection of the formal-legal approach often favoured by traditional institutionalism.[2] Formal institutions are no longer treated as the basic independent variables that determine political behaviour (Rhodes, 1997: 67), because 'political institutions form only part of the explanation whatever the theory under scrutiny' (Rhodes, 1997: 80). The development of new institutionalism can thus be understood as an attempt to address the dialectical relationship between structure and agency. In particular, it tries to synthesise the structuralist elements of traditional institutionalism with the individualist emphasis of behaviouralism (Lowndes, 2002: 91, 107).

Historical institutionalism and path dependency

To assist in the development of the theoretical and methodological principles underpinning this study, it is helpful to consider Peters' (1999: 63–77) analysis of historical institutionalism in which he discerns a number of complex analytical questions behind the apparently straightforward concept of path dependency:

1. Are initial choices 'institutional'?
2. What is the definition of 'path dependency'?

3. What is the potential for and causes of subsequent institutional change?

With regard to the first question, Peters (1999: 64–6) suggests that institutions are defined by routinised standard operating procedures and ideas. Thus, path dependency is deeply dependent upon initial policy decisions in these areas. Garner's implicit invocation of path dependency places over-riding emphasis on perceived aspects of the embryonic policy process that coincided with the assent of the 1876 Act. He characterises these aspects in broad terms: the idea that the Act was set up to regulate animal experimentation, and the perception that the Home Office did not have a prior structural relationship with the research community. However, it is debatable how the initial political circumstances in 1876 comprised rules and ideas that were sufficiently 'institutional' to provide the starting point for path dependency in animal research policy. Furthermore, the observation regarding the AAMR's policy role from 1882 brings into focus the problem for historical institutionalism of defining the point of institutional formation. Peters (1999: 67) contends that this task is 'crucial for making the case that those initial patterns will persist and shape subsequent policies in the policy area'. Thus, if 1882 (rather than 1876) is the starting point for the institutionalisation of this policy area, then the expected developmental path may be quite different, for it would appear to emerge from a situation of institutionalised relationships between the Home Office and animal researchers.[3]

This brings the discussion to Peters' second point regarding what is meant by 'path dependency'. The assumption of a clear, linear direction of development for this policy area reflects what Hay and Wincott (1998: 953) see as new institutionalism's hitherto 'characteristic "creational" bias and its emphasis, subsequently, on institutional inertia'. However, Peters (1999: 65) notes that historical institutionalism has developed more dynamic approaches to path dependency, where evolution can occur, for example, through actors' perceptions of the need to correct initial institutional choices that were later perceived to be dysfunctional. Thus, path dependency does not necessarily imply a direction that has been pre-determined by formative institutional choices, it simply acknowledges that all decisions emerge from within a context that includes inherited institutions and structures that may privilege certain actions and outcomes over others. Different conjunctions of institutions, exogenous structures and actors can lead to variations in path direction. In addition, assumptions of institutional stability are unsafe in the light of Lowndes' (2002: 99) comment on the contribution of policy network analysis to this issue: 'Those adopting a network perspective emphasise that institutional stability is dependent upon a continuing process of consensus and coalition-building among actors, within a continually changing environment.'

Thus, Peters' third point concerning historical institutionalism's problems with accounting for institutional change highlights aspects of animal research public policy that require further analysis. In the first instance, Peters (1999: 71) emphasises the need to study the link between the institutional constraints and individual decision, which is a key source of change. Furthermore, Peters' observation of the following tendency in historical institutionalism needs be borne in mind when understanding path dependency in this policy area:

> Indeed, there is a certain sense of *deus ex machina* in the historical institutionalist approach, with decisions taken at one time appearing to endure on auto-pilot, with individual behaviour being shaped by the decisions made by members of an institution some years earlier. (Peters, 1999: 71)

Likewise, Hay and Wincott (1998: 952–4) identify an implicit structuralism as a significant element in the historical institutionalist canon, while attempts to overcome this problem tend to vacillate between the structuralist and intentionalist extremes, rather than transcend this dualism.

In addition to these three problem areas related to historical institutionalism, Peters' (1999: 123) observation that network approaches normally conceptualise institutional change in 'organic' terms reinforces the need to be wary of assumptions of persistent institutional homeostasis.

This is particularly relevant to animal research policy as the 1876 institutions displayed a relative *absence* of structure, in the form of an issue network, which one might expect to be less likely to initiate a stable evolutionary path. This observation perhaps highlights a useful analytical contribution of policy network analysis to new institutionalist theory, in that it identifies variability in institutions that may help to explain different patterns of change and stability. In other words, as discussed in Chapter 2 under the heading of 'Endogenous network factors', it proposes that some institutional settings, characterised as issue networks, are less stable or 'institutionalised' than others because of the relative absence of institutionalised relationships or a hegemonic ideology to constrain and regularise individual behaviour in the policy network or 'institution'. A stable issue network that may initiate strong path-dependent evolution could be explained by examining a 'higher' level of rules above the policy network: 'constitutional rules (the rules that govern the rules!)' (Lowndes, 2002: 101). This study will offer such an analysis of the broader socio-political context in which the proposed 1876 issue network was embedded.

Structure and agency: the need for reconciliation, not vacillation[4]

The discussion of path dependency has thrown into relief the role of ontological assumptions regarding structure and agency in shaping conceptualisations of how institutions and policies evolve over time. This is because, on the one hand, an explanatory emphasis on structure will privilege the constraining and enabling role of institutions to the neglect of actors in the policy process, while on the other hand, a focus on agency will tend to downplay the role of institutional contexts and emphasise the intentions of the participating actors as the fundamental driver of political events (Hay, 2002: 102, 109–10). However, Lowndes (2002: 107) argues that one key aspect of new institutionalism is a more reflexive and explicit approach to the ontological question of structure and agency because it is fundamental to a more adequate understanding of political processes, and therefore influences how research is conducted. What Lowndes recommends is the rejection of both a prioritisation of either concept and the conflation of structural and intentional explanations. Instead, analysis must be informed by an attempt to transcend the structure/agency dualism. Despite their aforementioned criticisms of some aspects of the approach, Hay and Wincott (1998: 953–4) discern the beginnings of such an attempt in the historical institutionalist literature and aim to develop it further. The task, as they see it, is to establish: '...a theory of institutional innovation, evolution and transformation...[based on] the relationship between...institutional 'architects', institutionalised subjects and institutional environments'.

The evolution of institutions and policy occurs through these interrelationships in conjunction with their outcomes and actors' perceptions of those outcomes. Hence, policy evolution is path-dependent in the sense that:

> the order in which things happen affects how they happen; the trajectory of change up to a certain point itself constrains the trajectory after that point; and the strategic choices made at a particular moment eliminate whole ranges of possibilities from later choices

while serving as the very condition of existence for others. (Hay and Wincott, 1998: 955)

It is essential to be reflexive concerning the dangers of adopting a conflationary or vacillatory approach to the structure/agency problem. On the one hand, the assumption of a persistent form of institutional structure for animal research policy-making appears structuralist and deterministic. Yet conversely, applications of policy network analysis may sometimes incorporate intentionalist, individualist assumptions, for example, by relying heavily on data from questionnaires and interviews with political actors to ascertain the relative quality of access of interest groups to government policy-makers as the primary determinant of the network's position on the Marsh/Rhodes typology. But unless the relationship between these variable modes of access and network power relationships is explicated, there is a danger that this type of approach implicitly favours interpersonal, rather than structural, network links (Rhodes, 1997: 36). Similarly, Hay (1995: 195) notes the association between the methodological individualism of intentionalism and an emphasis on 'accounts which tend to take issues of social and political interaction largely at face value, constructing explanations out of the direct intentions, motivations and self-understandings of the actors involved....' This naive approach, often implicit in journalism and the utterances of political actors themselves, appears to embrace:

> distinctive features of behaviouralism. These include a focus on power as decision-making and a tendency to assume that an analysis of inputs into the political system, such as the pressure exerted by interest groups upon the state, is sufficient to account adequately for political outcomes. (Hay, 2002: 10)

Peters (1999: 14) characterises this behaviouralist trait as 'inputism', where the political process is conceived in linear fashion, emphasising individualistic 'inputs' into an institutional 'black box' that converts inputs into policy 'outputs'. This can be contrasted with the more systemic and interactive dialectical policy network model postulated by Marsh and Smith (2000) and reproduced at page 24. Furthermore, behaviouralism is also said to be 'reductionist': reducing those 'inputs' to their individual components, with collective entities merely a product of the interaction of individuals, rather than social or political structures having any impact on individual preferences or behaviour (Peters, 1999: 16). This reveals individualistic assumptions: 'individuals are autonomous, with their preferences and actions unconstrained by institutions' (Peters, 1999: 1). Thus, it can be seen that in the behavioural approach, institutions are downplayed in favour of an analysis of the inputs, as if institutions did not matter (Peters, 1999: 14). One example of 'inputism' would be an approach which assumed that the Animals (Scientific Procedures) Act 1986 and the cost-benefit assessment of animal research projects were implemented in a transparent, pluralistic manner, instead of withholding judgement pending exploration of the way institutionalised networks can mediate and thus affect policy outcomes. However, it is important to recognise that this flawed approach is understandable, given the legally-enshrined confidentiality surrounding the regulatory process in this policy area. The danger is that, when analysts are unable to describe what goes on in the 'black box' at the heart of policy processes, they become reliant on making inferences from other data that depend on their underlying methodological position (Hill, 1997: 24).[5] This observation reinforces the desirability of scrutinising the methodological assumptions that underpin the interpretation of data.

These basic methodological issues concerning the relationship between structure and agency are expressed in policy network terms by Marsh and Stoker (1995: 292–3), when they argue that it focuses on the meso-level of analysis of structured relationships between actors in a particular policy area. Thus, in order to explain policy outcomes, they suggest that what is required is an *explicit* integration with both micro-level theories and descriptions of individual behaviour, together with macro-level theories of power distribution across a nation-state and beyond (that act as constraining 'superstructures' in relation to policy networks). This means that in order to deepen our understanding of the genesis and implications of the 1876 Cruelty to Animals Act, it is essential to address relevant contextual questions, such as whether professions enjoyed macro-level structural advantages at the time, or, relatedly, whether the new institutions for regulating animal research adopted an established template that may itself have embodied structural power relationships (Hay, 2002: 105).

New institutionalism and critical realist epistemology

The preceding discussion indicates how new institutionalism, and in particular the variant of 'historical institutionalism', which focuses on the distribution of power in a polity and power dependence, offers a robust basis for exploring animal research policy. As Marsh and Stoker (2002: 313) observe, the dominant epistemology underpinning historical institutionalism is realism, which in its contemporary 'critical' form recognises that reflexive actors' interpretations of structures affect their behaviour and hence outcomes, and that those interpretations are influenced by social constructions of reality. Hay (1995: 199–202) conceptualises this core, interactive aspect of the critical realist approach in terms of a dialectical relationship between structure and agency, reflecting the dialectical policy network model utilised in this study. Thus, layered structures represent a nested hierarchy of action settings (Hay, 1995: 200) that are strategically selective, in that they favour certain strategies and actors over others. They also reflect at any point in time the embodiment of past actions, which were themselves constrained or enabled by strategically-selective structures. Outcomes and structures evolve through the intended and unintended consequences of strategic agency. Importantly, critical realism's ontology parallels new institutionalism insofar as it allows for the possibility of non-directly observable, informal institutions or structures shaping the nature of power relationships (Hay, 2002: 186–7; Marsh and Furlong, 2002: 20).

Thus, it can be seen that a critical realist philosophy has underpinned the previous discussions of policy network analysis and Garner's application of this tool to animal research policy. It has generated a number of outstanding research questions that need to be addressed in order to furnish a more comprehensive understanding of this policy area. Those questions are both empirical and theoretical in nature, in the sense that they pertain to the need both to describe processes and outcomes in animal research policy, and also to examine phenomena and dynamics that, within the policy network conceptual framework, are said to affect evolving policy outcomes. One of the tasks that is crucial to understanding the evolution of the animal research policy network is to describe and compare the resources of relevant actors in this policy area, and thus infer deeper social, economic and political structures that cannot be directly perceived.[6] Such deeper structures are key aspects of both the dialectical network model and the critical realist epistemology that pays attention to 'the structural constraints within which individual operate...the most important [of which] are generally the impact of differential allocation of resources....' (Marsh and Furlong, 2002: 38). This raises questions regarding what methods are appropriate for the examination of research questions framed within a critical realist epistemology.

Methodological implications of critical realism

The insight that certain important causal factors – unobservable structures – are not amenable to quantification (Hay, 2002: 252) leads critical realist epistemology to require empirical evidence to be related to a theoretical narrative. Here, 'theory' is understood in terms of a heuristic guide to empirical analysis that highlights particular relationships and phenomena that are held as relevant to policy processes (Hay, 2002: 45–7). In the present case, the aim is to elucidate the dynamic relationships between strategically-selective structures and strategic actors. Thus, as Burnham et al. (2004: 28) comment:

> In terms of methods, it should be clear that critical realists will not rest content with simply recounting 'actor's views' or with bald statistical presentation; instead, the aim must be to reject surface explanation and, wherever possible through the use of primary sources (particularly documentary material and elite interviews), reconstruct and reinterpret the events under investigation.

Therefore, the critical realist approach points to the suitability of a broadly qualitative method of research that can go beyond ephemeral perceptions of documentary data towards an elucidation of such data's meaning in relation to both deeper unobservable power structures and the interpretative activity of the participating actors (Marsh and Smith, 2001: 529). In other words, critical realism eschews a method that predominantly relies on the systematic collection of quantitative data from a wide range of cases in the hope of developing formal explanatory models because, on the contrary, 'social reality is complex and involves reflexive agents' (Marsh and Smith, 2001: 533). So, critical realism is associated with qualitative analysis that reflects an epistemological position: 'that stresses the dynamic, constructed and evolving nature of social reality' (Devine, 2002: 201). Thus, critical realism's emphasis on understanding specific, inherently unique processes rather than seeking universal, regular patterns suggests that an in-depth, qualitative analysis of the historical evolution of animal research policy would be an appropriate method of providing new narratives of this policy area.

One important corollary of the critical realist epistemology and its mainly qualitative method is the recognition of limits to the kinds of claims that can be made by this approach to political research. As realists, it is accepted that at a basic or immediate level, knowledge has a universal character: Abraham (1995: 30) gives the example of being able to agree on the distinction between a live animal and a dead one as a foundation for the analysis of the much more complex issue of corporate bias in scientific testing. However, beyond that basic level, the immense complexity and inherent irregularity of the social and political world[7] means that political theories can provide, at best, a limited and provisional explanation of events, but they cannot be predictive in the sense of laws in natural science, because no two situations are identical across time and space (Burnham et al., 2004: 27–8). Furthermore, the various standpoints of analysts mean that interpretations of complex systems and the relationship between unobservable structures and empirical observation are likely to be contested, although agreement on basic empirical points provides a bridgehead for some degree of comparison and adjudication between different theories or models (Hay, 2002: 252).

Interestingly, the recognition of the essential role of empirical analysis in critical realism is consistent with Read and Marsh's (2002: 232–5) observation that although different ontological and epistemological positions tend towards either qualitative or quantitative methods, there is no simple, determinant link. In particular, critical realists may 'wish to make claims that have a quantitative basis' (Read and Marsh, 2002: 233). Similarly, Hay (2002: 252), who proposes an analytical strategy similar to critical realism, comments that 'while empirical evidence alone is never enough it is an important and necessary starting point'. Therefore, elements of quantitative analysis may be appropriate in this study to describe policy outcomes and compare them with certain regulatory requirements, for example, compliance with severity limits that are conditions of licences to conduct specified procedures on animals. However, the broader requirement to conduct a cost-benefit assessment of animal research projects, which was introduced by the Animals (Scientific Procedures) Act 1986, is clearly a subjective, value-laden decision-making process. Therefore, in order to derive a richer understanding of this particular process, a qualitative analysis of the way in which the harms and benefits of a proposed animal research programme are predicted, balanced and monitored, offers the most appropriate overall methodological approach (Devine, 2002: 205).

The role of case studies

These considerations imply that a case study design would be a suitable component of this present work, as indicated by Burnham et al. (2004: 53): 'While both quantitative and qualitative data can be generated by case study design, the approach has more of a qualitative feel to it as it generates a wealth of data relating to one specific case'. Interestingly, Abraham's (1995) study of pharmaceutical regulation in Britain identified similar obstacles to those experienced by Garner's analysis of animal research policy in that 'the data handled by regulatory authorities are highly confidential' (Abraham, 1995: ix). Consequently, the limited primary data, particularly in documentary form, meant that pre-existing studies lacked empirical depth and instead attempted to judge regulatory performance by employing crude comparative benchmarks, e.g. the number of drugs approved relative to number of submissions and withdrawals (Abraham, 1995: ix). Likewise, Garner's study attempts to provide broad comparative indicators of the *relative* balance of power in the animal research policy process between the US and UK, rather than more specifically analysing the *actual* nature of power in the black box of the UK animal research policy process. As Rhodes (1997: 79–80) comments, this type of situation points to the desirability of including traditional descriptive analysis in this study, because so little is currently known about how this policy process works and its outcomes. This, in turn, indicates the case study approach as an appropriate method that may facilitate deeper understanding of the operation of the policy network, as embodied in the cost-benefit assessment, in order to test the present policy community hypothesis:

> The attractiveness of case studies is that data on a wide range of variables can be collected on a single group, institution or policy area. A relatively complete account of the phenomenon can thus be achieved. This enables the researcher to argue convincingly about the relationships between the variables and present causal explanations for events and processes.... (Burnham et al., 2004: 55)

The aim is to shed more light on the 'black box' of animal research policy-making, although it is important to recognise, as the term 'black box' indicates, that traditional confidentiality constraints have meant that detailed case studies have not been undertaken in animal research. However, Abraham (1995: x) suggests that this type of obstacle may be overcome through choosing controversial cases that emerge into the public domain, which therefore represent the most practical route into examining normally confidential policy-making processes. Therefore, this situation provides a particularly clear example of the notion that the choice of 'case studies...must be largely governed by arbitrary or practical, rather than logical, considerations' (Eckstein, 1979; quoted by Rhodes, 1997: 81). Nevertheless, in order for such a case to be useful in terms of testing and developing theory (Rhodes, 1997: 82), one important benchmark against which to assess its utility is the extent to which it matches up to the requirements of a *heuristic case study*:

> for 'discerning important general problems and possible theoretical solutions'. Such case studies are directly concerned with theory building

and 'can be conducted seriatim, by the so-called building-block technique, in order to construct increasingly plausible and less fortuitous regularity statements'. (Rhodes, 1997: 81; quoting Eckstein, 1979)

Therefore, considerations such as the *generalisability* of a case study are relevant to this assessment of case study utility and are discussed below. This then raises the question of what appropriate primary data and related case(s) might be available.

Data collection

In order to test the hypothesis that forms the focus of this study – that the nature of UK animal research policy-making reflects a persistent policy community rather than an issue network – it is necessary to address the four research questions outlined in the previous chapter that underpin this hypothesis:

1. Which group(s) interests were served by the assent of the Cruelty to Animals Act 1876?
2. Did the policy network that emerged during the passage of the 1876 Act evolve into a policy community in the subsequent years?
3. Did the passage of the Animals (Scientific Procedures) Act 1986 signify a core change in policy or an example of dynamic conservatism?
4. Does the implementation of the Animals (Scientific Procedures) Act 1986 reflect an issue network or a policy community?

In order to respond to these questions, data are required that describes the group actors in this policy area in terms of their resources, ideologies and strategic actions. The data must also provide evidence of the evolving institutional framework for animal research policy, the relationships between groups and state actors, and who benefits from the policy outcomes. At the same time, it is essential to incorporate insights into the structural context of this policy process. By applying this information to the four research questions above (each of which must be disaggregated into a group of underlying questions derived from the dimensions of the Marsh/Rhodes typology and the policy network dynamics table), the nature of the evolving policy network can be explored.

Constraints on collecting data

However, obtaining such information is not straightforward. Thus, it was noted at the beginning of the previous chapter that the extant literature relating to the evolution of animal experimentation policy has tended to lack empirical detail regarding the policy process and outcomes in this policy area. This is due to the fact that detailed public records relating to animal research policy implementation and outcomes are normally unobtainable due to government restrictions,[8] legal barriers to disclosure, and commercial confidentiality. In particular, Home Office documents relating to this policy area have been subject to a hundred-year restriction (French, 1975: 178n8), while the Animals (Scientific Procedures) Act 1986 prohibits the unauthorised disclosure of any such information, except for the purposes of discharging functions under the Act such as publishing official reports and guidance. Furthermore, the Freedom of Information (FoI) Act 2000, which did not come into full effect until 1st January 2005, does not formally allow access to information as it is trumped by the secrecy clause in the 1986 legislation.[9] The range of accessible data is limited to government reports, published scientific papers and tertiary[10] historical accounts. In fact, only French's 1975 study has had the opportunity to triangulate this information with more reliable primary documentation – dating from the late 19th century – that was generated specifically for the purposes of regulating animal procedures and was not intended for public disclosure. Thus, French was able to reconstruct the early administration of the 1876 Cruelty to Animals Act and provide a more reliable insight into the relationship between the Home Office and the relevant interest groups and individual actors.

Data relating to the historical evolution of animal research policy

The tradition of secrecy surrounding animal experimentation means that virtually all of the work is based on secondary sources, such as government legislation and policy statements, pressure group reports, official inquiries and parliamentary debates. The general absence of primary data concerning policy implementation and outcomes, combined with the extensive scope of discretion afforded front-line regulators in the interpretation of vague rules, means that the reliability of such sources in this respect is questionable. Nevertheless, these secondary sources, and the studies that rely upon them, can provide some useful evidence relating to the formal structure of policy-making, as well as the broader political conflict over animal experiments, including the ideologies and strategic actions of participants in the public debate and open arenas of policy formulation.

Therefore, in order to address the first, second and third research questions concerning the evolution of the animal research policy network in the most reliable way possible, this study will analyse and compare these various historical accounts, looking for points of convergence and difference, thereby trying to reconstruct the evolution of the network and its context by triangulating the various sources of information. One of the most reliable available sources is French's (1975) study of the vivisection controversy in the final quarter of the nineteenth century, which was based on unique, discretionary access granted by the Home Office to public records that were normally subject to a hundred-year restriction. French's analysis will be particularly useful when addressing the first and second research questions concerning the genesis of the policy network and its early evolution following the Cruelty to Animals Act 1876. But this still leaves a serious absence of primary data with which to construct a reliable analysis of this policy area.

Obtaining primary data

Overcoming this problem raises the issue of utilising elite interviews, which are one of the primary sources suggested by Burnham et al. (2004: 28) as appropriate for a critical realist approach. Moreover, intensive interviews with participants in policy processes are a common qualitative technique as they aim 'to explore people's subjective experiences and the meanings they attach to those experiences' (Devine, 2002: 199) However, there are three significant drawbacks or limitations to this particular method in this policy area. Firstly, at a general level, the confidentiality constraints regarding the internal workings and outcomes of this policy area clearly affect the degree of access to many of the key 'insider' actors. Secondly, the problem of access is compounded by the writer's own involvement with a lobby group operating in this area. Devine (2002: 205–6) states that in order to collect useful empirical materials through interviews, 'The relationship cannot be distant if confidential personal information is to be revealed or when sensitive topics are discussed'. In this case, it is therefore important to recognise that it is not feasible to achieve a consistent and representative level of access to the salient actors involved. Thirdly, it could also be argued that such interviews would be, *in relation to the empirical task of assessing policy outcomes in terms of the implementation of the cost-benefit assessment*, more properly classified as secondary sources than primary because they would be retrospective, indirect accounts of such events.

The emergence of primary documents

There is, however, a set of relevant primary material that this study can draw on, derived from the writer's own involvement in the emergence into the public domain of documentary data relating to a recent (1995–2000) programme of animal experimentation. This primary material comprises confidential material that came into the writer's possession through two unauthorised disclosures: from within Imutran Ltd in spring 2000, and then the Home Office in October 2002. The documents were sent anonymously and unsolicited to the writer in his capacity as an activist and lobbyist. In April 2003, a Court Order permitted the writer and the animal protection organisation of which he was Director,[11] to publish in redacted form over one thousand pages of these confidential documents, together with the report, entitled 'Diaries of Despair' (Lyons, 2003), based on the first set of leaked documents.[12] At this point, it is appropriate to note that this data represent the type of unusual, controversial case that occasionally enters into the public domain and which Abraham (1995: x) contends can potentially facilitate an advance on previous studies of this policy area (i.e. a heuristic case study, see above). In fact, to the best of the writer's knowledge, no other similar set of documents has ever come into the public domain.

The Imutran documents relate to the company's programme of pig-to-primate organ transplantation research, a technique known as 'xenotransplantation',[13] which they commissioned the contract research company Huntingdon Life Sciences (HLS) to carry out between 1995 and 2000. The documents had originally included thirty-nine final draft study reports that detailed the design, materials, methods and results of various xenotransplantation procedures on 49 baboons (*Papio anubis*) and 424 cynomolgus monkeys (*Macaca fascicularis*, also commonly known as 'crab-eating macaques'). Other documents included correspondence with HLS, suppliers and Home Office Inspectors, and also meeting minutes, feasibility studies and internal reports concerning many aspects of the conduct of, and plans for, xenotransplantation research. The Home Office documents comprise correspondence between Imutran and both the Home Office Animals (Scientific Procedures) Inspectorate (ASPI) and the APC, reports submitted by Imutran to both those bodies in support of their licence applications, and actual project licence authorities.

The permitted publication of the confidential materials took place following a two-and-a-half-year legal battle. Having initially sought outright suppression of the documents, following mediation pursuant to a court order, a settlement was reached between the parties. The settlement agreement required the claimants to abandon their original claim for damages and costs in respect of breach of confidentiality and copyright and gave the defendants the right to publish an 'agreed bundle' of redacted[14] confidential documents, together with the 'Diaries of Despair' report (Townsend, 2003).[15] These documents together comprised the majority of those listed by the defendants as demonstrating the key public interest elements stated in the Defence (Bean and Afeeva, 2000: 5):

13.2 (c)   The Home Office inspectors whose duty it is to monitor the activities carried out by or on behalf of the Claimant have a relationship with the Claimant that is too indulgent, in that the monitoring undertaken is ineffective and fails to ensure that the primates are protected as the law requires;

13.2 (d)   The Claimant had distorted the truth in its public statements concerning the success of its research and the welfare of the primates on which it experiments.

The question now arises: What is the specific relevance of this documentation and the case study to this study?

The Imutran xenotransplantation research case study

This section considers the validity of the primary case study data (Uncaged Campaigns, 2003) through the criteria of authenticity, credibility, representativeness and meaning (Burnham et al., 2004: 184–6), alongside the relevance of the related secondary and tertiary resources.

The documents include key sections of the thirty-two xenotransplantation study reports: the 'Study Design' and 'Surgical procedure' sections of the 'Experimental Procedure' chapter, and the appendix for each report listing the 'clinical signs' of the primates following the transplant procedures. Relevant details (i.e. those considered in the public interest) from the sections of the study reports that remain confidential have, however, been disclosed within the 'Diaries of Despair' report (Lyons, 2003). These study reports represent highly reliable evidence; for example, they were intended to be compliant with the principles of Good Laboratory Practice (GLP). The Good Laboratory Practice Regulations 1999 (Department of Health, 1999) are a Health and Safety Statutory Instrument administered by a Department of Health body known as the Good Laboratory Practice Monitoring Authority (GLPMA). GLP compliance is intended to ensure that data generated by tests of technologies and products are as reliable and valid as possible when submitted to the relevant regulatory authorities for approval.

The information in the study reports is complemented by various internal documents produced by Imutran as part of the development of their research programme. These offer additional analysis of the conduct and results of the research, in terms of both the 'benefits' accrued and the 'costs' to animals in terms of pain, suffering, distress and death.

The interpretation of these documents with respect to the policy process and policy outcomes is facilitated by the information contained in the confidential documents obtained from the Home Office, none of which would normally be in the public domain. The most important of these are the project licences, which form the legal basis upon which animal research is licensed. The anatomy and purpose of project licences, and their components such as severity limits, will be discussed in more detail in the case study chapter below. But here, it is helpful to note that project licence applications are supposed to be detailed, contain all relevant information, and hence form the basis of the Home Secretary's decision to grant a licence. The following extract taken from a Home Office (2002: 11–14) report confirms this:

> A.40 The severity limit for each protocol is determined by the upper limit of the expected adverse effects that may be encountered by a protected animal, taking into account the measures specified in the licence for avoiding and controlling adverse effects.For the purposes of the statutory cost/benefit assessment the precise animal welfare 'costs' considered are derived from detailed narrative descriptions of the nature, incidence and severity of the likely adverse effects (and the measures to be taken to prevent, identify and ameliorate the adverse effects). These are set out on the form of [project licence] application or provided as supplements to an application judgements of animal welfare costs, the level of suffering that may be produced, and the humane endpoints to be applied are determined by the detailed narrative descriptions on the form of application and licence.

Project licences also require a description of the objectives and expected benefits of the research programme. The Home Office has stated that animal research is permitted on the basis that it is 'likely to achieve the stated objectives' (O'Brien, 2000a). Therefore, the project licence is a vital document for policy analysis, as expressed in this Home Office (O'Brien, 2000a) statement:

> In deciding whether to grant a licence for any regulated procedure, the 1986 Act requires that the likely benefits of the programme be weighed against the likely adverse effects on the animals concerned (the cost/benefit assessment).

Comparing the project licence with the results of the licensed research not only permits an assessment of the adequacy of the initial cost/benefit assessment and the Home Office's scrutiny of the application, but also provides a benchmark by which to assess the ongoing implementation of the regulatory framework and the compliance of researchers with conditions such as severity limits. This is demonstrated by the Chief Inspector's 'Note on the Cost/benefit Assessment' (Home Office, 1998: 50–9), where he states that the assessment of benefits should be an ongoing process rather than a one-off event at the initial licence application stage: 'The cost/benefit assessment is a process rather than an event. Licensed work is scrutinised to determine that the benefits are being realised in practice and that the costs cannot be further reduced'.

Therefore, it can be seen that by examining and comparing official government publications, confidential project licence applications and related regulatory documentation, and primary information about the actual conduct of the research and its costs and benefits, it is possible to compile a unique picture of how this policy area is operationalised and implemented. This contrasts with Garner's heavily constrained analysis of the formulation and passage of the Animals (Scientific Procedures) Act 1986. For example, it was noted above that Garner's analysis of policy implementation was hindered because the definitions of the severity bands attributed to project licences are vague and cover different procedures with different severity. Furthermore, there was a lack of evidence to indicate what these severity bands meant in practice in terms of animal suffering. Thus, the primary and other documentation presented here relating to the xenotransplantation experiments and their regulation overcomes these fundamental data constraints.

The confidential documentation also includes papers relating to the workings of the APC, a factor relevant to the fourth research question concerning the policy impact of the Animals (Scientific Procedures) Act 1986. Combining these papers with published material on the role of the APC, the Committee's observations on this case, and its advice to the Home Secretary, will help to address this question.

In relation to the question of the quality of access to state policy-makers enjoyed by the different interest groups, the primary documentation provides fresh information in the form of correspondence between regulators and researchers, and internal company reports on relationships with regulators. Once again, this can be usefully augmented with material related to the Home Office's response to allegations of maladministration, including correspondence with the various interested parties and public statements such as Written Answers to Parliament, letters to MPs, media statements and submissions to bodies considering the case, such as the House of Commons Home Affairs Select Committee and the Parliamentary Ombudsman. Together, this data can help in understanding the ongoing processes of re-negotiation among network actors that drives the network's evolution (see Lowndes, 2002: 97).

This material will also inform analysis of the operation of grievance procedures and constraints on the arbitrary exercise of power by policy-makers, which is germane to the question of whether the 'model of justice' in animal research policy implementation reflects an élitist 'professional treatment' model or a pluralistic 'moral judgment' model.

Overall, this case study, comprising primary sources together with related secondary and tertiary documents, allows the construction of a narrative designed to test the fourth set of research questions regarding the recent character of the policy network. This will, in turn, illuminate persistent uncertainties regarding the degree of policy change introduced by the Animals (Scientific Procedures) Act 1986. Therefore, the areas addressed will include:

- the operation of the statutory cost/benefit assessment of animal research projects
- the impact of the Animal Procedures Committee in terms of public accountability and policy outcomes
- the relative quality of access of interest groups to policy-makers.

The more detailed and reliable description of these empirical issues provided by this case study facilitates a re-appraisal of the extent to which the various interest groups have achieved their goals in this policy area, and thus the balance of power in the policy network and its position on the Marsh/Rhodes typology. In other words, it becomes possible to address this study's central hypothesis: UK animal research policy is best characterised by a policy community-type network.

Furthermore, the fact that the availability of primary documentation is limited to this case study means that the main focus of this book is the fourth research question which examines the balance of power and policy network characteristics in contemporary animal research policy. However, this may also have some relevance to an understanding of the history of the policy area. For, to the extent that policy-making is path-dependent. then, by the same token, present-day policy-making is a partial manifestation of that historical development and may offer some indicators as to earlier policy dynamics that remain obscure due to the lack of primary evidence. As Burnham et al. (2004: 19) observe, path dependency involves understanding the '... key historical choices made by states. Critical moments create branching points from which historical development moves on to a new path and once that new path has been taken it is difficult to change track'. At the very least, this approach will have implications concerning the claim that modern animal research policy-making occurs in an issue network which reflects the nature of the original network structure and institutional choices made in 1876.

Having considered the relevance of the data to this study, it is now necessary to consider its limitations.

The limits of the methodology

It will be clear from the earlier discussion of the derivation of the primary documentation that this author approaches the issue of animal research and its regulation from a critical perspective. However, this does not automatically preclude the writer's pursuit of informative and relevant knowledge in this area of study. Indeed, it could be argued that there is no such thing as an entirely objective observer. As Burnham et al. (2004: 41) point out, '... all researchers will have expectations about the kind of results the project is likely to generate and this may influence their analysis'. The important task for a researcher is to be self-conscious about their commitments.

The second point is more practical: The primary data presented here would not be in the public domain were it not for the original lobbying activities of the writer. Consequently, a detailed case study has emerged which provides an opportunity to utilise primary data to test the research hypothesis concerning the balance of power in UK animal research policy.

In relation to the data themselves, limitations arise as to what can be claimed in this study. First, the necessary reliance on secondary and tertiary data regarding the historical evolution of the policy area raises questions of reliability, because they themselves tend to be hindered by a lack of primary data and are often written from a committed viewpoint. It is hoped, however, that by examining a range of tertiary studies and secondary sources in the shape of a departmental report, for example, a reasonably accurate view of the historical development of the policy network can be ascertained. However, as Burnham et al. (2004: 172) state, 'secondary and tertiary sources in political science are most effectively employed in combination with elite interviewing and/or with the analysis of primary documents'.

Those primary documents relate to the Imutran xenotransplantation case. It must, however, be acknowledged that this data do not represent the entirety of the information generated in the course of the xenotransplantation research programme or its regulation. Whether this introduces bias into the data is a difficult question to answer with absolute confidence. However, there are a number of factors that suggest that the data are reliable for their purpose.

The outcome of the legal proceedings means it is reasonable to infer that the data are significantly reliable. Both Imutran (during the proceedings) and the Home Office (2004) have referred to the existence of additional unpublished observations of the experimental animals, such as those contained in surgeons' logs, in an attempt to cast doubt on the reliability of the primary data presented here. However, the observations contained in these appendices are particularly informative, as the following excerpt from a confidential study report reveals:

> The clinical signs presented in this Appendix are only the first and last observations reported for each day the animal survived. Due to the frequent and numerous procedures performed on the animals throughout the course of each day, the signs displayed by them other than first thing in the morning or last thing in the evening, were considered to be unrepresentative of the underlying clinical condition of each animal.[16]

It may also be significant that, having referred to such additional data in their pleadings, Imutran refused to disclose the data to the Court. Moreover, the writer applied to the court for an Order requiring Imutran to disclose such documentation. This application was contested by Imutran, and the hearing was cancelled following the settlement agreement between the parties that confirmed the abandonment of Imutran's claim and the right of the defendants to publish the confidential documents. Similarly, the Home Office (2004) has referred to such additional data in defending its actions, while simultaneously stating that they did not have possession of such information and thus could not disclose it. Apart from anything else, such actions offer interesting data in regard to the strategies of the two actors. Comparison with other sources, such as analysis of the data and the case by other actors, will help to triangulate the primary data.

Furthermore, incomplete information about a research project is not necessarily a barrier to examining the adequacy of implementation.

Thus, in relation to the assessment and enforcement of severity limits on procedures, the Home Office's 'Guidance on the Operation of the Animals (Scientific Procedures) Act 1986' (1990: 10) explained:

> 4.8 Such an assessment should reflect the maximum severity expected to be experienced by any animal. It should not take into account the numbers of animals which might experience the maximum severity or the proportion of the animal's lifetime for which it might experience severe effects.

In other words, the severity limit should reflect the worst-case scenario for any single animal. Therefore, evidence of just one procedure exceeding the severity limit raises legitimate questions about the assessment of severity and the enforcement of severity limit conditions. Clearly, However, the implication of any such instances also depends on the reactions of licence holders and regulators.

While this case study may have the advantage of facilitating a relatively complete account of the various phenomena involved in this research programme, this raises questions concerning the generalisability of the case to the rest of the policy area. In answer to this question, it should be noted that if the hypothesis of a policy community-type policy area is supported by the data, then this case study is particularly useful for drawing wider generalizations about animal research as it represents a 'critical' case most likely to prove the contrary hypothesis (Burnham et al., 2002: 54) of an issue network model of policy-making. This is because experimentation on higher primates, according to Section 5(6) of the Animals (Scientific Procedures) Act 1986, requires special justification. This was especially the case in this instance, as illustrated by the following comment on the Imutran research by the APC in their 1996 Annual Report:

> The speed of development of this work and its sensitivity makes it essential that the Sub-Committee and, indeed, the full Committee keeps fully appraised of the progress of this work and its direction. It is also essential that the work is carefully and closely controlled. We accept that this will place extra regulatory burdens on those undertaking such work and that work may be delayed as a result. We do not apologise for this. (APC, 1997: 10)

In terms of its representativeness across time, as will become clear, there have been no significant, core-level changes to this policy area since the research programme and the publication of the primary documents, while the amended 2012 regulations retain all the key measures in force during this case study.

Conclusion

This chapter has outlined a critical realist epistemological and methodological position that both raises questions about existing analysis of animal research policy and provides a foundation for this book. This critical realism consciously attempts to reconcile the structure/agency dualism in a way that is consistent with the dialectical model of policy networks, discussed in Chapter 2, which forms the analytical framework for this study. Therefore, the key methodological issues outlined in this chapter are:

- Structures do not determine action; instead they constrain and facilitate action with some types of strategy and actors being favoured over others.
- Actors are not unconstrained, though they are reflexive, and their actions affect both outcomes and structures.
- Political processes and events must be explained in terms of the dialectical interaction between strategic actors and strategically selective structures or contexts, a conception also captured in the dialectical policy network model outlined in Chapter 2.
- Key structures, e.g. power structures, can be unobservable and therefore unquantifiable.
- Political processes are complex and irregular.
- Qualitative methods are required to analyse the complex, interactive processes that include non-observable structures, and are thus appropriate to addressing the research questions which are derived from a dialectical policy network framework.
- Quantitative methods may have a role in ascertaining basic empirical facts.
- Case studies are an appropriate way of examining political processes in detail.
- The high level of confidentiality in animal research policy points to the desirability of utilising controversial cases in the public domain.
- The available data facilitates the use of secondary and tertiary data to examine evolution of animal research policy, with the primary data providing a basis for a case study of recent practices.

- The case study and associated data satisfy criteria of authenticity, credibility, representativeness and meaning.

Having established the key methodological principles that inform this study, the next task is to begin applying these to the analysis of the data in relation to the four research questions that underpin the present hypothesis.

5
The 1876 Cruelty to Animals Act: Protection for Animals or Animal Researchers?

Introduction

The previous three chapters have provided an overview and analysis of the policy network literature, its application to animal research policy and underlying methodological issues. This has generated a series of research questions that underpin this study's hypothesis. The aim of this chapter is to address the first research question concerning the circumstances surrounding the formation of the animal research policy network: Which group(s) interests were served by the assent of the Cruelty to Animals Act 1876? This issue must be addressed in order to provide an analytically useful understanding of this allegedly critical point in the historic path of animal research policy.

Answering this question within a critical realist methodological framework necessitates the detailed reconstruction of the political process up to this point in time. Therefore, the available tertiary sources and the 1876 Act itself will be analysed in order to build a chronological narrative of the interactions between the participating actors and their structural contexts. When state actors began to communicate with groups with a view to formulating animal research policy (as eventually expressed in the 1876 Act), then from that point that structural context will have included an embryonic policy network. Applying policy network analysis to the narrative of this period will involve describing:

- the relevant group and possibly any individual actors in terms of their core and secondary ideologies and their resources[1]
- relations between policy actors

- the rules of the game and strategies employed
- policy outcomes (in this case, the 1876 Act).

Thus, it not only will be possible to estimate the distribution of benefits to the relevant interest groups implied by the 1876 Act. but it will also reveal the key dynamics and relations that explain the distribution of benefits and potentially influence the future trajectory of the network.

The emergence of animal experimentation on to the political agenda

To start, this chapter discusses the emergence of animal experimentation on to the political agenda, leading to the assent of legislation in 1876. At this juncture, a note on terminology is appropriate: animal experiments during this period consisted mostly of procedures involving surgical incisions into live animals, which is the literal definition of the term 'vivisection' (Turner, 1980: 166n10). Hence, as will become apparent, discussions of this period in particular tend to use the term 'vivisection' (among others) to cover all experiments on living animals, and that same terminology is followed here.

The philosophical, religious and cultural roots of vivisectionist and anti-vivisectionist thought

Most of the literature that deals with the emergence of vivisection acknowledges that Cartesian philosophy played a central role (Brooman and Legge, 1997: 8–10; Radford, 2001: 17–19; Ryder, 1989: 55–8; Garner, 1993: 10–2; Monamy, 2000: 10–1). Descartes (1595–1650) provided an ontological, epistemological and normative rationale for vivisection as a scientific method of physiological research. He did this, firstly, by extending the mechanistic theories being developed in mathematics and physics to the biological realm, arguing that human and other animal bodies could also be thought of as conforming to laws of nature (Monamy, 2000: 10; Maehle and Trohler, 1987: 25). Secondly, Descartes posited that human animals were unique in being divinely-endowed with a soul. Consequently, non-human animals were viewed as non-conscious. Monamy (2000: 11) describes the significant ideological implications of Descartes' position:

> This concept of 'beast-machine' was critical to the way in which scientists viewed other animals. It provided a convenient ideology for early vivisectionists: how could animals suffer real pain if none

had a soul? How could animals suffer real pain if none had real consciousness? In Descartes' writings was found a reason to discount the behavioural responses of animals to vivisection (which would be described a symptomatic of pain in humans) as the mere mechanical reactions of robots.

Consequently, from the mid-1600s onwards, Descartes' theories were used by experimenters (and others) '...to deny that even the most minimal of humane considerations should be applied to non-human animals' (Brooman and Legge, 1997: 10). This absence of moral concern was reinforced by the dominant religious and philosophical notion that non-human animals existed essentially for the purpose of serving the interests of humans (Radford, 2001: 17).

Radford (2001: 19–22) notes that isolated thinkers throughout history have consistently questioned this mainstream anthropocentric position on the relationship between humans and other animals. However, the development of a vocal and popular animal welfare lobby in Britain is widely considered to have been significantly influenced by the utilitarian political philosopher Jeremy Bentham, who dismissed the absolute distinction between humans and (other) animals drawn by the Cartesians (Brooman and Legge, 1997: 13; Monamy, 2000: 19; Radford, 2001: 25–6, 30–1; Ryder, 1989: 75–7). In his 1789 work *An Introduction to the Principles and Morals of Legislation,* Bentham asserted that animals could, indeed, suffer, and that this capacity was the most important factor in determining how they should be treated, rather than the possession of the 'rational soul' postulated by the Cartesian position. According to Monamy (2000: 19), Bentham's work meant that 'The anthropocentric world view was being challenged by a more holistic notion that animals ought to be protected for their own sake'. In addition to this explicit normative critique, the following cultural factors also played a significant role in underpinning subsequent opposition to vivisection:

- a general development of compassionate, humanitarian sensibility towards others, starting with other humans (the most prominent example being the anti-slavery movement)
- scientific thinking about the similarities between humans and other animals
- certain religious conceptions that 'the good life' required benevolent action
- the notion that pain was a form of evil

- an elevated conception of animals in the divine order of nature (Quakers being an example of the growth of such tendencies)
- indirect moral and social arguments against animal cruelty (i.e. that it reflected a lack of virtue and led to cruelty to humans) (Turner, 1980: 4–14).

Nevertheless, it is widely accepted that Cartesian philosophy was a major influence on nineteenth-century French physiologists such as Magendie (1783–1855) and Bernard (1813–78), who later played a significant role in the routinisation of animal experimentation in Britain and elsewhere (Monamy, 2000: 11–2; Ryder, 1983: 125). Significant anti-vivisection sentiment in Britain is said to have arisen from public animal experiments performed in London by Magendie in 1824 (Monamy, 2000: 20). These lectures provoked strong condemnation, some of which came from the scientific and medical communities, signifying their ambiguous attitude towards vivisection at that time (French[2], 1975: 18–21).

The ideology of the French physiologists is illustrated by Monamy (2000: 12) when he quotes Bernard's 'powerful philosophical rationale' for vivisection taken from his 'landmark' 1865 publication, *An Introduction to the Study of Experimental Medicine*:

> Have we the right to make experiments on animals and vivisect them?...I think we have this right, wholly and absolutely....No hesitation is possible; the science of life can be established only through experiment, and we can save living beings from death only after sacrificing others.

In the same vein, Ryder (1983: 123) cites a particularly famous passage of Bernard, who, echoing Cartesian doctrine, described the motivations of the vivisector thus:

> The physiologist is not an ordinary man: he is a scientist, possessed and absorbed by the scientific idea that he pursues. He doesn't hear the cries of animals, he does not see their flowing blood, he sees nothing but his idea, and is aware of nothing but organisms which conceal from him the problems he is wishing to resolve.

Bernard's core position entailed a zero-sum ontology of human flourishing in relation to other animal species and a radical prioritisation of the pursuit of scientific knowledge over considerations of animal welfare. Both Smith and Boyd (1991: 3) and Garner (1998: 176) assert that the work of Bernard and Magendie stimulated public antipathy, which eventually resulted in the 1876 Act.

The birth of vivisection and anti-vivisection as British social movements

French (1975: 20) cites evidence that during the early Victorian period, a significant strand of medical and scientific opinion in Britain was critical of Magendie's vivisection lectures as 'unnecessary torture'. In general, however, medical and scientific bodies were supportive of animal experiments in principle and concerned about parliamentary attention that threatened to result in restrictive legislation. In a similar vein, Monamy (2000: 20–1) suggests that the British experimenters, who until 1860 were few in number (Turner, 1980: 83), were perhaps more sensitive to animal welfare and/or public suspicion of vivisection than their French counterparts, possibly reflecting a cultural difference between Britain and the Continent. Thus, physiologist Marshall Hall is said by Monamy (2000: 21) to have pioneered the discipline's response to 'increasing societal abhorrence of animal cruelty, including painful vivisection'. Reflecting the reaction to Magendie, Hall is said to have acknowledged that 'unnecessary, inept, and cruel experiments sometimes took place and felt that control and prevention of these would remove the stigma attached to vivisection' (French, 1975: 21). Thus, in 1831, Hall recommended the formation of 'a society for physiological research' (Hall, 1831; cited in Paton, 1993: 2).

The stated intention of this proposed society was the enforcement of principles designed to prevent perceived gratuitous infliction of pain on animals, while still giving ultimate priority to the pursuit of knowledge as directed by researchers.[3] Nevertheless, Smith and Boyd (1991: 249) note that Hall's fellow physiologists did not take up his call for explicit self-regulation.

Rising concern for animals and condemnation of cruelty became institutionalised in Britain during the first half of the 19th century (French, 1975: 25). In 1824, two years after the first limited anti-cruelty statute was passed, the Society for the Prevention of Cruelty to Animals was formed at a meeting in London by public figures with reputations for social reform, including several MPs and clergymen (Ryder, 1989: 90). The Society grew and established links with the aristocracy, attracting the interest of the soon-to-be Queen Victoria in 1835 and subsequently earning the 'Royal' prefix in 1840 (French, 1975: 27; Ryder, 1989: 90–2). Ryder (1989: 91–2) suggests that this signalled that the Royal Society for the Prevention of Cruelty to Animals (RSPCA), and the cause of animal protection in general, had become a recognised and esteemed social movement. However, in the process, the Society had become more conservative as the more radical, active figures who had helped establish the Society were marginalized in favour of a more aristocratic composition and outlook.

The RSPCA was first stimulated to concerted agitation against vivisection by revelations that horses had been subjected to the practice for surgical training purposes in French veterinary colleges, reports which also attracted criticism from the British medical press and the mass media (Turner, 1980: 85; French, 1975: 30). The significance for the politicisation of the issue in Britain was 'that the RSPCA's initiative stimulated wide discussion of experiments on living animals on both sides of the channel, much of it colored by the undoubted brutality of the French veterinary schools' (French, 1975: 31).

The RSPCA went on to investigate vivisection in Britain, and formulated their crucial policy in 1863 that expressed opposition to any *painful* vivisection: i.e. without anaesthesia[4] (French, 1975: 32). During the 1860s and early 1870s, medical journals were not completely opposed to the RSPCA's general approach, although they defended vivisection for experimental purposes rather than surgical technique and demonstration purposes, countenanced the infliction of pain 'in rare cases' and, significantly, opposed government regulation in favour of formal professional self-regulation (French, 1975: 33).

The politicisation of the vivisection debate

The interaction between vivisection and animal protection

Both Smith and Boyd (1991: 249) and Rupke (1987a: 5) argue that the decisive factor behind the birth of an organized anti-vivisection movement in Britain was the emergence of experimental physiology as an academic discipline in the 1860s. Rupke (1987a: 6–8) avers that this development can be explained by changes in British scientists' beliefs about knowledge and related perceptions of professional interests. In the first instance, having trained in continental vivisection laboratories, influential British physiologists followed Bernard in according experimentation a higher status than observation as a means of advancing knowledge. Thus, from the physiologists' perspective, vivisection, as an experimental method, was important as a defining feature of proper science. Therefore, animal experiments were seen as facilitating the professionalisation and institutionalisation of physiology and 'could be used as a legitimising factor for the professional and social ambitions of certain scientists' (Rupke, 1987a: 6).

At the same time, scientific research was becoming more closely integrated in medical practice as a result of the aspiration of medicine to elevate its status from art to science, and simultaneously elevate the status of its practitioners. Furthermore, most of the British medical profession perceived that they were falling behind their counterparts elsewhere in Europe. Therefore, it was seen as desirable to increase physiological education and research and recognise the embryonic profession of experimental science whose focus was the testing of physiological hypotheses through vivisection (French, 1975: 10–1, 39–44). As a consequence, vivisection began to be institutionalised. In 1870, leading medical institutions such as the Royal College of Surgeons indicated their desire to increase animal research activity to French and German levels and to expand the use of animals in training (Brooman and Legge, 1997: 125). Also that year, British vivisectionists were appointed to prestigious new positions at universities and other establishments (French, 1975: 41–4). However, the patient-centred, observation tradition partially persisted in British medicine, which explains the residual divisions in medical opinion over the vivisection issue in this period.

French (1975: 35) comments that, until this time, British medical scientists had been able to dampen public consciousness of the issue by emphasising the small scale of vivisection and the concern of experimenters to minimise pain, relative to continental researchers. However, this distinction lost credibility after 1870, '...thereby allowing anti-vivisectionists to argue with some plausibility that British experimental medicine in the seventies was neither more modest in its demands nor more ethical in its conduct than the French or German varieties' (French, 1975: 35).

The interaction between public opinion and vivisectionists: self-regulation as attempted depoliticisation

The institutionalisation of vivisection and an awareness of public concern led the scientific community to respond to calls in the medical press for explicit self-regulation (French, 1975: 44–6). Thus, in 1871, the British Association for the Advancement of Science (BAAS) published guidelines recommending that anaesthetics be used whenever deemed possible; painful experiments to demonstrate known facts were not justifiable; and that painful experiments to advance knowledge only be conducted by skilled persons with adequate equipment 'in order that sufferings inflicted may not be wasted' (quoted by Smith and Boyd, 1991: 250).

However, Smith and Boyd (1991: 250) relate two events that undermined the BAAS's aim to reassure the public with this code of practice (see also French, 1975: 46). Firstly, in 1873, the *Handbook for the Physiological Laboratory* was published. According to Richards (1987: 127), it was written by four of the leading medical scientists of the time, led by editor and co-author John Scott Burdon Sanderson, who had been a student of Bernard's at Paris and was at the time Professor of Practical Physiology and Histology at University College, London. This first English language manual in experimental physiology aimed to instruct new British physiologists in the methods of the authors' French and German teachers, and gave an unprecedented insight into the vivisectionists' methods (Smith and Boyd, 1991: 250). Richards (1987: 127) asserts that its publication signalled a 'watershed in the transmission of Continental methods to British laboratories'. The perception of 'suspicious' foreign practices being imported to Britain is said by Turner (1980: 89) to have: 'profoundly disturbed many of those who monitored Britain's mind and morals for the newspapers and reviews'.

Sanderson's *Handbook* is said to have fuelled the controversy because it appeared to indicate the physiologists' 'apparent indifference to pain, evidenced by its failure to specify the need to use anaesthetics' (Radford, 2001: 68). Furthermore, the BAAS's guidelines were not even mentioned, despite the fact that Sanderson had been on the committee that had recently drawn them up. Rupke (1987a: 9) argues that this indicates that the BAAS's attempt at encouraging voluntary self-regulation by experimental physiologists, like Hall's before it, was ineffective. Underlying the failure of self-regulation was a belief system where animal welfare considerations were overridden by the fundamental priorities of 'scientific curiosity and professional ambition' (Rupke: 1987a: 9)[5]. Although Sanderson later claimed that anaesthetics were not mentioned because their use was routine and therefore taken for granted, French (1975: 48–50) observes:

> The importance of the publication of the *Handbook* at this time can scarcely be overestimated.... At last British antivivisectionists had a clear mandate to scrutinize the institutions and literature of British physiology for cruelty in the new and growing practice of animal experiment.... The *Handbook* provided ample cause for vigilance on the part of the animal protection movement.

The second event that is said by Smith and Boyd (1991: 248) to have undermined public faith in vivisectionists occurred in August 1874 and involved the abandonment of the annual meeting of the British Medical Association (BMA) in Norwich. An outcry had ensued over the public vivisection of two dogs by the French physiologist Eugene

Magnan. The RSPCA attempted to prosecute in December 1874 under existing anti-cruelty legislation, but the move was unsuccessful because Magnan had fled to France, and the three Norwich doctors who had organised the demonstration had not actually conducted the vivisection procedure. However, a rare partial success was achieved because the magistrates declined to award defence costs as they felt that the RSPCA had been justified in bringing the case.

This series of events, particularly Magnan's attempted demonstration in Norwich and the ensuing trial, intensified public concern over vivisection. Turner (1980: 91) asserts, 'the RSPCA had stoked to white-hot intensity the public outrage against vivisection'. French (1975: 57–8) relates that 1874 saw steadily increasing public interest in vivisection, and the Norwich trial received national coverage, while editorial opinion 'was extremely suspicious of the practice of vivisection'. In fact, publications which were later to defend vivisection were highly critical of the practice at this point. In the face of a widespread lack of trust in physiologists, 'Experimental medicine seems to have been in retreat before an increasingly negative public opinion' (French, 1975: 57–8).

Professional groups: strategic actions stemming from perceptions of interests and structural constraints

French (1975: 52) helps to illuminate the evolving response of the scientific and medical communities to this situation when he observes that in 1874, parts of the medical press, including the *British Medical Journal* 'closed ranks in explicitly criticizing medical professionals who attacked vivisection in order to exert professional pressure upon them'. Henceforth, such critics were labelled 'eccentric', 'ignorant' and 'publicity-seeking'. Meanwhile, other periodicals, such as the *Lancet*, were said to be more circumspect, while one or two were highly critical of animal experimentation. Thus, 'Medical opinion was as divided as public opinion, if the medical press can be taken as any indication' (French, 1975: 53). However, in the *Lancet's* opinion, the justification of experiments was best left to scientific experts as the only people with sufficient expertise, rather than 'such ignorant bodies as the Norwich Bench of Magistrates' (French, 1975: 59–60).

This evidence raises the key question of the attitude of the medical and science lobby to legislation on the matter, given the potential application of existing law to vivisection as highlighted by the awarding of costs to the RSPCA following the Norwich trial. Indeed, the ambiguous legality of animal experiments was also hinted at by a motion passed at a convention of animal protection societies in 1874 (French, 1975: 54).

The *Lancet* thus recommended the establishment of a regulatory body that institutionalised scientific control over the statutory regulation of vivisection (French, 1975: 59–60, 64). The stated aim of such a mechanism would be to permit 'justified' experiments, defined as those where: 'experimenters were skilled persons with adequate scientific objectives' (French, 1975: 59–60). In other words, as Radford (2001: 68) puts it:

> [T]he scientific community, fearful that the legality of vivisection would fall to be determined by magistrates, as a result of their interpretation of the anti-cruelty legislation, were moving towards the view that legislation *specifically permitting* the practice might be required.

This point relates to the research question addressed in this chapter: Who benefited from the assent of the Cruelty to Animals Act 1876? Interestingly, the observation that researchers themselves were, to some extent, favourable towards the principle of legislation is absent from interpretations of the 1876 Act as a compromise between the opposing factions. It also raises question marks over the pivotal assertion that the purpose of the 1876 Act was to protect animals. Alternatively, the Act may have been designed to defend scientists' freedom to experiment on animals, in which case the institutional framework for animal research policy may have incorporated structural links with research interests from the point of the Act's assent. Ascertaining the veracity of these two interpretations requires further evidence regarding the political processes leading up to the assent of the legislation.

Animal protection and anti-vivisection: strategic action in the midst of structural opportunities and constraints

For the anti-vivisectionist cause, the Norwich case brought unprecedented publicity and sympathy and 'permitted the evolution of an explicitly anti-vivisection movement' (French, 1975: 61). Significantly, it attracted the attention of the journalist, women's rights and anti-poverty campaigner Frances Power Cobbe, who was to become the most high-profile anti-vivisectionist of the late Victorian era (Ryder, 1989: 109). She published the first anti-vivisection pamphlets in Britain at the time of the Norwich trial, interpreting the failure to secure convictions as an indication that existing legislation was inadequate to deal with the cruelty of vivisection (French, 1975: 62). Thus, Cobbe perceived the necessity of legislation for precisely the opposite reasons as stated by the *Lancet*: '...[to] remov[e] such decisions from the sphere of medical and scientific self-interest' (French, 1975: 64).

Perceiving the crucial role of the RSPCA's organisational resources, Cobbe set about drawing up a 'memorial' – an open letter – addressed to the Society in attempt to stimulate it to Act against vivisection. The memorial cited the Norwich incident, expressed concern about the expansion of vivisection in Britain, and called on the Society to test existing law and hence, if necessary, initiate restrictive legislation that required experiments involving pain to be conducted by licensed individuals at certified places (French, 1975: 65). Cobbe's high public profile and connections with prestigious literary, religious and political figures assisted her efforts to endow the movement with credibility (French, 1975: 62). Furthermore, the support she received for her memorial indicates the strength of anti-vivisection sentiment among the elites of Victorian society. According to Ryder (1989: 111):

> This memorial was signed by seventy-eight medical practitioners, by many peers and bishops, and by such illustrious Victorians as Cardinal Manning, Lord Shaftesbury, W. E. H. Lecky, the Reverend B. Jowett, John Bright, Major-General Sir Garnet Wolseley, Thomas Carlyle, Alfred Tennyson, John Ruskin and Robert Browning.

One important advocate for the anti-vivisection cause was Queen Victoria. While Cobbe was assembling her memorial, the RSPCA received a letter and donation expressing Victoria's concern about cruelty to animals in vivisection. In the coming years, she would also communicate her opposition to vivisection to leading figures such as the eminent surgeon Joseph Lister and Prime Ministers Disraeli and Gladstone (Ryder, 1989: 111). As Ryder (1983: 134) explains, 'After nearly forty years on the throne, Queen Victoria was a deeply respected figure and it would be rash to underestimate the force of her influence upon respectable opinion'.

In contrast to Victoria's consistent and deeply-held concern, the RSPCA was perceived as equivocal on the question of vivisection (Ryder, 1989: 116). The RSPCA's stance was crucially important because it was by far the pre-eminent animal protection group in this period: 'By the seventies, the RSPCA was a large, wealthy, powerful, and prestigious organization' (French, 1975: 81). The opportunities stemming from the RSPCA's structurally-enhanced legitimacy resources are indicated by Radford (2001: 79–82) when he explains how Victoria's support for the Society:

> gave the cause of animal protection a degree of status and influence out of all proportion to its relative novelty, especially in the early years [of her reign].... In an age of deference, the RSPCA's association with

the monarchy attracted the support of the rich and powerful...which alone made it a force to be reckoned with.

The Society had also accumulated significant organisational resources as it had become professional and adept at pressure group activities such as public education, lobbying and legal challenges. Furthermore, the Society 'had nurtured a growing relationship of mutual trust with government, slowly building an impressive body of legislation for animal protection' (French, 1975: 81). However, French (1975: 81) suggests that the RSPCA's political resources were essentially the consequence of a moderate 'modus operandi'. The implication is that such a style of activity had become institutionalised and perceived as beneficial. Therefore, in respect of Cobbe's memorial that was presented in January 1875, the RSPCA is said to have 'dithered when it came to decisive action' (Ryder, 1989: 112).

Radford (2001: 68) suggests one additional factor behind the Society's response was its ambiguous position on the ethics and preferred legal framework for regulating animal experiments. French (1975: 64, 66) opines that the Society's restrained reaction may have been due to the variety of opinion on the issue therein and its links with influential supporters of animal experimentation, including in government. Both Radford (2001: 83–4) and Rupke (1987a: 3–5) explain the RSPCA's ambivalence in terms of a class bias in Victorian animal protection activity, which meant that working-class cruelty was targeted, while élite practices such as hunting and, to a certain extent, vivisection, were relatively ignored. French (1975: 61) also notes that the new anti-vivisection movement which came to the fore after Norwich had a constituency that was far from identical to the RSPCA's and that there was something of a love-hate relationship between the two camps. He explains this friction in terms of the anti-vivisection movement's much deeper suspicion towards science and technology (French, 1975: 83). The conduct of the RSPCA in keeping anti-vivisectionists at arms length is also said to have been, to some extent, forced upon the Society as a result of 'the hysteria and sensationalism that were [also] to discredit the movement in the eyes of a significant proportion of the press and the public' (French, 1975: 82). Nevertheless, French (1975: 66) suggests that the RSPCA's actions at this juncture may have had far-reaching consequences for the strength of the anti-vivisection movement:

> The response of the RSPCA to the proposals of the memorial was crucial. Really decisive action by the society at this point, of the kind

Cobbe envisioned it taking, might have maintained for it an exclusive franchise on effective action from the humanitarian side and an undivided constituency in anti-vivisectionist opinion. (French, 1975: 66)

However, the publication of a letter from Dr George Hoggan in the *Morning Post* on 2 February 1875, which amounted to a powerful indictment of Bernard's vivisection demonstrations that he had witnessed, had a momentous impact in further inflaming public opinion against the practice (French, 1975: 68). As a consequence, public concern was so intense that the anti-vivisectionists, led by Cobbe, 'no longer needed the wealth and power of the RSPCA to gain access to Parliament' (French, 1975; 68).

Vivisection enters the parliamentary arena: private members' bills

Cobbe seized this window of opportunity and utilised her access to Parliamentarians to organise the introduction of a bill to regulate vivisection through Lord Henniker in the House of Lords in May 1875. As Hampson (1987: 314) argues, at this stage, rather than seeking total abolition, anti-vivisectionists were mainly 'reformers, believing vivisection could be controlled through legislation to prevent animal pain and ensure public accountability'. There was, however, a considerable amount of public support for outright abolition, which meant that Cobbe's initiative was perceived as credible (French, 1975: 69). The main provisions of the Henniker Bill were: firstly, that experiments should only be performed at inspected, annually-registered locations; and secondly, all procedures must be performed under anaesthetic, unless a six-month licence were granted to an individual researcher at the Home Secretary's discretion.

French (1975: 71) observes that, having been informed of Cobbe's memorial to the RSPCA at the beginning of 1875, a group of physiologists and other sympathetic scientists, including the influential figure of Charles Darwin, had anticipated the forthcoming political battle and came together in: '... the preparation of a legislative measure to take the wind out of the enemy's sails.... From this group, a scientists' lobby to protect experimental medicine was to emerge'. Hence, within a week of the presentation of Henniker's Bill, a rival bill had been tabled by the MP Lyon Playfair which, according to Ryder (1989: 112), was 'instigated by scientists such as John Burdon Sanderson and T. H. Huxley, who wished to maintain almost complete freedom of research'. Playfair was an established parliamentary representative of the interests of the medical profession (French, 1975: 73). Radford (2001: 68–9) explains that this alternative bill sought to enshrine the automatic legality of 'painless' vivisection, and allow the Home Secretary to issue licences (for a period of five years (French, 1975: 75)) to researchers to permit them to conduct painful experiments.

At this point in the debate, with the issue relatively novel and complex, the position of some of the prominent actors within the conflicting groups was changeable (French, 1975: 79). This may have contributed to the situation as perceived by French (1975: 75):

> The very significant differences revealed by closer examination of the bills should not obscure the fact that the parties were very close. The differences in question were more the product of two independent attacks on the same complex question, without benefit of precedent or any extensive experience with similar legislative problems, than they were of deep seated irreconcilable viewpoints on what was or was not permissible in the way of experiments on living animals. After all, both bills allowed for painful experiments under appropriate conditions.

However, in addition to the proposed duration of licences to perform painful experiments, a major divergence between the bills occurred around the issue of inspection. This may reflect the sharp aforementioned distinction between the position of the *Lancet* and Cobbe over where the authority to make licensing decisions should reside: a fundamental question of power distribution. For, unlike the anti-vivisection proposal, the scientists' bill contained no provision for inspection, reflecting professional resentment at being supervised by others deemed to be of equal or lesser status (French, 1975: 77). Combined with the proposal to allow complete freedom for painless experiments, anti-vivisectionists feared that, in the absence of inspection, researchers could escape any control of their research simply by categorising it as painless. It is therefore possible that French underestimates the difficulty in reaching a compromise between the two lobbies, although, to a certain extent, both appear to have been responding in a pragmatic fashion to structural constraints and a shared perception of the desirability of some form of legislation. This would also account for attempts by both sides to appear moderate and reasonable by proffering licensing regimes and networking with a broad range of opinion.[6]

Ultimately, both bills failed due to a combination of poor drafting and lack of parliamentary time (Brooman and Legge, 1997: 125). But with the controversy unabating, the government moved to instigate a Royal Commission to investigate vivisection and make recommendations on legislation (Smith and Boyd, 1991: 250–1; Monamy, 2000: 22–3). Thus, vivisection had finally been forced upon the government as a policy issue.

Characterising the embryonic policy network

With the government's establishment of a Royal Commission, the contours of a policy network were beginning to take shape as state actors started to take an active interest in the vivisection controversy. In order to understand this process, it is useful to analyse the empirical evidence discussed thus far in terms of the Marsh/Rhodes typology.[7] During the earlier discussion of the respective traits of issue networks and policy communities, it was noted that nascent policy networks tend towards the issue network end of the spectrum (Smith, 1993a: 10; Hay and Richards, 2000: 7). It should not be surprising, therefore, to find a number of issue network characteristics at this stage.

Firstly, in terms of the *membership* dimension, there were numerous active participants[8] in the politics of vivisection from many shades of opinion, and, furthermore, Parliament was involved. Secondly, *integration* was low in the network, especially between groups and government policy-makers, who did not appear to have definite preferences regarding the outcome, judging by the manner in which the Royal Commission was established. There was also a degree of fluidity in the policy stances of many actors. Although many parties agreed on the need for legislation, there was conflict about its content. In terms of those élites engaged in policy formulation at this stage, such as the aristocracy, the clergy and the professions, overall there was clearly significant dissensus.

Thirdly, the *resource* dimensions of the network also indicate an issue network: the government's lack of intrinsic interest and the apparent absence of institutionalised relationships with actors signifies its perception of the groups' resources as of limited, though perhaps growing, value. Meanwhile, members of groups on both sides were not acting in a uniform manner, with internal schisms over values and strategy. And fourthly, at this stage it would be difficult to discern the balance of *power*, but certainly, in many respects, this was a zero-sum game. This is due to the fact that the key issue between the lobbies was not whether there should be legislation but rather, and more fundamentally, whom the legislation would empower to make decisions regarding the licensing of vivisection. This is signified by the fact that even those actors on the scientific and medical side, such as the *Lancet*, who had reservations over the morality of vivisection in certain circumstances, were strongly committed to the principles of professional status, autonomy and, thus, self-regulation (see Table 5.1).

This analysis of the animal research policy process identifies an embryonic issue network. However, an adequate understanding of this (or any) policy process requires attention to the dialectical interactions over time among the network, the actors and their structural context. Nevertheless, this preliminary characterisation helps to provide a base level from whence the subsequent dynamic process can be described and analysed in the remainder of this chapter.

Vivisection as a policy issue

As state actors began to interact with group actors with a view to formulating animal research policy, vivisection had finally arrived as a public policy issue with a corresponding nascent policy network that, at this stage, was engaged in formulation rather than implementation. The rest of this chapter therefore traces the development of the debate and the network up until the assent of the 1876 Act.

The Royal Commission on vivisection, 1875–6

According to French (1975: 92), in appointing members of the Commission, the then Home Secretary Richard Cross 'attempted a balance of scientific expertise, practical medical experience, judicial wisdom and humanitarian zeal'. The access of anti-vivisectionists to government and the policy formulation network at the time is demonstrated by the fact that Cross was advised on this matter by another Cabinet Minister, the Earl of Carnarvon, who was an anti-vivisectionist sympathiser (though the only one in the Cabinet) (French, 1975: 92). However, Carnarvon appears to have perceived the need for an effective Commission that could command widespread confidence, and thus recommended that most of the appointments not be committed participants in the controversy. French's (1975: 92–6) detailed discussion of the Commission's composition implies that it was reasonably balanced, with each side of the debate being represented by committed advocates, together with four further members with apparently more moderate positions.

The Commission heard witnesses through the latter half of 1875. For the anti-vivisection lobby, the RSPCA gave the pre-eminent submission (French, 1975: 102). True to its official policy, the RSPCA presented draft legislation that would abolish painful experimentation, with a licensing regime for premises and individuals engaged in painless experiments. However, although the RSPCA generally impressed the Commission, the Commission determined that such an Act would have prevented some useful experiments. Other anti-vivisectionist witnesses either made little impression or, in the case of the Society for the Abolition of

Vivisection, succeeded in antagonising the Commission. Cobbe, who may have had a beneficial impact on behalf of anti-vivisection, did not testify at all because, according to Turner (1980: 91), she lacked confidence in the Commission. The variable quality of the relations between anti-vivisectionists and this policy formulation body could be a useful indicator of the movement's potential effectiveness.

An important issue of contention concerned the distinction between vivisection aimed directly at medical advances and that aimed at the production of abstract or 'pure' knowledge (French, 1975: 101–3). Anti-vivisectionists were particularly critical of vivisection for the purpose of generating abstract knowledge, despite physiologists' assertions that the ultimate benefits of pure research were unpredictable. Indeed, this particular suspicion of pure research was shared by some prestigious medical professionals who otherwise supported vivisection. Nevertheless, the evidence of several scientific and medical professionals in defence of the ethics and utility of vivisection is said by French (1975: 103) to have been more persuasive to the Commission.

But the greatest impact on the Commission is said to have arisen from the professed indifference to animal suffering of some of the pro-vivisection witnesses (Radford, 2001: 69). By far the most significant of those was Dr Emanuel Klein, who had contributed to the *Handbook*: both Dr Klein's evidence and the *Handbook* appeared to indicate that Cartesian ideology remained prevalent among the vivisection community. His admission that he deliberately ignored animal suffering during experiments, together with his assertion of similarity between the attitudes of British and Continent physiologists, had the effect of uniting the Commission in convincing them of the need for regulation, for they deemed that vivisection was intrinsically liable to abuse, even when practiced by 'eminent' scientists (Ryder, 1989: 113; Radford, 2001: 70–1). Meanwhile, Cross's subsequent diaries recorded the public outrage in response to the publication of Klein's testimony (French, 1975: 106).

The Commission published its report in January 1876, proposing a law to regulate animal experimentation, the stated intention of which was 'to reconcile the needs of science with the just claims of humanity' or 'to restrict experimentation within what society deems to be acceptable limits while causing least harm to free scientific enquiry' (Hampson, 1987: 219). Thus, the final report ruled out the abolition of vivisection on the grounds that:

- 'it "would not be reasonable" since "the greatest mitigations of human suffering have been in part derived from such experiments"' (Radford, 2001: 70).

- there was a possibility of law-breaking or vivisection being transferred abroad which would 'certainly result in no change favourable to animals' (quoted by Radford, 2001: 70; Ryder, 1983: 134).
- the expert use of anaesthetics could greatly mitigate pain (Radford, 2001: 70; French, 1975: 107).

But the Commission recommended the avoidance of severe and protracted suffering and the infliction of what it deemed *unnecessary pain* due to unskilled or 'inhuman' practitioners (Radford, 2001: 70). The report went on to express confidence that its proposals for a system of licensing would be supported by 'the most distinguished physiologists and the most eminent surgeons and physicians' (quoted by Radford, 2001: 70). Although the recommended licensing regime partly borrowed from the RSPCA's proposals, the Commission further countenanced:

- painful experiments on animals under licence
- experiments solely for the advance of knowledge
- experiments for demonstration purposes at medical colleges, conditional on the use of anaesthetics to prevent any pain (French, 1975: 103, 108).

French (1975: 106) describes the report as: '...a curious and somewhat equivocal document, reflecting differences of opinion between the commissioners'. However, according to French (1975: 110), most members of the opposing lobbies were content with the Commission's recommendations, though for differing reasons. Anti-vivisectionists interpreted the report as recommending legislation because of the recognised potential for abuse (though they dissented on the Commission's conclusion that no actual abuse had been uncovered). Meanwhile, the physiologists' supporters rebutted the evidence of cruelty by British practitioners and claimed that regulation was being proposed as a necessary evil to appease ill-informed public distrust (French, 1975: 110). However, it can be argued that a pro-vivisection leaning in the report is indicated by the fact that the only dissenting voice on the Commission came from the anti-vivisectionists' main representative, Richard Hutton. He described the report as a partial whitewash of vivisection and appended a minority report calling for an absolute ban on the use of cats and dogs (French, 1975: 108–9). Perhaps most significantly, the RSPCA complained about the lack of rigour in the hearings, having preferred evidence to be given publicly and under oath, with cross-examination assisted by counsel. Furthermore, the RSPCA would not back proposals to allow painful vivisection.

During the period of the Commission, in November 1875, Cobbe had responded to her frustration with the RSPCA's apparent inaction by forming a new society that was dedicated to restricting vivisection (French, 1975: 85; Ryder, 1989: 114). She successfully attracted patronage from prestigious figures such as Lord Shaftesbury and others from Parliament, the aristocracy, literature, and the church, as a perceived essential means of garnering public attention and support. Consequently, the group, which later came to be known as the Victoria Street Society (VSS), dominated the anti-vivisection movement (French, 1975: 88). As the issue became increasingly politicised, and anti-vivisection disillusion with the RSPCA grew, other societies were founded, such as the London Anti-Vivisection Society, the Society for the Abolition of Vivisection and the International Association for the Total Suppression of Vivisection. However, these organisations were relatively small, prone to infighting, and more uncompromising in their demand for total and immediate abolition (French, 1975: 89–90). Paradoxically, unlike the RSPCA, the VSS supported the Royal Commission report and lobbied for Government legislation in line with its recommendations (French, 1975: 112). This suggests an interesting strategy on the part of the VSS, who, ostensibly, were more radical than the RSPCA, and indicates that their strategic action was affected by their perception of the constraints posed by their structural context. Thus the anti-vivisectionists' disenchantment with the RSPCA may have been more to do with their strategy and inaction than ideological differences.

The passage of the Cruelty to Animals Act 1876

In the first two months following the report of the Royal Commission, the Government was non-committal regarding its response, including the question of legislation (French, 1975: 113). Then, in March 1876, support from prestigious figures such as Lord Shaftesbury and Cardinal Manning enabled the VSS to be favourably received by the Home Secretary, and they were invited to submit suggestions for legislation (Ryder, 1983: 135). At this juncture, the anti-vivisection lobby (excluding the RSPCA) appear to have been seizing the initiative. The anti-vivisectionist minister Carnarvon, with whom the VSS had close connections, introduced a government bill into the Lords in May.

Carnarvon's original bill

In introducing the bill's second reading, Carnarvon encapsulated the conflict of interests and values involved in the vivisection controversy and indicated where the balance lay in the initial version of the bill:

'His legislation attempted to "reconcile the high laws of modern science with the *still higher* laws of morality and religion"' (French, 1975: 114; emphasis added here). The bill was similar, in certain respects, to that proposed by both the RSPCA and the Royal Commission, particularly in terms of the licensing and inspection arrangements, though it reflected the Commission's recommendations in envisaging painful experiments under certain circumstances 'if certified as necessary by certain specified individuals, such as the presidents of various scientific and medical bodies' (French, 1975: 115).[9]

However, both French (1975: 115) and Smith and Boyd (1991: 251) identify the main controversy over the bill as surrounding those provisions that went further than the Royal Commission in proposing to:

- ban any and all experiments on dogs and cats (a key feature of Hutton's minority report following the Royal Commission)
- prohibit vivisection for the sole purpose of advancing knowledge.

With legislation imminent, Balls reports (1986: 6) that this point saw the first sign of organised lobbying from research interests. This was the crucial phase of political activity that gives an insight into 'the interplay of interest groups on an issue of relatively small importance to Disraeli's government' (French, 1975: 116).

Group reaction to the bill

Radical abolitionists from the animal protection and anti-vivisection side rejected the bill, but those with a restrictionist strategy, such as the VSS, welcomed the measure. For the RSPCA's part, French (1975: 118) notes the Society's failure to anticipate the emergence of the bill, which was perhaps an indication of its inconsistent attention to this issue at the time. While happy with the stricter welfare measures compared with the Royal Commission report, the RSPCA protested at the provision for painful experiments. Interestingly, they proposed to Carnarvon an amendment to the effect that certificates to allow painful experiments should be conditional on a positive character reference by a *non-scientific* individual of high standing, although this suggestion does not appear to have been adopted. In fact, this proposal echoed a broader anti-vivisection concern about the potential inability of the Home Secretary to justify any restrictions on animal experiments due to a lack of organizational resources within the Home Office and the claims to hegemonic knowledge-based authority by the scientific profession (French, 1975:

178). However, following the Royal Commission, the RSPCA's participation in the formulation of legislation was generally reactive and limited (French, 1975: 111).

While the bill in its original form is said to have been generally supported by the mainstream press (Ryder, 1989: 115; French, 1975: 117–8), the medical and scientific press condemned it vociferously and mobilised against it. The pro-vivisection lobby focussed particularly on those two points that went beyond the Royal Commission's report and in pursuit of a further amendment that would have required any prosecutions under the Act to be approved by the Home Secretary, thereby preventing anti-vivisectionists from launching prosecutions against researchers (French, 1975: 118–20). According to Monamy (2000: 23), 'Many [scientists] claimed the right to use any animal species for any purpose'.

French (1975: 121–2) identifies two factors that had a major impact on the trajectory of the bill at this point. Firstly, Carnarvon's mother died, and he had to leave London, resulting in a delay to the passage of the Bill. This gave the scientific lobby crucial time to marshall their case and organise support. In the context of a divided and uncertain cabinet,[10] the absence of the bill's sponsor weakened the position of the bill as it stood in the eyes of the government. Secondly, the points where Carnarvon's bill was more restrictive than the Royal Commission report provided an opportunity to discredit the bill.

The impact of the medical profession

It was in this context that the relatively weak experimenters' lobby (French, 1975: 151) successfully engaged their colleagues in the medical profession in pressing the government for amendments. French's (1975: 150) overview of the status of the British medical profession at this time gives a vital insight into the considerable resources at its disposal and thus is essential to an adequate understanding of the emerging animal research policy network:

> British medicine in the seventies was a profession of increasing unity and power. It enjoyed a high degree of autonomy, which was ceded to its leadership by legislation, especially from the fifties onward, and was based upon an impressive array of institutions: hospitals, medical schools, Royal Colleges, the governing General Medical Council, and the mass membership British Medical Association. The profession had achieved a good deal of collective political experience, originating with the issues of public health and professional qualification in the forties and fifties.

The General Medical Council (GMC), for example, was piqued by the perceived failure of the government to consult it on a matter related to its legal responsibilities for medical education, and recommended amendments to the bill:

- allowing cats and dogs to be used in experiments
- the permissibility of experiments for the pursuit of knowledge
- the permissibility of experiments aimed at the advancement of veterinary medicine
- the requirement that inspectors be scientists (French, 1975: 124–5).[11]

In addition to the élite General Medical Council (GMC), the mass-membership British Medical Association (BMA), the Royal Colleges of Surgeons and Physicians, as well as scientific bodies such as the Royal Society and the Linnean Society were involved in pressing ministers for amendments on behalf of the physiologists. The question of professional pride and status appears to have been at least part of the BMA's motivation for supporting the GMC's stance (French, 1975: 125), and the BMA submitted a 3,000-strong petition of medical professionals to the Home Secretary (Ryder, 1983: 135). Thus, as Hampson (1987: 315) remarks, Carnarvon's Bill was opposed by 'a small core of experimental physiologists [who] succeeded in mobilising almost the entire medical profession'. Similarly, Radford (2001: 71) comments that the medical establishment was to the fore of the scientists' response.

The potential effects of this deployment of political strength can be understood when placed in the historical context of a period 'of professional independence and deference from politicians' (French, 1975: 11). Consequently, with several speakers referring to medical opposition to the bill, by the time the measure had passed through the Lords, Carnarvon had been forced to agree to amendments that allowed experiments for the advancement of knowledge, for veterinary medicine, and on cats and dogs if a certificate endorsed by scientific and medical bodies were obtained. French (1975: 127) comments: 'there can be no doubt that the medical and scientific interest had gained a considerable victory'.

With the bill about to enter the House of Commons, French (1975: 128) notes that the intervention of the medical profession (especially the GMC and the BMA) during the delay caused by Carnarvon's bereavement had been decisive. Now, rather than being satisfied with their gains in the Lords, the pro-vivisection lobby sought to press home their advantage, with 'spectacular' lobbying of the Home Secretary Cross during July 1876. The BMA agitated its members across the country to write to their MPs and local press. The display of 'professional power and unity' also had the effect of attracting support from the national press, many of whom had backed the original Bill (French, 1975: 132–3). Indeed, a university MP closely linked with the research lobby, and the *Times* newspaper, both criticised the notion that research should be at all constrained by a Home Secretary lacking scientific expertise (French, 1975: 178). In general, the perception of vivisection had shifted from that of a dubious continental practice to something that was integral to British medical and scientific prestige.

The impact of the medical profession is also revealed by the fact that, by this time, the RSPCA and the VSS had changed their position from tentative aspirations for a stricter bill than Carnarvon's original version, to one based on not conceding any further amendments. Meanwhile, several new anti-vivisection societies were springing up demanding total abolition and intensely petitioning the House of Commons to that effect (French, 1975: 129). Conversely, 'The objectives of the research and medical communities were to render the Bill 'innocuous' so that it might serve the purpose of soothing the agitated public while imposing no real restrictions on fundamental or medical research' (Hampson, 1987: 315).

In respect of this power struggle, French (1975: 133) observes: 'The Victoria Street Society, whatever the influence of its individual representatives, scarcely counterbalanced the BMA'. Indeed, such was the power of the pro-research lobby that, despite warnings from the RSPCA of public unrest, and their objections to what they characterised as legislation that would protect experimenters rather than animals, the Home Secretary informed their deputation: 'Of course, you must be aware that at this period of the session it would be absolutely impossible to carry a bill of this kind in the face of opposition from the medical profession and scientific world' (quoted by French, 1975: 133–4).

Thus, despite reactive counter-lobbying by anti-vivisectionists, including the RSPCA (Ryder, 1989: 116), the political context meant that the government was persuaded at a private meeting with key members of the scientific and medical lobby that their requested amendments had to be met for any legislation to be passed (French, 1975: 138). The government's principle motivation for persisting with the bill was, according to French's (1975: 150) interpretation, 'simply to reduce the general level of public concern over the issue by using the Act to spike the guns of the anti-vivisectionists'. Thus, on 9 August 1876, the Home Secretary published and introduced a bill in the Commons that contained all but one of the amendments agreed with the scientists' lobby[12] (French, 1975: 140).

Hampson (1987: 315) describes these amendments as: 'so substantial as to change its fundamental nature'. Not surprisingly, according to Brooman and Legge (1997: 126), these 'provisions...were much more favourable to the research community than the provisions of the original Bill, and difficult to enforce...'. Indeed, previously implacable medical opponents to any legislation, on the grounds of its perceived slur on their honour, were now recommending the acceptance of a 'harmless bill' (French, 1975: 140). The amendments won by the pro-vivisection lobby included:

- occasionally allowing experiments to take place at private addresses
- certificates would be required for procedures on cats, dogs and equidae only when anaesthesia was omitted
- prosecutions could only take place with the consent of the Home Secretary
- only vertebrate animals would be covered by the Act (French, 1975: 138–9).

According to Balls (1986: 6), the researchers' amendments had rendered the bill 'much weaker' to the extent that it deviated further from the Commission's recommendations than Carnarvon's original bill. By this stage, many of the leading anti-vivisection figures from the RSPCA and VSS would have preferred the bill to have been thrown out in its final reading in the Lords on 11 August. However, Lord Shaftesbury, then president of the VSS, decided to permit the bill to proceed because of his belief that it was better to have some piece of legislation which could then be built upon in the future (French, 1975: 141–2). Hence, the bill was given Royal Assent on 15 August, and is characterised by Ritvo (1984: 59) as: 'representing a legislative victory for scientists'. However, French (1975: 142–3) suggests that the government was the most satisfied of all the concerned parties, having dealt with a controversial issue of no perceived benefit to itself. Meanwhile, the partisans remained uncertain as to the practical consequences of the Act once implemented.

The provisions of the Cruelty to Animals Act 1876

The uncertainty regarding the impact of the new statute stemmed from what French (1975: 177) describes as 'the extraordinary degree of discretionary power that the Act conferred upon the Secretary of State for the Home Office'. Those decision-making powers concerned:

- the granting of licenses and certificates to permit persons to conduct painful experiments on vertebrate animals
- the duration, revocation and conditions of licenses and certificates, consistent with the Act
- the registration of addresses where experiments took place
- the form and content of reports submitted by licensees regarding experimental results
- the appointment and activity of Inspectors
- permission for prosecutions.

There were absolute bans on:

- experiments aimed at improving surgical skills
- any public performance of painful animal experiments.

But, in general, the text of the Act reflects the shifting balance of political power during its passage. Many of the restrictive clauses were subsequently attenuated by provisions for the issue of certificates of exemption. Thus, the permitted purposes of vivisection were:

- 'the advancement by new discovery of physiological knowledge'
- 'knowledge which will be useful for saving or prolonging life or alleviating suffering'
- (with a certificate) if judged 'absolutely necessary' for the teaching of physiological or medical knowledge
- (with a certificate) if judged 'absolutely necessary' to repeat a previous experiment for the sake of advancing knowledge.

The impact of certification on the Act's initial stipulations on the use of anaesthetics and the protection of favoured species was particularly pronounced:

- Animals had to be anaesthetised throughout experiments to prevent them feeling pain, but a certificate permitted anaesthesia to be withheld if it interfered with the research objectives.
- Animals had to be killed before recovery from anaesthetic if they were likely to be in pain or had suffered serious injury, but once again a certificate was available to permit the researcher to keep the animal alive until the object of the experiment had been attained.
- No unanaesthetised cats, dogs, horses, asses or mules were to be used in experiments, unless a certificate were obtained on grounds that

anaesthesia and the use of another species would frustrate the object of the experiment.

The Act also required that licence or certificate applications be approved by one of the Royal Societies, Colleges or the GMC. Furthermore, a licence or certificate application was also to be signed by a university professor of medicine or biological science, such as physiology or anatomy, unless the applicant were such a professor. Considerable practical authority appears to have been granted to these scientific bodies, as this section (11) also stated:

> A certificate under this section may be given for such time or for such series of experiments as the person or persons signing the certificate may think expedient. A copy of any certificate under this section shall be forwarded by the applicant to the Secretary of State, but shall not be available until one week after a copy has been so forwarded.

However, the final decision, at least formally, appears to have rested with the Secretary of State, who 'may at any time disallow or suspend any certificate given under this section'.

Characterising the animal research policy network in 1876

This narrative of the genesis of the 1876 Act has sought to scrutinise Garner's two key claims underpinning his perception of a nascent animal research issue network:

- The purpose of the 1876 Act was to regulate animal experimentation.
- The Home Office did not have a pre-existing structural relationship with the research community.

In addressing these issues, this analysis adopts the dialectic model of policy networks discussed in Chapter 2 in order to provide a detailed description of the purposes of the Act, the relations between government and interest groups, and other relevant features of the evolution of the animal research policy network, including its structural context and the actors within it. Thus, in addition to a characterisation of the network and the balance of power as expressed in the Act, phenomena with the potential to affect the trajectory of this policy area are also elucidated.

The purpose of the 1876 Act and government-vivisectionist relations

Firstly, Garner's two key claims will be discussed. In relation to the question of the purpose of the 1876 Act, in strictly formal terms, it could be said to have introduced a regulatory framework in the guise of a regime of licensing, certification and inspection. However, this chapter has demonstrated that the most influential actors on both sides of the debate believed that statutory regulation was desirable, though for differing reasons. In fact, the focus of dispute was on the extent of any absolute restrictions contained in the Act and the question of who would be empowered in its administration. It appears that the scientists' lobby achieved the vast majority of their aims in both these respects, and so it is reasonable to suggest that the potential impact of the Act would be one of facilitating, rather than restricting, vivisection. Indeed, the perceptions of most actors on either side of the debate appear to confirm this as the most plausible interpretation of the Act, albeit partially conditional on expectations of the Home Secretary's exercise of discretion.

In understanding the purpose of the Act and this policy process in general, it is also necessary to consider its political context, particularly the pattern of relationships between professions and the state, and pre-existing institutional models that may have affected the creation of new institutions. Thus, as French (1975: 148–9) observes, it is instructive to note the parallels between the passage of the 1876 Act and the broader pattern of administrative innovations in Victorian Britain. French (1975: 149) states that, at first, '...legislation was usually ineffective because of the lack of practical experience with the problem and the lobbying of the endangered interests before passage through Parliament'.

Similarly, Moran (2003: 42) discusses the development and replication of a cooperative style of regulation in the Victorian period, where newly organized professions (in the present case, experimental scientists or physiologists) modelled their political goals on established professions (e.g. medicine). These activities were based on '...a powerful ideology of self-regulation: legal authorization went with a light touch from the state, an emphasis on cooperation within the profession itself and a distaste for the imposition of sanctions' (Moran, 2003: 42). Thus, inspectorates, which were the established mechanism for formal regulatory enforcement in Victorian government, in practice eschewed the imposition of legal penalties and instead employed a collegial or 'gentlemanly' approach that resulted in informal self-regulation. This was an approach rooted in 'the wider culture of club government' (Moran, 2003: 61–3).

140 *The Politics of Animal Experimentation*

In the case of medicine, Moran (2003: 49) notes the existence of established, close relationships between medical institutions (the same ones who lobbied on behalf of the physiologists and were increasingly engaged in vivisection) and government predating the Medical Act of 1858, which had reorganized and empowered the medical profession. This observation raises doubts over the second proposition of an absence of institutionalised relationships between the Home Office and the research lobby prior to the 1876 Act. Strictly speaking, although the Home Office may not have had formal relations with animal researchers, the advocates of vivisection had institutionalised relations with government in other networks that represented both a considerable resource in the conflict over the passage of the Act, as well as a potential network-on-network constraint on the new issue network.

Animal research policy network dimensions at the assent of the 1876 Act

In the following analysis of its evolution so far, the network will be initially characterised in the terms of the Marsh/Rhodes typology, which will incorporate a discussion of the exogenous constraints and dynamics, as well as the conduct of strategic actors. The previous analysis of the period up to the Royal Commission identified a number of factors that combined to form an issue network pattern of relations. With the subsequent intensification of political activity up to the point of the assent of the 1876 Act, it will be interesting to trace the dynamics of this process and relate them to any changes in policy network dimensions and potential future trajectories.

The first task involves ascertaining the boundary and hence membership of the policy network rather than the broader policy universe. In this case, insufficient time has elapsed for any consistent pattern of policy outcomes to emerge that would allow assessment of whether groups were authentic members of the network or merely peripheral insiders. Therefore, membership will have to be decided on the basis of observed interactions over this period of policy formulation.

However, in relation to the network's *membership* (see Table 5.1 below), there appear to have been some subtle shifts over the previous year. In some respects, the pattern of membership for the anti-vivisection lobby was particularly sporadic. For example, while the VSS did not directly participate in the Royal Commission, they were able to achieve some quality access for a subsequent period, as reflected in the formulation and presentation of Carnarvon's original Bill. The RSPCA, on the other hand, appears to have been partially marginalized following the Royal

Commission report because of its formal opposition to the permissibility of painful experiments and its inability to quickly adapt to the situation where such experiments were inevitably going to be made possible by future legislation. Similarly, following the Royal Commission, more radical abolitionist groups enjoyed significant public support and continued to lobby Parliament but were excluded from other arenas of policy discussion, particularly with the government.

In general, anti-vivisection ideology comprised the following beliefs, arranged from core to secondary in descending order:

- a metaphysical rejection of the medical utility of vivisection on the grounds that no good can come of a perceived evil
- a normative opposition to the infliction of pain
- a prioritisation of animal welfare over the pursuit of knowledge
- a distrust, to varying degrees, of reductionist science and medicine
- an empirical rejection of the medical utility of vivisection in favour of clinical observation and anatomy as the basis of medicine
- a lack of faith regarding physiologists' capacity for effective self-regulation and a perception of sadistic cruelty in vivisectionist practice
- a need for legislation that enshrined public accountability into any regulatory regime.

The pro-vivisection lobby, consisting of a broad range of professional groups from experimental science, other areas of science and the medical profession, moved quickly to intensify their access to both Parliament and the government during the passage of the bill. Interestingly, although Parliament was a key player in the network at this time, insofar as its opinions (which were themselves strongly affected by pro-vivisection lobbying) constrained the government, the final provisions of the bill were privately negotiated by the government and the scientific lobby, and then passed by Parliament with relatively little discussion and scrutiny. However, in summary, the membership dimension of the policy network continued to indicate an issue network, though in some respects the network was evolving away from this end of the typology. In general, pro-vivisection ideology comprised the following beliefs, once again arranged from core to secondary in descending order:

- an ontology that perceived biology in similar mechanistic and reductionist terms as other natural sciences

- an epistemology that perceived the strong validity of extrapolation of knowledge from animal models to human medicine
- a metaphysical position that saw nonhumans as instruments of humans, and humans and nonhumans in a zero-sum relationship
- a normative prioritisation of pursuit of knowledge for its own sake[13] over animal welfare
- professional autonomy and self-regulation, accompanied by a perception of opponents, especially non-scientific critics, as 'ignorant'
- a need for legislation to institutionalise scientific control over animal experiments while assuaging lay agitation.

Likewise, the network's *integration* dimension maintained a broadly issue network form. Over this period, access for conflicting groups fluctuated in frequency and intensity, corresponding to major shifts in the policy position in terms of the relative prioritisation between the values of animal welfare and the pursuit of knowledge. The entry of both the VSS and then the medical lobby into the network was a causal factor in such changes. However, the provision in the Act for the participation of scientific and medical bodies in the licensing and certification of researchers appears to signal a shift from consultative relations to implementation relations for these actors. Once again, this indicates a possible change in this aspect of network integration towards the policy community end of the spectrum.

In order to analyse the *resource* dimension of the network, it is worth initially noting a straightforward resource advantage for the scientific lobby stemming from the particular political circumstances of the passage of the bill. Namely, that beyond the broad model of inspectorate-based regulation: '... suggestions for legislation had to be developed de novo and with only the most grudging assistance from the group most capable of providing it' (French, 1975: 145).

Setting this factor aside, it is appropriate to employ the dialectical network model (see Figure 2.1) to help understand how actors' resources are related to the network and its structural context, and affect policy outcomes through strategic action. For example, the establishment of an Inspectorate appears to have represented a relatively conservative example of policy transfer from other sectors, and this institutional template may thus have had the potential to reproduce the growing structural power of professional elites in Victorian Britain in the form of self-regulatory, policy community-type arrangements. Meanwhile, the ideological and cultural context was relevant to the composition of the opposing lobbies (Rupke, 1987a: 7) and, relatedly, the political legitimacy resources of the participating actors. Firstly, the ideology of vivisection was attractive to the medical profession, who generally perceived it as a means of further increasing its effectiveness and, hence, credibility by endowing medical practice with an apparently more scientific foundation (Rupke, 1987a: 6–7; French, 1975: 411). At the same time, scientific expertise was generally being increasingly invoked as the basis of political and moral authority (Ryder, 1989: 117; Elston, 1987: 274). In conjunction with a significant culture of public and political deference to the medical profession, then it is clear that the pro-vivisection lobby enjoyed considerable political legitimacy.

However, 1876 was a time of considerable cultural conflict. Interestingly, many commentators perceive animal welfare as merely one, albeit significant, aspect of anti-vivisection sentiment that was essentially intertwined with other cultural and ideological currents that sought to elevate perceived moral values over scientific imperatives. Indeed, the the RSPCA, and particularly the VSS, enjoyed support from the Church of England, the aristocracy, the judiciary and literary figures because of wider concerns about the growing institutionalisation and power of science in Victorian society, at the expense of more traditional sources of authority. The support for anti-vivisection from elite groups is said by Rupke (1987a: 6) to have been a potentially decisive factor in the emergence of anti-vivisection onto the political agenda. It helped to attract public support for animal protection and anti-vivisection, which in turn was vital to fundraising for such groups, thereby providing financial and organisation resources, especially in the case of the RSPCA. Furthermore, elite support facilitated access to government, another valuable resource.

Nevertheless, the research lobby's success in amending the bill to produce an Act that they judged would not be inimical to their interests might be thought to signify an uneven resource distribution (in their favour) within the network, symptomatic of an issue network. However, there is no determinate relationship between resource distribution and policy outcome, because the effectiveness of resources depends, to an extent, on the deployment of actors' skills in the network and their strategic learning based on perceptions of what is possible and desirable, as well as chance events.

The scientific and medical lobbies were inevitably composed of highly-educated individuals with considerable expertise and skills. However, the testimony of Klein to the Royal Commission demonstrates that technical knowledge does not always translate to political acumen. Nevertheless, the medical lobby were able to put the skills learned in

Table 5.1 Politically-active groups 1874–6

Pro-vivisection lobby		Anti-vivisection (AV) lobby		
Overall character: Initial disagreements over tactics; commitment to vivisection and professional status of medicine and experimental science; science and advancement of knowledge seen as ultimate value.		Overall character: Divided, active in inverse proportion to endowment with resources. Suspicious of reductionist materialism embodied by vivisection. Morality more important than science. Religious idea that benefits cannot be derived from cruel acts.		
Actors engaged in vivisection	Other medical and scientific bodies	RSPCA	Victoria Street Society	Other AV groups[15]
Composition: Scientists at institutions such as University College London, the Brown Institute; Cambridge University; Edinburgh University. Physiological Society.	Composition: Eminent scientists such as Darwin, Huxley, Simon (Medical Officer of Local Government Board). Medical institutions: General Medical Council; British Medical Association; Royal College of Surgeons; Royal College of Physicians; Royal Society; Linnean Society. Press (during final stages of passage of bill): Lancet, British Medical Journal, Nature, the Times, the Standard, Pall Mall Gazette, Punch.			
Policy aims: Differences over desirability of legislation, but on the whole preferred some legislation if it protected their interests; research should be permitted on all species and for purposes of advancing scientific knowledge and veterinary medicine as well as promoting and protecting human health; research permitted at non-registered premises; prosecutions require permission of Home Secretary; inspectors must be scientists.	Policy aims: Prepared publicly to oppose legislation; commitment to professional autonomy of British physiology and experimental medicine; opposition to lay involvement in regulation and inspection.	Policy Aim: Officially, abolition of painful experiments, permissive of painless experiments under anaesthesia, but see below; involvement of lay authority in licensing.	Policy Aim: Restriction of vivisection through legislation involving registration of premises, licensing of all researchers at discretion of Home Secretary; inspection; no vivisection for pure knowledge; no vivisection of cats and dogs.	Policy Aim: Total abolition of vivisection.
		Other ideological features: As under 'overall character', but internal differences of opinion on vivisection; not universally hostile to 'science'; reluctant to criticise élite cruelty; fairly conservative and excluded women from leading positions.	Other ideological features: As under 'overall character', and critical of medical utility of vivisection; linked with other progressive causes; women heavily involved.	Other ideological features: As under 'overall character', though particularly suspicious of science.

Other ideological features: Mechanistic view of life; animals of principally instrumental value to humans; pursuit of scientific knowledge paramount.	*Other ideological features*: Pursuit of scientific knowledge paramount; resentment at legislation interfering with scientific and medical matters.	*History*: Born of frustration with perceived weak response to Cobbe's 1875 Memorial. Pre-eminent AV society.	*History*: Born of frustration with RSPCA and VSS 'moderation'.
Resources: Connections with medical profession, leading to support; connections with government and key MPs allowed them to control policy formulation; expert authority; highly-educated; some members politically clumsy.	*Resources*: Skilled and committed leaders able to mobilise profession and persuade politicians; professional power and unity; BMA – members and branches across country; possible veto power over legislation.	*Resources*: Links with rich and powerful Victorian élites, especially clergy, literary figures, aristocracy and Queen Victoria, and some medical figures; considerable financial and organisational resources; established, close relations with government (insiders); perceived as respectable, credible, and in tune with public opinion.	*Resources*: Good links with sections of Victorian élites, particularly clergy and literary figures; sporadic access to government; access to resources required for drafting and tabling of legislation.
Key strategic choices: Enlisted help of medical profession; exploited more extreme position of powerful medical profession while marginalizing them from policy discussion, thus wringing concessions out of government who wished for some legislation.	*Key strategic choices*: Used expertise and resources to lobby Ministers, Lords and MPs vigorously; proposed amendments while threatening to prevent any legislation; finally assented to legislation having won amendments.	*Key strategic choices*: Prudence and moderation led to widespread support; hesitant on specific policy and legislative goals regarding regulation of vivisection; distant relationship with dedicated AVs; possibly failed to seize window of opportunity in 1875, possibly because of internal division and/or professed links with pro-vivisection actors.	*Key strategic choices*: Took advantage of public outrage in 1875 to bypass RSPCA to gain degree of access to Parliament and Government; perceived that public did not support total abolition.
			Resources: Commitment and energy; politically naive and radical relative to public opinion; patchy credibility with press and public; outsiders in relation to government.
			Key strategic choices: Fractious relations with each other, the VSS, consistent public criticism of RSPCA; tendency towards hysteria and sensationalism; continued to pursue ultimate goal of abolition as short-term objective.

earlier political activities to effective use in their decisive agitation against the bill. Consequently:

> The ultimately dominant group within the scientists' lobby...chose to exploit the altered balance of power to achieve legislation that would protect experimental medicine from anti-vivisectionist harassment.... Thus, the nascent professional group – the scientists – achieved their ends by using a private, conciliatory style of politics that enabled them to utilise the power of the established profession. (French, 1975: 158)

For the animal protection side, the RSPCA in particular had cultivated considerable public education and lobbying skills. However, the deployment of these skills in this particular situation was somewhat hampered by a certain tentativeness and slowness to react to a rapidly changing political environment. This was perhaps a disadvantage of the cautious style that had also garnered them considerable prestige and allowed them to grow into a formidable organisation. Another potentially constraining factor on the RSPCA (which remains true today) was its broad animal welfare remit, which would create dilemmas for the Society regarding how it should prioritisation of its resources. This apparent inertia may also have been exacerbated by the paradoxical combination of a relatively strong policy against any painful experiments, together with a support base that was equivocal on the issue of vivisection. Conversely, the rather weaker VSS were in a better position to seize opportunities to act, but perhaps made a strategic mistake with the obvious restrictiveness of their proposed bill that was tabled by Carnarvon, as well as lacking the resources to win a set-piece political battle. Nevertheless, were it not for the chance event of the death of Carnarvon's mother, it is quite possible that their swift and agile strategic action may have paid greater dividends in the form of a more stringent Act.

The 'immediate abolitionist' wing of the anti-vivisection movement offered the resources of energy and commitment but tended to be politically naïve and prone to alienating moderate opinion in the press and the public. Their often tactless radicalism would certainly have put them at a disadvantage in lobbying government actors compared to the scientific and medical lobby who had considerable experience of the culture and 'rules of the game' of the metropolitan policy-making elites. Furthermore, the abolitionists were also wont to criticise those anti-vivisectionists who perceived restriction as a more feasible strategy (and vice versa). Conversely, although the scientific and medical lobby were not entirely united on all questions arising from the vivisection issue, there was a greater degree of unity, discipline and tactical flexibility on their part. Thus there was a varied and variable distribution of the resource of organisational unity in the network (generally to the advantage of the research lobby), another indicator of an issue network.

Finally, the *power* dimension of this network also exhibits issue network characteristics: the members of the network were often involved in highly conflictual, zero-sum power games in terms of many of the crucial provisions of the Act relating to permissible purposes of vivisection, which species could be used, and which bodies were empowered to participate in inspection and the approval of licences and certificates. However, the private negotiations between the government and scientists in the latter stages of this period to find an acceptable bill, appears to suggest power relationships indicative of a policy community, with the actors engaging in positive-sum, exchange interactions. Nevertheless, given that anti-vivisectionists can generally be considered as members of the network during this period, it seems reasonable to conclude that power was generally unequal in the network, if measured by the fact that the scientists' lobby won most of the policy battles.

Conclusion

This chapter has established that, up to the point of the 1876 Act, the animal research policy area most closely resembled that of an issue network. However, the purpose of the Act was equivocal at best, and in fact most evidence suggests that it was more protective of vivisectionists than animals. Furthermore, the Act appeared to potentiate institutionalised relations between the Home Office and those scientific and medical bodies formally empowered to participate in its administration, which would undermine the assertion that a previous absence of such relationships determined the evolution of this policy network.

The earlier review of the policy network literature generated the proposition that issue networks are particularly vulnerable to change due to exogenous pressure.[14] Similarly, it was proposed in Chapter 4 discussion of historical institutionalism and path dependency that the relatively weak degree of institutionalisation in issue networks may indicate that they are unlikely to persist without some 'higher' institutional framework to stabilise their structure. Given that the resource and power distributions in the network at this point appear to favour the pro-vivisection lobby, it seems plausible to suggest that the most likely trajectory from this point would have been outcomes (in terms of both licensing and certificate decisions as well as those affecting the decision-making structure) that tended towards the evolution of a policy community dominated by the scientific and medical communities.

However, the Home Secretary was granted extensive discretion in the operation of the Act, and the anti-vivisection movement's resources were not insignificant. This means that despite the apparent structural advantages enjoyed by the pro-vivisection lobby at this point, it would be unjustified to presume any future trajectory for policy outcomes. The next chapter addresses the second research question that asks whether the animal research issue network evolved into a policy community following the assent of the 1876 Act.

6
The Evolution of the Animal Research Policy Network: 1876–1950

Introduction

At the time of its assent, the implications of the 1876 Cruelty to Animals Act for the animal research policy network were uncertain. Despite the gains made by the animal research lobby during the passage of the bill, the legislation came into being in the midst of an issue network policy-making environment, and its impact was dependent on the extensive discretionary power afforded the Home Secretary. The future evolution of the network and the distribution of benefits represented by the implementation of the Act were of an unpredictable trajectory.

However, on a theoretical level, issue networks are relatively unstable policy-making environments. Furthermore, the existing literature indicates the development of close relations between animal research groups and the Home Office from 1882 following the incorporation into the administration of the Act of the Association for the Advancement of Medicine by Research (AAMR). This raises the possibility of a post-1876 network transformation towards a policy community model. A comparison with another Victorian Inspectorate regime – in air pollution policy, where a policy community also emerged – emphasised the plausibility of such a trajectory.

These considerations gave rise to the second research question, which asks whether the animal research policy network evolved into a policy community after 1876. Therefore, this chapter, like the previous one, utilises the available tertiary sources in order to present a narrative of this policy network's dialectical evolution from 1876 until the beginning of a new wave of politicisation following World War Two. The chapter is split into three main sections: the first deals with approximately the first six years of the Act's operation prior to the intervention of the AAMR; the second section then focuses on the impact of the AAMR; while the third part surveys the remainder of the period in question for evidence of significant events with the potential to change the network and policy outcomes.

What happened to the 1876 issue network? The early administration of the Cruelty to Animals Act

It has already been argued that the 1876 Act came into being in the context of an issue network. This proposition is reinforced by Rupke' (1987b: 188) observation that the legislation was a 'traumatic development in the still informal and largely untried relationship between scientists, the government and the lay public'. In other words, it was the relative absence of *institutionalised* relationships between policy actors in 1876 that indicate an issue network. This raises the fundamental question of how these relationships subsequently developed, a process which, according to French (1975: 179), provides 'a case study in the more general issue of the interaction of professional expertise and political power'.

Early Home Office implementation

There are two broad areas of the implementation of the 1876 Act that are relevant to an assessment of the animal research policy network:

- decision-making in connection with licence and certificate applications
- the detection and punishment of infringements.

Before these issues are examined, it is necessary to note that the appointment of inspectors to advise the Home Secretary and visit registered premises was a crucial feature affecting both of these factors. This process appears to have been significantly influenced by the perceived power of the pro-vivisection lobby, as the first inspector chosen by Home Secretary Cross is said to have been associated with leading proponents of animal experimentation (French, 1975: 179). In addition to the inspector's advice, Cross occasionally consulted London-based medical experts who were also supporters of vivisection.

Applications

In general, there is a consensus in the literature that the initial enforcement of the Act resulted in some limited restraint upon animal experimentation (French, 1975: 188; Ryder, 1989: 120). This is perhaps more noticeable in relation to the first aspect of the Act's implementation: the scrutiny of applications. The first two ministers after 1876 exercised 'remarkably' firm control according to Ryder (1989: 120), with 15 per cent of applications rejected, mainly on the grounds of excessive pain or lack of 'utility'. French (1975: 188) avers that it is likely that such refusals would have been in the face of contrary advice from the inspector, other scientific figures and, of course, the presidents of leading scientific and medical institutions who endorsed applications.[1] Assuming French is correct, the stance of the inspector and others supports Hill's (1997: 165–77) aforementioned argument that implementing officials tend to assume the interests and ideologies of regulatees with whom they share social and professional relationships.

Therefore, it is interesting to note that Sir William Harcourt, who succeeded Cross in 1880, initially refused a higher proportion of applications than his predecessor despite his noted sympathy for animal experimentation (French, 1975: 184–8). French (1975: 206) attributes this paradox to Harcourt's apparent lack of interest and knowledge about the Act. However, he (1975: 188) also provides an insight into the decision-making criteria applied by the Home Secretary at this time when he recounts how, in one rejection letter, Harcourt stated that he:

> does not wish it to be taken for granted that the discovery of every new poison is to be the reason for instituting a set of vivisection experiments unless there is some particular prospect of the utility of such experiments.[2]

It appears, therefore, that Harcourt's decision-making reflected an ideological position which, while not opposed to animal experimentation in principle, believed that certain extra-scientific conditions needed to be fulfilled in order for a proposed set of experiments to be justifiable. Given the inspector's apparently predominant allegiance to his professional colleagues in science and medicine, it appears that the level of scrutiny and public accountability to which vivisection was subjected in these early years was mainly due to the willingness of the Home Secretary to exercise his power of discretion in determining applications. It may also be the case that the relatively small-scale of vivisection at this time – 277 animal experiments were conducted in 1877 – facilitated a level of personal ministerial involvement that appears anachronistic from a modern perspective.

Thus, the pro-vivisection belief in the ultimate value of the pursuit of knowledge did not consistently dominate licensing decisions to the exclusion of anti-vivisection claims for animal protection. To a degree, the Home Secretary's implementation of the Act bore some of the features of the 'moral judgment' model of justice whereby a balance between competing social values was sought, one of the hallmarks of a pluralistic issue network-type of policy process. In contrast, the pro-vivisection lobby sought a 'professional treatment' model where regulation of activity would be controlled by the professions themselves, based on claims to exclusive decision-making competence and professional autonomy, a process that would resemble a policy community-type of network.

An additional determining factor is revealed by French's (1975: 189) calculation that the group most likely to have their applications refused were younger scientists based outside London. This suggests that political power in this policy area flowed from status and personal contact with prominent figures in the metropolitan scientific and medical elite.

Infringements

The other important aspect of the operation of the 1876 Act concerns the detection and punishment of infringements, because, as mentioned in Chapter 2, the capacity of grievance procedures to provide accountability affects the type of network, in particular through its implications for power distribution and outcomes (Hill, 1997: 178). French (1975: 191) implies that Inspectorate visits to registered premises were co-operative rather than adversarial, in proportion to the status of the researcher. Regulatory responses to anti-vivisectionist accusations of infringements were more rigorous, with the licenses sometimes refused or revoked if infractions were confirmed (French, 1975: 191–2). However, prosecutions proved to be difficult, confirming the RSPCA's objection to the Act that it was 'very unsatisfactory as a means of discovering offences' (Ryder, 1989: 116).

Richards' (1987) study of the methods and attitudes of nineteenth-century physiologists provides further insight into the effects of the 1876 Act. Richards describes the licences and certificates issued from 1876 to E.A. Schafer, a researcher at University College London, for the investigation of nutrition. Schafer was issued with certificates exempting him from the requirement to kill animals before anaesthesia wears off, and waiving the 'prohibition' on the use of cats, dogs, horses and monkeys. His experiments included the removal of organs or parts thereof from cats and dogs, with subsequent 'recovery' from the procedure to investigate effects on nutrition.[3] Although the number of experiments was restricted by the Home Office, when Schafer exceeded the permitted number, the Home Secretary informed him that he had decided 'to consider that the three unauthorized experiments were performed through inadvertence in failing to remember the limitations' (Richards, 1987: 143). Schafer therefore escaped prosecution or any infringement action, apart from a request from the Home Secretary to familiarise himself with the licence conditions.

Richards' (1987: 143) proposes that this episode demonstrates that the Act was a 'fine example of British compromise'. However, if this description is meant to convey the idea of a balancing of interests between the opposing lobbies then this is surely a dubious proposition, particularly in the light of the following justification offered by Richards:

> the Act was implemented effectively by the Home Office and scrutinised with care. It was a genuine attempt to allay public misgivings by being seen to regulate vivisectional activities, an attempt that was sufficiently rigorous to irritate and even sometimes impede the physiologists, yet at the same time one that continued to allow the great majority of experiments to proceed unhindered.

Indeed, the primary objection from scientists was that 'while grudgingly admitting that it protected them from malicious prosecution, [they] resented being identified by the stigma of legal restriction' (Richards, 1987: 143). This reflects a common theme in pro-vivisection ideology: a sense that the mere concept of external, legal oversight was an affront to their social status.[4] Such resentment would have been even more acute in respect of the threat of criminal prosecution. Thus, state actors would have been especially reluctant to authorise prosecutions because of the potential for intense political conflict to occur, not only with physiologists, but with their powerful allies in the medical profession. Such an arms-length, formal relationship with researchers would also have substantially diverged from the general pattern of co-operative regulation that became established at this time (Moran, 2003). Consequently, the capacity of grievance procedures to punish and hence discourage infringements is likely to have been attenuated. In conclusion, therefore, the failure to prosecute Schafer and the relatively marginal constraints on animal experimentation suggest, *pace* Richards, that the Act may not have quite represented a true compromise on the part of the researchers' lobby.[5]

Furthermore, Richards' reference to the Home Office's intention to *be seen* to oversee vivisection is revealing. It indicates that the desire for symbolic public reassurance was a significant factor underpinning the Act's administration. The stance of the Home Office is further revealed by French (1975: 206) when he comments that, contrary to contemporary complaints from the pro-vivisection lobby, 'the only evidence I have run across as to partisan feeling on the part of civil servants has indicated their sympathy with experimental medicine'.

In summary, when licensing and infringement action are taken into account, Brooman and Legge's (1997: 126–7) assertion that this period of implementation signified a major success for the anti-vivisection lobby should be viewed as something of an exaggeration. Instead, it would be marginally more (but not entirely) accurate to say that the Cruelty to Animals Act:

> interfered significantly with research in experimental medicine in Britain between 1876 and 1882.... To an important minority of applicants under the Act, the right to perform such experiments was denied... [and] did curb to some degree the extent of experiments on living animals during this period. (French, 1975: 191)

Therefore, it can be concluded that the following aspects characterised the Home Office's administration of the 1876 Act in the first six years of its operation:

- Inspector, advisors and officials were broadly sympathetic to animal research
- Status and personal contacts were an advantage for applicants
- Ministerial decisions introduced broader social values, such as the requirement to prove the utility of experiments beyond the 'pursuit of knowledge'
- Therefore, a small minority of license and certificate applications were refused
- However, infringement action was weak.

Pro-vivisection strategy and action

The pro-vivisection lobby also played an important role in the administration of the 1876 Act through the requirement for applications to be endorsed by representatives of leading scientific and medical bodies. In fact, refusals to endorse applications were a rare occurrence, which French (1975: 184) implies signified a general failure of such institutions to scrutinise applications to the extent expected by the Act.

Most importantly, the pro-vivisection lobby strongly attacked the notion that the recommendations of scientific and medical bodies should be questioned by political appointees such as the Home

Secretary (French, 1975: 192; Rupke, 1987b: 204). Indeed, numerous scientists' statements cited by French and Rupke support Richards' assertion above that opposition to lay and regulatory interference remained a core feature of pro-vivisection ideology. For example, an 1882 article by Professor of Physiology G.F. Yeo indicates that it had been assumed by animal researchers that the Home Office, which was perceived to lack the required expertise, would abide by the certification decisions made by the named scientific and medical authorities (French, 1975: 195n53).

Thus, the organized pro-vivisection lobby of 1876 had dissipated following the assent of the Cruelty to Animals Act (Rupke, 1987b: 188–9), presumably because they assumed the Act would not interfere with their activities. However, in 1877 the Physiological Society[6] presented a memorandum to the GMC asking for assistance in lobbying the Home Secretary against unexpected refusals and delays in the issuing of licences (French, 1975: 196). In fact, the GMC postponed action while it monitored developments, and the scarcity of refusals in 1878 and 1879, the Physiological Society's relative lack of autonomous power, and a strategy to avoid any public conflict with anti-vivisectionists that might politicise the issue, meant that there was little concerted action on the part of the pro-vivisection lobby until 1881 (French, 1975: 197).

However, the unexpected increase in the rate of refusals under the new Liberal Home Secretary Harcourt, combined with the increasingly aggressive agitation of anti-vivisectionists, spurred the Physiological Society into action. Thus, in 1881, the group produced another memorandum alleging obstruction of research by the Home Office's implementation of the Act and reversed its policy of non-confrontation by stimulating pro-vivisection columns in the media (French, 1975: 197–8). But it was the International Medical Congress in London in August 1881 that was to mark the birth of significant and organised pro-vivisection activity involving virtually the whole scientific and medical establishment (Rupke, 1987b: 188–9; French, 1975: 198–200). The impact of these actions on the policy network is discussed later.

Anti-vivisection politics: 1876–82

When the Act was first passed, the predominant perception within both the anti-vivisection movement and the RSPCA was that the legislation authorised researchers to inflict severe pain upon animals – actions that had otherwise been criminalised by previous legislation (Richards, 1987: 125; Ryder, 1989: 116; Ritvo, 1987: 160). This led anti-vivisectionists such as the VSS to become disillusioned with formal political institutions, and instead their activities became more focussed on direct emotional appeals to the public, which included angry denunciations of the scientific and medical establishment (Ritvo, 1987: 161–2). However, Ritvo (1987: 160–1) argues that while anti-vivisectionists intensified their political activities, the RSPCA's ambiguity towards middle- and upper-class cruelty led it to significantly reduce its attention to the matter, focusing its remaining efforts in this area on the complex task of trying to ensure enforcement of the new Act. At the same time, according to French (1975: 217), by 1881–2, public interest in the issue had somewhat diminished since 1876.

Nevertheless, anti-vivisection sympathisers such as Lord Carnarvon and, most saliently, Queen Victoria continued to lobby the Home Secretary over the 1876 Act's implementation, thereby contributing significantly to the minor but perceptible degree of oversight detailed above (Ryder, 1989: 111; French, 1975: 181). The VSS themselves did not entirely abandon legal and parliamentary activities, and their close monitoring of vivisection also encouraged the Home Office to scrutinise applications and investigate their allegations of non-compliance (French, 1975: 176, 191–2).

However, published data about the administration of the Act was relatively uninformative. Anti-vivisectionists were therefore unaware that a small proportion of applications were being refused, although, on the other hand, it could be discerned that the number of experiments increased every year (French, 1975: 176, 191). Together with their repeated failure to pass amendments to tighten up the Act (Ryder, 1989: 117), this set of circumstances appears to have fostered the perception within the anti-vivisection movement that the legal framework could not provide any meaningful protection for animals. It seems reasonable to argue that this view may have been exacerbated by the fact that there was no formal structure to enable anti-vivisectionists to participate in the policy process, thereby promoting a sense of exclusion and alienation among the movement. Hence, the anti-vivisectionists drew the conclusion that a more radical stance was necessary to even partially achieve their goals (Hampson, 1987: 314–5; Rupke, 1987b: 190), although whether the macro-level structural context enabled or constrained the realisation of their goals through this type of strategic action would remain to be seen.

The animal research policy network, 1876–82

The evidence presented so far indicates that, until 1882, neither of the opposing lobbies completely dominated animal research policy in terms of the consideration of applications and the policing of licensees and procedures. With 15 per cent of applications refused, despite the endorsement of influential scientific and medical figureheads, it can be seen that anti-vivisectionists did not consistently lose policy battles. The implication of this for the question of network boundaries means that anti-vivisection groups may be considered members of the policy network.

In assessing the evolution of the membership dimension of the policy network between late 1876 and 1882, it will be useful to compare this with the analysis of groups for the start of this period provided in Table 5.1. For the pro-vivisection side, the 'actors engaged in vivisection' acted as individuals and through the Physiological Society. Although other scientific and medical bodies were involved in signing licence and certificate applications, in some respects their relationship with the physiologists was less intense than had been the case during the passage of the Act. On the anti-vivisection side, despite the diminution of the RSPCA's engagement with the network and the VSS's sense of alienation, some policy impact was achieved. In general, therefore, it would seem appropriate to conceive of the policy network in terms that have at least some resonance with the issue network model. The network included a range of interests, although professional research interests were generally the most powerful.

The network's integration dimension also displays some issue network characteristics. Interactions between state actors and the interest groups varied in intensity. Scientific and medical groups, together with researchers themselves, generally had closer contact with the Home Office than anti-vivisectionists. However, their relationship was probably more distant in the case of applications and in response to anti-vivisectionist allegations of non-compliance, than the inspection of premises. The prestige of the applicant was a further source of variation in integration between regulators and researchers. On the anti-vivisection side, there was some access to government for prominent sympathisers and distant, but not entirely ineffectual, relations between the VSS and the Home Office.

The relatively loose integration is also evidenced by the fluctuations in the comparative salience of the competing values in this policy area. Thus, instead of consistent elite consensus, conflict sporadically ensued as a result of application refusals and delays caused by scrutiny of applications. The Home Office did not entirely share the ideology of absolute professional autonomy and the fundamental prioritisation of the pursuit of knowledge. However, the dissensus with anti-vivisection groups was, of course, more intense.

Policy outcomes were clearly affected by the uneven pattern of resource distribution in the network in favour of the experimenters' lobby, many structural sources of which, such as the considerable political resources of the medical profession, were described in the previous chapter. Furthermore, the form of the 1876 Act, particularly the lack of significant specific restrictions therein, was a considerable legal resource for the vivisectionists. In addition, the involvement of the scientific and medical élites in endorsing applications provided them with additional political and legitimacy resources. However, before 1882, there was no organised lobbying of the Home Secretary by the scientific and medical establishment. On the other hand, the opposing anti-vivisection lobby did enjoy limited political and legitimacy resources stemming from support from prominent figures such as the Queen, as well as a fair degree of public support. Therefore, in the absence of any pre-existing 'rules of the game' or dominant, institutionalised ideology, the degree of discretion afforded the Home Secretary meant that claims for the need to protect animals could be considered. The anti-vivisectionists were relatively weak but not completely excluded from the network. Another way of putting this is to note that the key aspect of the resources dimension is how groups' resources are perceived by government. During this period, it must be said that the resources of all groups were seen as limited by the government, a perception perhaps revealed by the lack of interest shown by the second Home Secretary, Harcourt, prior to the AAMR's intervention.

Meanwhile, the distribution of resources between participating organisations within each lobby was somewhat variable. On the pro-vivisection side, the group most eager to influence the direction of policy, the Physiological Society, was relatively weak, compared to more established medical bodies, and they could not agree on a direct approach to the Home Office over its operation of the Act. On the other side, relations between the various shades of anti-vivisection opinion remained fractious. Finally, power was certainly unequal, with pro-vivisectionists enjoying a significant, if not entirely decisive, advantage over their opponents in a zero-sum game. It does, however, appear that a positive-sum game was possible between the Home Office and compliant licensees, who both had an interest in stymieing anti-vivisection agitation.

In summary, then, it seems reasonable to conclude that the animal research policy network during this period possessed some characteristics of an issue network (see Table 6.1 below). However, the degree of openness that existed in the network rested largely on the Home Secretary's operation of his discretionary powers. The persistence or otherwise of pluralistic tendencies in the network would therefore appear to have rested on the stability of this arrangement.

Pro-vivisection strategic action: an issue network under pressure?

A number of changes in the network and its environment took place during 1881 and 1882 that were also associated with the strategic actions of policy actors, a process that will be analysed using the dialectical policy network model and critical realist methodology outlined in Chapters 2 and 4, respectively.

The International Medical Congress: the initiation of concerted pro-vivisection activity

The 1881 International Medical Congress (IMC) in London was 'arguably the largest and grandest medical congress ever held' (Rupke, 1987b: 190). The Physiological Society took the opportunity to make common cause with the prestigious and well-publicised gathering and its attendant medical professionals (French, 1975: 199). As a result, the defence of vivisection dominated the congress, culminating in a unanimous declaration that asserted the essential role of animal experiments to medical progress and 'that it is not desirable to restrict competent persons in the performance of such experiments' (Rupke, 1987b: 191–2). French (1975: 200) avers:

> There can be no doubt that the presentation of its case by the International Medical Congress was a substantial propaganda coup for the Physiological Society and it sympathizers. The wide publicity given the proceedings of the congress reached a tremendous audience....

In the terms of the dialectical policy network model, the physiologists had successfully applied their skills within a favourable, strategically-selective structural context. The context included the sympathy of the influential medical profession that enabled them to advance a highly credible pro-vivisection message to the public. Thus, the congress is said to have given great impetus to the animal research lobby, which resulted in increasingly hyperbolic claims for the utility of animal experimentation and virulent protestations against any restrictions on experimenters' activities (Rupke, 1987b: 195). Richards (1987: 125) argues that premature assertions of the medical benefits of vivisection served the deeper purpose of 'establishing physiology as independent science justified by the pursuit of knowledge for its own sake'. Thus Rupke (1987b: 203) observes:

> that the authority of science was at stake in the vivisection controversy is also apparent from repeated claims by scientific authors that only they are proper judges of the right or wrong of animal experimentation, not the lay public.

Indeed, pro-vivisection pronouncements in 1881 and 1882 proposed complete freedom for researchers and complained of the need to petition the Home Secretary for permission to experiment. According to Rupke (1987b: 204), they believed that 'any sort of restriction or supervision represented an infringement and a slur on an honourable class of men'.

There was one further consequence of the 1881 congress that is said by French (1975: 202) to have had a deep impact on the pro-vivisection lobby. The VSS had compared the transcript of the congress with information relating to the licenses and certificates that was published by the Home Office. Thus, a prosecution for unlicensed research was brought in late 1881 by the VSS against the 'very eminent' researcher Professor David Ferrier (French, 1975: 201).[7] On the basis of the congress accounts, and reports in the medical press, Ferrier stood accused of performing severe and invasive brain procedures on monkeys without the necessary licence. However, due to testimony from an attending physiologist, Ferrier persuaded the court that the procedures were, contrary to the published reports, performed by the licensed Dr Yeo (French, 1975: 201–2).

The failed prosecution of Ferrier in 1881 further undermined the VSS's faith in the potential for legal regulation of animal experiments and led them to adopt a total abolitionist policy. From the opposite perspective, despite the failure of the VSS's prosecution, this attack on Ferrier, combined with the restrictions being imposed by the Act, led animal researchers to perceive the urgent need for co-ordinated action to defend their interests (Rupke, 1987b: 188; Balls, 1986: 6; Ryder, 1989: 120; French, 1975: 203; Turner, 1980: 107–8). To this end, the powerful pro-vivisection coalition brought together by the IMC provided a valuable resource for animal researchers to utilise.

The impact of the Association for the Advancement of Medicine by Research

Thus, with the assistance of the profits generated from the IMC, the Association for the Advancement of Medicine by Research (AAMR) was formed in March 1882 at a meeting convened by the Presidents of the Royal Colleges of Physicians and Surgeons (Rupke, 1987b: 192). Indeed, the leading medical and scientific institutions were formally represented on the governing Council of the AAMR, indicating how once again the physiologists had harnessed the considerable resources of the scientific and medical professions in support of pro-vivisection aims (French, 1975: 204, 218). The composition of the IMC and the AAMR in the early 1880s revealed an unprecedented unity among the entire community of medical and biological scientists, even when the majority had little direct connection with vivisection:

> The new unanimity of the 1880s showed how much during the preceding decade the practitioners of the biomedical sciences had developed a sense of professional identity, closing ranks when outsiders demanded public accountability. (Rupke, 1987a: 202)

Despite the broad aim suggested by its title, the AAMR's fundamental goal was the removal of the restrictions on animal research caused by the implementation of the 1876 Act (Balls, 1986: 6; Rupke, 1987b: 192; French, 1975: 204). The Association conducted two types of activity in pursuit of that goal: public propaganda and private lobbying of policy-makers, particularly the Home Secretary.

Rupke's (1987b: 191) analysis of media coverage indicates that interest in vivisection had subsided markedly after 1876, and then re-emerged in 1881 and 1882 in response to the activities of the pro-vivisection lobby through the IMC and the AAMR. However, this re-ignition of the debate appears to have favoured the pro-vivisection argument that animal experiments provided major medical benefits, while on the other hand the anti-vivisectionists were perceived as increasingly extreme (Rupke, 1987b: 190–7; Ritvo, 1987: 162). The relative decline of the anti-vivisection lobby was exacerbated by their inability to attract public support from sympathetic scientists and doctors (French, 1975: 217).

The researchers' publicity campaigns appear to have complemented their lobbying efforts. Thus, French (1975: 217) suggests that the decreased level of public concern about vivisection in 1882[8] facilitated the Home Secretary's positive response to the AAMR's confidential offer to advise him on the administration of the 1876 Act. For it was public support for anti-vivisection groups that had enabled them to push non-scientific political actors to interfere with the freedom of animal researchers (French, 1975: 217-8). This relative susceptibility to public opinion is consistent with the issue network policy-making environment found in this area up until this point.

However, in contrast to the notion of policy process influenced by public opinion, the AAMR's approach to the Home Secretary was couched in terms of their allegedly unique possession of the technical knowledge required to judge the legitimacy of research proposals. The influence of this ideology was strongly related to the resource distribution between the actors. In the middle of the nineteenth century, powers of professional self-regulation had been legally bestowed upon leaders of the medical profession who were on the AAMR Council, resulting in a hierarchical profession where the elite could control their membership. This helped to ensure unanimity or discipline among the profession with regard to their position on vivisection, despite significant reservations amongst many scientists and physicians (Turner, 1980: 109). Consequently, the anti-vivisection movement was deprived of valuable scientific and medical support, while the pro-vivisectionists magnified their influence over government actors (French, 1975: 216; see also Moran, 2003: 49).

With the Home Secretary inclined to be sympathetic to the claims of scientific expertise, this proposal from a united scientific and medical establishment, with whom the Inspectorate had close relationships, was irresistible in the absence of any significant dissenting expert opinion (French, 1975: 217). Furthermore, the discretion afforded the Home Secretary by the 1876 Act meant there was no formal obstacle to such alterations in its implementation structure (French, 1975: 218). In summary, the political, organisational, information and knowledge resources of the professional groups were a major constraint on government action. Leading scientists and medical professionals had:

> mobilized the power and prestige of the [medical] profession through bodies such as the British Medical Association and the General Medical Council. As a result, the antivivisection movement and the Home Secretary ultimately found themselves confronting an established profession rather than a handful of scientists. (French, 1975: 215-6)

The essential change made is encapsulated by the Home Secretary's order that all applications under the Act should first be submitted to the AAMR for their comments before being considered by the Inspectorate (French, 1975: 207). The AAMR had thus succeeded in assuming the 'self-appointed role of advisory body to the Home Office in the administration of the Act' (Rupke, 1987b: 192–3). Importantly, whereas the Home Secretary had applied utilitarian criteria to the determination of applications, 'the AAMR rarely applied any binding criteria whatsoever to applications that came before it' (French, 1975: 219).

The pro-vivisection lobby's new, dominant role in the implementation of the Act undeniably represented a more integrated relationship with the Home Office. This new 'standard operating procedure' (French, 1975: 207) led to major policy changes that expressed a shift in the network's core values or policy paradigm:

- Expert knowledge alone had come to be seen as sufficiently competent to the task of dealing with the policy problem of determining licence applications
- Lay participation was seen as a hindrance to the policy process
- Professional autonomy was prioritised
- The goal of the pursuit of knowledge automatically took precedence over animal protection.

This ideological consensus among the scientific and medical profession (or so it appeared), and the Inspectorate, with which the Home Secretary tended to agree and in any case acquiesced, helped to stabilise the network and produce continuity in outcomes. This dominant appreciative system in turn affected how the groups' resources were perceived within the network, thereby helping to exclude groups with aims that conflict with those of the pro-vivisection lobby. Consequently, this arrangement remained in place for thirty years until the report of the second Royal Commission in 1913 (Turner, 1980: 109; French, 1975: 207; Balls, 1986: 7).

The outcome of this new situation is described by French (1975: 207–8) when he refers to Home Office records that reveal an unequivocal shift in policy outcomes whereby applications were almost invariably granted:

> Under the new procedures, experimental medicine in Britain enjoyed spectacular growth; the number of licenses increased from 42 in 1882 to 613 in 1913. The substantial alteration in the mode of

administering the Act, which effectively transferred decision-making on applications from the Home Office to the AAMR, was responsible for greatly facilitating the licensure of would be researchers.

Indeed, the Littlewood Report (1965: 7) states that the number of experiments rose from 270 in 1877 to about 800 in 1885. By 1895, the figure had risen further to 4,679. 1900 saw over 10,000 experiments, while in 1905 nearly 38,000 experiments occurred, mostly without anaesthesia. Thus, Brooman and Legge (1997: 127) conclude that the role of the AAMR 'radically transformed the Act's operation, helping to bring about a rapid growth in the use of animals in research'. Likewise, Balls comments (1986: 7): 'all applications for licenses and certificates were recommended by the Council of the AAMR, so rejection by the Home Office became a very rare occurrence'. When the Home Secretary did occasionally threaten to reject an application, the AAMR's control of the decision-making process and semi-official role put it in a powerful position to confront the Minister (French, 1975: 208–9). The AAMR also weakened the Act in other ways, for example, by advising applicants and complaining about bureaucratic delays within the Home Office (French, 1975: 209). Not surprisingly, infringement action remained slight, with no prosecutions in this period.

Ryder (1989: 120) provides the following summary of the result of these changes: '...the AAMR entered into a clandestine liaison with the Home Office, which in effect allowed the scientists themselves to control the administration of the Act'. Given the prominent role of active vivisectors in the AAMR's Council (French, 1975: 209), this implication of self-regulation and professional autonomy seems particularly apt. Ryder is, however, a committed critic of animal experimentation, so it is therefore striking that a commentator sympathetic to animal researchers, Turner (1980: 108), reaches a similar conclusion regarding the effect of the AAMR:

> Within months the AAMR had negotiated a comfortable *modus vivendi* with the Home Secretary. The AAMR Council became the vetting agency for all licence applications; its support meant virtually automatic approval. The physiologists could hardly have asked for more.

In fact, no commentator provides a detailed or specific refutation of this analysis.

Having achieved their primary aim, the AAMR's propaganda role was rapidly mothballed during 1884–5 (Rupke, 1987b: 192; Turner, 1980: 108) partly because, according to French (1975: 211), their '...quick accession to a crucial position in the administration of the Act, in what was at best a somewhat questionable arrangement, made further publicity positively undesirable'.

The anti-vivisection movement interpreted these changes as removing the degree of protection offered to animals by lay involvement in decision-making and attacked the AAMR for what they perceived as an underhand and secretive approach to achieving goals that could not be openly pursued (French, 1975: 212–3). However, the AAMR refused to be drawn into public debate, and the Home Secretary remained unmoved by anti-vivisectionist lobbying against the policy role of the AAMR (Turner, 1980: 108). As a consequence, anti-vivisectionist trust in the 1876 Act, the scientific and medical community and the administrative process dwindled yet further (French, 1975: 212–3). In this increasingly polarised atmosphere, anti-vivisectionists proclaimed science in general as dangerous and evil, a position that intensified their isolation from the policy network and the scientific and medical community, and eroded public confidence in their cause (Ritvo, 1987: 162–4). Furthermore, attacks on physicians' pro-vivisection claims lacked credibility because of public trust in the profession, a resource that had been harnessed by the physiologists who had enlisted their support (Turner, 1980: 111).

French (1975: 217) suggests that one of the major reasons for the pro-vivisection lobby's desire to prevent public debate was because of the controversial differences between the argument it presented to the public and, on the other hand, conflicting aspects of their ideology that they uttered in bureaucratic domains. In particular, one core aspect of the pro-vivisection lobby's belief system, as successfully conveyed to the Home Secretary, was that it was inappropriate for non-scientists to play a role in animal research policy decisions. However, this argument tended to be excluded from the public domain:

> The basic purpose of the AAMR was to allow medical scientists to choose their problems and carry out their research as they wished, without the application by nonprofessionals of criteria of acceptability deemed irrelevant by the investigators themselves....Even as the AAMR flaunted the utilitarian potential of experimental medicine before the public at large, it worked to remove the application of

utilitarian or any other non-scientific criteria to the research actually being carried out. (French, 1975: 219)

Citing French, Ryder (1989: 117) summarises the implications of the events of 1882 and 1883:

[I]t is, as French puts it, that the politicians revealed an 'awe of science' and a deference towards it which resulted in 'a measure ultimately administered to protect experimental medicine rather than restrict it, under which research upon living animals prospered as never before'.

Understanding the dialectical process of network evolution

Having described the changes in animal research policy-making that occurred in the 1880s, it is now appropriate to analyse this evolution in terms of the dialectical network model that elucidates the interactions among the macro-level structure, the policy network, the strategic actors and policy outcomes (see Figure 2.1). This will allow the second research question to be addressed: Did this policy network evolve into a policy community, and if so, how?

One of the key dynamics in the dialectical network model is agents' perceptions of policy outcomes, and their interpretative mediation of exogenous and macro-structural factors in the context of their network structure. This is how strategic action arises. It is clear that the strategic action of the physiologists and, subsequently, the leaders of the medical profession and other branches of science, involved the successful deployment of their skills and resources in the public domain.

This helped set the context for strategic action that affected the dialectical processes within the network: in particular, the structures and interactions within the network that affect outcomes.[9] The absence of institutionalised rules that characterised the early issue network (and may, if they had existed, have helped preserve such a structure) meant that there was considerable scope for *meta*-policy transformation – i.e. policy concerning decision-making structures that can, in turn, lead to fundamental changes in policy outcomes. Assisted by a beneficial structural context, pro-vivisection interests discursively constructed a resource interdependency – based on technical expertise – that favoured themselves over the Home Office and other actors. This affected the Home Office's perceptions and behaviour, leading to changes in the network structure and hence the pattern of policy outcomes. Thus, united 'meta-policy' action from the scientific and medical professions that was designed to

Table 6.1 Comparing the policy network before and after the advent of the AAMR

Dimension	1876–81	1882–
Membership:		
Number of participants	Fairly broad membership. Parliament *partially* involved through role (albeit attenuated) in passing of initial legislation. Public opinion had some impact.	Limited, anti-vivisection groups now excluded. Parliament and public bypassed during entry of AAMR.
Type of interest	Both pro- and anti-vivisection interests involved.	Only professional interests involved.
Integration:		
Frequency of interaction	Access variable for different interests and at different stages of process. Home Secretary relatively independent from both lobbies.	Regular interaction between pro-vivisection interests (such as AAMR) and inspectors over implementation, particularly applications.
Continuity	Fluctuating policy outcomes. Including, eventually, through pro-vivisection action, which changed structure of network.	Membership, values and outcomes now consistently pro-vivisection.
Consensus	High conflict between lobbies and in state–anti-vivisection relations. Instances of dissensus between Home Secretary and pro-vivisection interests.	Home Office and researchers now shared pro-vivisection ideology. Episodes of dissensus rare and consistently resolved in favour of group interests.
Resources:		
Distribution of resources within network	Uneven. Group resources perceived as of limited value by government. Pro-vivisection lobby had greater resources than anti-vivisection movement.	Scientists and pro-vivisection lobby now perceived by government as possessing all necessary resources valuable for policy-making.
Distribution of resources within participating lobbies	Pro-vivisection fairly united, except regarding tactics on lobbying Home Secretary. Anti-vivisection relatively split and ill-disciplined.	Hierarchical and united pro-vivisection lobby maintained discipline among members, facilitating influence over government while depriving outsider groups of valuable resources.
Power:	Unequal, involving zero-sum games between opposing lobbies, some positive-sum games between Home Office and pro-vivisection lobby.	Stable positive-sum power relationship between dominant pro-vivisection lobby and government.

alter the 'rules of the game' in this policy area had the effect of essentially usurping the decision-making powers of the Home Secretary.

Meanwhile, anti-vivisectionists were hampered in their strategic action by a lack of reliable information about the policy process, a lack of explicit support from scientific and medical professionals, and a tendency towards political naivety and radicalism that undermined their public credibility. This process may have been exacerbated by the RSPCA's disengagement from the controversy.

Instead of the inconsistent policy outcomes and conflict associated with the early issue network, policy decisions began to follow a consistent pattern, and the conflict within the network virtually disappeared. A core change in policy had taken place through the ejection of animal protection values from the policy process. The ideological structure of the network tightened around a consensus that fundamentally prioritised the pursuit of knowledge, as defined by research scientists.

From this analysis, it can be concluded that a major network transformation took place in the early 1880s whereby animal research policy came to be made in an environment more akin to the policy community model. These changes in the dimensions of the Marsh/Rhodes policy network typology are summarised above in Table 6.1. These findings also generate a number of analytical questions regarding the extant animal research policy literature and policy network analysis that merit further discussion.

Implications for animal research policy literature and policy network theory

New insights into animal research policy

Firstly, the foregoing analysis dismantles the case for a persistent issue network in this policy area by shining new light on the effects of the 1876 Act.

The perception in the literature that the 1876 Act had a noticeable impact on the activities of animal researchers appears to rest on Sperling's (1988: 45–7) work. However, an analysis of Sperling's text reveals that her conclusions regarding the *overall* impact of the Act rest on French's evidence of application refusals during *only* the first six years of its operation. There is no attempt to disaggregate chronologically the different phases and evolution of the Act's administration.

Specifications of the interests, ideologies and policy goals of the opposing lobbies and how these related to the evolving policy process are also too vague in preceding research. However, this study has demonstrated that, contrary to Sperling's account, the AAMR could not be said to have prevented the abolition of vivisection because that was not a realistic policy outcome in the early 1880s. Instead, the AAMR freed researchers from public accountability and hence the potential for significant restrictions on their activities.

The Act's prevention of 'outright abuses' is also said to indicate the significant incorporation of animal protection values in the regulatory system (Garner, 1998: 177). However, despite some of the concerns of the anti-vivisection movement raised by the indifference to animal pain professed by experimenters such as Klein,[10] the primary battleground was not over 'outright abuses' (which is not defined by Garner or Sperling, but presumably refers to deliberately sadistic or gratuitous cruelty) committed by vivisectionists. Instead, the policy debate concerned what the permitted purposes of animal experimentation should be, the degree of pain and suffering allowed in pursuit of researchers' experimental aims, and the processes whereby policy decisions were made. Thus, the prevention of 'outright abuses' is not really germane to the question of the nature of this policy network.

Furthermore, erroneous assertions of a consistent 'balancing of interests' by the Home Office are based not only on insufficient empirical analysis, but also on the postulation that animal researchers did not benefit from structural resource and power advantages. On the contrary, the present analysis has shown how vivisectionists made common cause with the powerful medical profession, who, compared to the anti-vivisection lobby, enjoyed decisive socio-economic privileges. Thus, the case for a persistent issue network overlooks the role of the medical profession and assumes a 'Westminster model' of power in the British polity. The Westminster model implicitly emphasises the role of Parliament in policy-making, the prioritisation of constituents' interests over expert claims, and bureaucratic insulation from group pressure. But in the case of the AAMR's seizure of power, it is hard to think of a clearer refutation of the Westminster model. Instead, the present study provides strong evidence for an élitist, asymmetric power distribution and structural inequality, where professional groups enjoy significant political and knowledge resources that act as a major constraint on government action and facilitate the exclusion of opposing interests from policy domains.

Implications for the policy network analytical framework

This study also raises the interesting question of the dynamics behind issue network-to-policy community transformations. In fact, one of the notable features of the variable policy network dynamics Table 2.2, which was derived from existing policy network literature, is that there has been a paucity of analysis concerning structural transformations (as opposed to ongoing fluctuations in policy outcomes) in issue networks. This may reflect the policy network literature's emphasis on the question of how the relative stability of *policy communities* comes under challenge.[11] Nevertheless, it may be possible to infer issue network dynamics from the scarce observations that have been made about the existence of such a network-type:

- Issue networks tend to exist in emerging policy areas where networks are forming and before group-state relationships become institutionalised (Smith, 1993a: 10; Hay and Richards, 2000: 7)
- Issue networks exist in policy areas where no threat exists to economic or professional groups (Marsh and Rhodes, 1992b: 254)
- The persistence of issue networks is facilitated by the institutionalisation of an open network structure, which results from a broader political culture and power distribution that values pluralism (inferred from Bomberg, 1998).

If these propositions are combined with the findings of the present study, it is reasonable to hypothesise that:

1. issue networks may transform towards the policy community ideal network type *soon after their formation* if:
2. the interests of economic or professional groups are threatened and...
3. in the absence of institutions or structures that ensure broad access to the policy-making process and some degree of state neutrality.

The lack of a structural foundation for a pluralistic, issue network policy process could be said to have removed an important political resource from both the anti-vivisection lobby and state actors. This, in turn, would have contributed to the observed hegemony of pro-vivisection interests in the network, which raises interesting questions about how the role of political authority and state interests is conceived in the policy network literature (see, for example, Marsh and Smith, 2000: 8; Marsh, 1998b: 189; Smith, 1993a: 10; Marsh and Rhodes, 1992b: 254). This policy area does not appear to have been intrinsically important to the government in the early 1880s, which is a scenario normally associated with the presence of issue network-type policy-making. However, the resources of the pro-vivisection lobby and the government's general desire to avoid political controversy helped to persuade the Home Office of the advantages of a stable and closed policy process centred around the AAMR. It therefore suggests a qualification to the proposition in the policy network literature that state actors are the main drivers of policy community formation. For, if government actors are not dominant in a network, policy communities may still emerge if the resource distribution is sufficiently advantageous for one set of interests. Thus, even apparently peripheral policy issues may not conform to the issue network model.

However, although policy communities are conceived as relatively stable entities, this was not, of course, the end of the dialectical policy process. As noted in Chapter 2, this type of secretive, one-sided policy has the potential to provoke a critical response due to perceptions of illegitimacy in both processes and outcomes. This is one source of potential destabilisation of the newly-established policy community. Indeed, it was mentioned above that the arrangement between the AAMR and the Home Office remained in place until the report of the second Royal Commission in 1913. The next section therefore examines the subsequent evolution of the policy network from the beginnings of the agitations that led to the second Royal Commission in the early 1900s, through to the beginning of the modern era of animal research politics shortly after World War Two.

The animal research policy network: 1900–50

Most of the pre-existing commentaries on animal research politics in the first half of the twentieth century imply that no significant changes occurred in the network during this period. Indeed, those authors who undertake a detailed examination of the impact of the AAMR point to their entry into the network as the crucial starting point for path-dependent policy-making. For example, French (1975: 215) cites pro-vivisection literature from the late 1950s and the 1965 report of the Home Office's Departmental Committee on Experiments on Animals (the 'Littlewood Report') to assert: 'From the late [eighteen] eighties on, the medical and scientific interest seems to have been generally satisfied with the administration of the Act. I believe this holds true to the present day'. If it is assumed, for the time being, that the ideologies of the pro- and anti-vivisection lobbies remained roughly similar until the time of French's work in the early 1970s, then this indicates a stable policy network until that point. Similarly, both Ryder (1989: 120) and Brooman and Legge (1997: 127–8) argue that these close, exclusive relationships between the animal research lobby and the Government remained in place at least until the Animals (Scientific Procedures) Act 1986.

However, it was noted in the previous section that the AAMR-Home Office relationship was disturbed in 1913 following the report of the second Royal Commission. In order to examine the subsequent effects of the policy network, the next section examines the dialectical processes surrounding the origins and impact of the second Royal Commission.

The second Royal Commission, 1906–12
Policy community success breeds external opposition

This lengthy inquiry is said to have been established by the Government in reaction to renewed public concern about animal experimentation (Radford, 2001: 71–2; Ryder, 1989: 139–40; Littlewood, 1965: 7). Ironically, as discussed in Chapter 2, this appears to have been a case where the very success of a policy community appears to have stimulated potentially destabilising activity from excluded interest groups (Richardson, 2000: 108). Mapping this onto the dialectical policy network model, anti-vivisectionist actors were able to learn about policy outcomes from Home Office statistics that showed relentless annual rises in the numbers of experiments conducted, most of which were conducted without anaesthetic (Littlewood, 1965: 7). This also affected the structural context, particularly public opinion, through both direct perceptions of the policy outcomes and vigorous public criticisms from anti-vivisectionist actors.

Kean's (1998: 139–41) account of this period of activity also indicates that anti-vivisectionists adopted the classic strategic activity of outsider groups, attempting to stimulate increased public opposition to vivisection as a form of external shock to the closed and stable policy community. In particular, two women, Louise Lind af Hageby and Liesa Schartau, perceived the need to obtain and publish first-hand testimony of vivisection. Thus, in 1903 they attended physiology lectures at one of the leading institutional centres for experiments on animals, University College London. The publication of their 'lurid' account, entitled *The Shambles of Science*, was 'a key moment in the anti-vivisection campaign' (Kean, 1998: 142).

Stephen Coleridge, the leader of the National Anti-Vivisection Society (NAVS, previously the VSS), subsequently quoted from the book the case of a brown dog and was successfully sued for defamation by one of the vivisectors involved (Ryder, 1989: 139). According to Kean (1998: 246), Coleridge was aware that this action might lead to a libel trial. As it turned out, despite Coleridge's defeat, the revelations of animal cruelty at the hearings outraged the public, encouraging anti-vivisectionists – with the support of Battersea Council – to erect a memorial statue of the small brown dog at the local Latchmere Recreation Ground in 1906 (Radford, 2001: 71–2; Kean, 1998: 152). The statue was the first to be erected to commemorate an animal killed by laboratory experimentation, bearing this 'provocative' inscription:

> In Memory of the Brown Terrier Dog Done to Death in the Laboratories of University College in February, 1903, after having endured Vivisection extending over more than Two Months and having been handed over from one Vivisector to Another Till Death came to his Release. Also in Memory of the 232 dogs Vivisected at the same place during the year 1902. Men and women of England, how long shall these Things be? (Kean, 1998: 153)

The 'brown dog' became a focal point for the controversy and a public order problem, with sometimes violent confrontations between medical and veterinary students attempting to remove it, and local residents defending the statue. According to Ryder (1989: 140): 'Parliament, the Battersea Borough Council, and public opinion generally, sided with the brown dog'. These events led to sufficient pressure on the Government for it to be persuaded to appoint a second Royal Commission of Inquiry into vivisection in 1906 (Radford, 2001: 72).[12]

The second Royal Commission

The general consensus in the literature discussing the impact of the second Royal Commission is that it did not result in significant changes to the administration of the Act. Interestingly, the second and third largest anti-vivisection societies refused to participate in the Commission, seemingly due to concerns about its composition and the likely dominance of a strongly pro-vivisection scientific lobby (Rogers, 1937: 7). Nevertheless Coleridge, representing the largest anti-vivisection group the NAVS, gave testimony that criticised the Home Office's 'improper confidential relations' with the pro-vivisectionist AAMR, and in his subsequent evidence, the head of the Home Office Inspectorate 'admitted that his department had been in continuous consultation with this organisation' (Ryder, 1989: 141). The president of the AAMR went on to admit to the Commission:

> that 'the object of the whole membership is favourable to the promotion of vivisection' and could not recall any occasion when the

association had finally refused to endorse an application for licensure. (French, 1975: 214)

In spite of this evidence, the Royal Commission dismissed accusations of Home Office pro-vivisection bias in its implementation of the Act and gave a general endorsement of the extant regulatory system (Littlewood, 1965: 8). An indication of the Royal Commission's satisfaction with scientific control over the administration of the Act is provided by French's (1975: 214) quote from the final report: '...all applications, coming as they do from and being recommended by competent persons, are granted. An absolute refusal is the very rarest occurrence'. The notion that it might be appropriate for the Home Secretary to overrule scientific judgements by introducing utilitarian and animal protection considerations had disappeared from the agenda. Furthermore, the Commission appears to have adopted pro-vivisection ideology by lauding the absence of restrictions on research (Radford, 2001: 72).

Thus, with the Government accepting the Commission's report, even the primary recommendations (Radford, 2001: 72; Littlewood, 1965: 8–9) were peripheral and had little impact. An examination of the three most salient recommendations demonstrates this. They related to:

1. the sources of advice to the Home Secretary
2. the Inspectorate
3. permitted pain levels.

Firstly, in relation to advice, although the AAMR was replaced, the new advisory committee (AC) was composed of individuals nominated by pro-vivisection scientific and medical bodies: the Royal Society and the Royal Colleges of Physicians and Surgeons. Furthermore, although *current* licence-holders were excluded, it is highly probable that previous licence-holders were appointed to the AC (Rogers, 1937: 54).[13] The AC was, like its predecessor, structurally pro-vivisectionist, albeit with slightly greater accountability than the AAMR.

In any case, the majority of the Commission rejected a minority report proposal that would have meant that this marginally more accountable body would be closely involved in decision-making (Littlewood, 1965: 10–1). Specifically, instead of the previous practice whereby the AAMR advised on every application, the role of the new AC was to give *general* advice on the administration of the Act (Littlewood, 1965: 9). In practice, the AC seldom met, considered on average only three (out of several hundred) research proposals each year, and in general had little impact (Balls, 1986: 7). Similarly, French (1975: 214) implies that the new structure made no discernible difference to the policy network and the pattern of policy outcomes:

> A perusal of the annual returns from the H.O. makes it clear that experimental medicine in Britain enjoyed tremendous growth under the AAMR arrangement and under the advisory arrangement that succeeded it... [I]t is hard to imagine such growth occurring under circumstances in which there was any significant interference with research.

French implies that the AAMR and then the AC were the most salient factors in the animal research policy network. However, the pre-AAMR situation, where the Secretary of State had refused applications, did not return following the passing of this body and the more peripheral role of its replacement. Therefore, other aspects of the evolution of this policy process need to be examined to help explain why the demise of the AAMR did not allow non-scientific values to re-enter the process. In other words, what was it about the distribution of resources in the network that facilitated its ongoing domination by the pro-vivisection lobby and its structuration by an ideology of professional autonomy?

A clue to this process can be found in a disagreement within the Commission. The minority report issued by three members indicates that, by the 1906–12 period, authority to approve applications for certificates had effectively shifted towards the scientific and medical institutions whose signatures were required, at the expense of the Home Secretary (Littlewood, 1965: 10–1). In the absence of formal rules, this must be interpreted as an informal institution or 'rule of the game'. The minority proposal to try to shift authority back to the Home Secretary, as had been the case prior to the advent of the AAMR, was rejected by the majority of the Commission. Thus the pro-vivisection lobby maximised their control over the administration of the Act by preventing the possibility of broader criteria being re-introduced into application determinations through the Home Secretary's exercise of discretionary power. Both the development of this informal rule, and the scientists' power to protect it in the arena of the Royal Commission, signify the dialectical relationship between the structural resource advantages enjoyed by the pro-vivisectionists and the ideological structure of the policy community that privileged scientific autonomy.

The constraining power of the animal research lobby's knowledge, expertise and social status resources is further demonstrated by the second relevant set of conclusions of the Royal Commission that concerned inspection. The Commission proposed that the number of inspectors should be increased from two part-time to four full-time to reassure the public that the 1876 Act was being adequately enforced (Littlewood, 1965: 8), although in fact only three full-time inspectors were subsequently appointed by the Home Office. Despite this enlargement of the Inspectorate, the rapid expansion of vivisection meant that the number of experiments per part-time inspector per year was many times more than had been the case in 1885, rising from 797 to 11,727 in 1920, and 112,705 by 1930 (derived from figures in Littlewood, 1965: 41, 253). Moreover, as noted in Chapter 3, extra resources for inspection (even if that had occurred in relation to the scale of the regulated activity) would only have had a significant impact on policy outcomes if they were accompanied by changes in the relationship between inspectors and researchers. That they did not is therefore related to the fact that the majority of the Commission rejected proposals from the RSPCA and the minority group to widen the eligibility criteria for appointment to the Inspectorate beyond medical professionals. Henceforth, the co-operative relationship between inspectors and researchers persisted (Rogers, 1937: 54). Indeed, the Littlewood Report's (1965: 77) observation that inspectors did not examine the merits of research proposals, a task which instead had become the domain of the scientific and medical bodies required to sign licence and certificate applications, indicates that professional autonomy remained pre-eminent.

The origin and implementation of the third main recommendations which gave rise to the Pain Condition on all licensed experiments, also demonstrate the ongoing dominance of experimental interests and their discretionary powers over policy outcomes. Once again, the majority Commission report rejected recommendations from the RSPCA and the minority group which would have placed significant restrictions on the pain caused by experiments (Littlewood, 1965: 8–11). Those proposals for reform included:

- a requirement that all potentially severely painful experiments be performed and terminated under complete anaesthesia and in the presence of an inspector (RSPCA)
- any experiment causing pain must be supervised by an Inspector (RSPCA)
- researchers must immediately kill animals in 'obvious' suffering (minority report).

The Commission's majority rejected these policies, in favour of pro-vivisection submissions that such measures would frustrate the object of experiments that might produce knowledge of potential utility in the alleviation of painful diseases (Littlewood, 1965: 10). Instead, the Commission proposed that:

1. Animals found in what researchers consider severe *or* enduring pain must be destroyed *if* the experiment has finished
2. Animals found in what researchers consider severe *and* enduring pain must be destroyed in any case
3. The Inspector has the power to order the killing of animals they believe to be in 'considerable' pain (Littlewood, 1965: 55).

The application of the first two clauses was dependent on the discretion of the experimenter and, as the Littlewood Report (1965: 56) averred, required complex subjective judgements on phenomena that were considered controversial and not always easy to detect. It is also noteworthy that the slightly greater degree of protection afforded animals by the first clause – that they be put down if pain is deemed severe *or* likely to endure, rather than both – is secondary to the researchers' interest in completing the experiment. Meanwhile, the third condition was entirely at the discretion of the Inspector, as there was no legal requirement for them to end an experiment in the event of considerable pain. Interestingly, the Commission defended its minor and attenuated recommendations on the limitation of pain by adopting another argument of the pro-vivisection lobby – that given public acquiescence to the infliction of severe and enduring pain on animals for the sake of sport, it would be perverse to go further in restricting pain in activities with the potential to reduce suffering.

This examination of the second Royal Commission has confirmed that this potentially destabilising event did not, in fact, cause significant disturbance to the animal research policy community that had been established by the AAMR-Home Office relationship in the 1880s. However, in order to understand this outcome, it is necessary to trace, as far as possible, the dialectical interactions among exogenous and endogenous factors and strategic agency that contributed to this policy community's resistance to exogenous political pressure.

The interaction between the animal research policy community and exogenous pressures

One account of how the scientific and medical institutions managed to maintain their dominance over animal research policy is offered by Ritvo

(1987: 162). She points to the growing perception of substantial medical benefits flowing from animal experimentation, such as the development of diphtheria antitoxin in 1894, which enhanced the authority and legitimacy of vivisectionists (see also Ryder, 1989: 145; Turner, 1980: 115) while undermining the credibility of the anti-vivisection movement. She goes on to assert that anti-vivisectionism became increasingly associated with radical new movements, and thus was marginalized from dominant social values and lost some public support:

> By the early years of the twentieth century antivivisectionism had become a fringe movement, appealing to an assortment of feminists, labor activists, vegetarians, spiritualists, and others who did not fit easily into the established order of society.

However, the exogenous politicisation of the vivisection issue that helped bring about the second Royal Commission does not sit easily with the notion of a marginalised anti-vivisection movement. For example, Kean (1998: 136) argues that the anti-vivisection movement's links with the groups identified by Ritvo were a source of political power rather than weakness:

> The optimism for the new century, fuelled by new socialist and feminist politics, spilled over into particular concerns about animals. In the early nineteenth century analogies had been made between the plight of animals and slaves; now links of a more complicated kind were being made: 'The same spirit of sympathy and fraternity that broke the black man's manacles and is to-day melting the white woman's chains, will tomorrow emancipate the working man and the ox.'[14]

Further on, Kean (1998: 163) comments: 'The close and growing links between anti-vivisection and animal rights issues generally and the suffrage cause had strong and mutual benefits on the respective campaigns'. The links between the feminist suffrage movements and anti-vivisection were further underlined by the attacks of pro-vivisection medical students on suffrage meetings during the brown dog controversy, which 'to some extent...had been a battle between the sexes and, more particularly, between machismo and feminism' (Ryder, 1989: 140).

However, the development of an arguably more consistent approach to political and ethical issues also had effects that potentially weakened the anti-vivisection movement. For example, it emphasised the arbitrary selectivity of many of the more aristocratic supporters of anti-vivisection who persisted with hunting and shooting, thereby creating yet more faultlines in the movement and opening them up to damaging charges of inconsistency and hypocrisy (Kean, 1998: 140–1). Furthermore, both Kean (1998: 144) and Ritvo (1987: 162–6) highlight an estrangement between the anti-vivisectionists and politically conservative animal welfare, or 'humane' organisations, exemplified by the RSPCA, who were perceived as 'pusillanimous' on the vivisection issue.

These contradictory effects of the development of the anti-vivisection movement point to an explanation for the apparent paradox of increased politicisation, combined with marginalisation. For, the militant and radical politics that helped to intensify anti-vivisectionist agitation simultaneously positioned the movement further from the centre of political power, both culturally and ideologically, reflecting broader socio-economic structural inequalities. The death of the anti-vivisectionist Queen Victoria in 1901 would have further weakened the cause's standing and ability to influence élites.

The close relationship with feminism and suffragism also helps to illustrate this process of marginalisation. In an era when women suffered intense discrimination and were denied the vote, the predominance of women in the anti-vivisection movement left the cause vulnerable to being denigrated as based on 'emotion' rather than reason and facts (Elston, 1987: 267). By the same token, 'male anti-vivisectionists had their masculinity impugned by critics. They too were sentimental, ignorant, easily led, falsely claiming moral superiority' (Elston 1987: 264). The association of anti-vivisection with the women's cause thus had the potential to undermine its credibility with significant sections of the public as well as the policy-making elite. This could be interpreted as a symptom of the structural inequality posited by the asymmetric power model where disadvantaged groups are excluded from policy-making. This situation was exacerbated by the lack of open support from medical professionals, as 'to do so at that time was to court censure from the profession' (Ryder, 1989: 142).

As the discussion of the policy network approach in Chapter 2 has argued, this would have weakened the ability of anti-vivisectionists to achieve policy change because the apparent monopoly of professional expertise enjoyed by the pro-vivisectionists meant their truth claims and interpretations of information were privileged over dissenting viewpoints. The comprehensive dismissal of the anti-vivisection case by the Royal Commission indicates that the pro-vivisection lobby's mediation of the network-to-context dynamic was particularly successful in that elite-composed body, compared to the general public. Indeed, the differing level of effectiveness enjoyed by the anti-vivisection movement is implied by Kean (1998: 143): 'Although the Cruelty to Animals Act of 1876 and subsequent royal commissions upheld the scientists' right to vivisect, the scientists themselves clearly felt that they had not won the moral argument'.

The prospects of the anti-vivisection movement having an impact on the Royal Commission and subsequent policy-making were probably not enhanced by the fact that they, like the socialist and feminist movements (Kean, 1998: 144), continued to be divided over strategy, undermining both their organisational resources and public credibility. For example, in 1897 Cobbe approved Stephen Coleridge's appointment as NAVS Secretary. However, he started to take a more moderate line than Cobbe, calling for restrictive legislation rather than immediate abolition. In the ensuing conflict, Cobbe and her allies were defeated and in 1898 left to form the more radical BUAV (Ryder, 1989: 138; Radford, 2001: 71–2; Kean, 1998: 144). Later in 1909, the rival camps organised two separate international anti-vivisection conferences, further emphasising the rupture between gradualists and total, immediate abolitionists: 'This rift in the anti-vivisection movement sapped much of its energies until the outbreak of the Second World War, and may have been responsible for alienating some people from the cause' (Ryder, 1989: 141). Indeed, French (1975: 406) avers that the anti-vivisectionists were caught in a 'catch-22' situation typical of reform movements:

> It was the potential for fragmentation of the movement that induced the obsession with total abolition as an acid test for anti-vivisectionist reliability. The movement was caught. It could not develop a larger constituency without transcending its public image of fanatical monomania, but any attempt to do so threatened its pre-existent power base.

The main goal of many supporters was the intrinsic satisfaction of expressing their moral values rather than achieving political results (French, 1975: 407).

The relative weakness of the anti-vivisection lobby is also indicated by the individualised and inconsistent nature of its links with socialist political groups: 'Individual support was strong; official endorsement by political organizations was weak' (Kean, 1998: 136). Ryder (1989: 145) notes that some anti-vivisection figures, such as Henry Salt and

George Bernard Shaw, were socialists and linked this with an ecological and animal rights philosophy. They were part of an influential intellectual network – Salt's work is said to have inspired Gandhi (Ryder, 1989: 127). But hopes that a socialist party would act against vivisection were later dashed. Although four future Labour cabinet ministers attended an anti-vivisection congress in 1909, they did not act to change animal experimentation policy when they obtained power twenty years later.

In contrast to the rather mixed evolution of anti-vivisection resources, the pro-vivisection lobby had grown more powerful. In the space of twenty years, British physiology had emerged from the shadows of its European counterparts to be considered as leading the world by 1900 (French, 1975: 402). The development of new treatments involving vivisection is said to have had a major effect on public opinion, rendering it much more favourable to the practice (French, 1975: 403, 405). At the same time, both the numbers of animal experiments and the institutional framework supporting the practice had expanded rapidly, with the disciplines of pharmacology and pathology now far outstripping physiology in their use of animals (French, 1975: 403). The animal research lobby effectively deployed these resources to dominate the policy process by advancing a discourse involving '...a confounding of "scientific freedom" with vague and insistent promise of a cornucopia of practical benefits' (French, 1975: 409). Having institutionalised cohesive, close relationships with the Home Office, who, in any case, increasingly lacked the resources to subject applications and research to scrutiny, the behaviour of both these sets of actors would be shaped, to some degree, by this institutional context. In particular, these apparently credible, authoritative actors were set to promote and defend this discourse at the Royal Commission, thereby lessening the possibility of dissenting exogenous perspectives from having significant purchase on the Royal Commission's findings. Thus. the institutionalised relationships and the hegemonic pro-vivisection ideology constrained not only the network but also its context.

Animal research policy, 1913–50

The combination of two world wars and the intervening economic crisis resulted in a reduction in support for anti-vivisectionism as the intensity of public concern for animal protection waned in the face of what were perceived to be more urgent human welfare challenges (Radford, 2001: 132; Ryder, 1989: 146–9). Ryder (1989: 148) suggests that from the First World War through to the 1950s, 'basic speciesism was accepted as common-sense necessity and dissent was dismissed as eccentricity'. This cultural context was associated with activity from the largest animal protection pressure group, the RSPCA, which sought minor, secondary policy changes (Ryder, 1989: 146–9). In a similar vein, the University of London Animal Welfare Society (later the Universities Federation for Animal Welfare), founded in 1926, avoided discussion of the ethics of animal experimentation in favour of issuing advice on the husbandry of animals in research laboratories (Ryder, 1989: 146–7). Meanwhile, the more radical anti-vivisection groups such as the BUAV continued pressing for a variety of reforms, ranging from the prohibition of experiments on dogs to the abolition of all animal research (Hopley, 1998: 36, 44). However, these moves foundered on the deference of ministers to pro-animal research expert opinion in the shape of the Medical Research Council and the predominance of the issue of national security on the political agenda (Hopley, 1998: 36, 44).

Meanwhile, the field of biomedical research continued to expand rapidly. New branches of medical science had emerged, including 'chemotherapy',[15] virology, endocrinology, radiobiology, medical genetics, immunology (leading to the development of antisera against tetanus and diptheria), biochemistry and nutrition. Furthermore, the emergence and growth of the pharmaceutical industry had stimulated research in applied pharmacology (Littlewood, 1965: 22). Veterinary research involving animal experimentation had also expanded due to an expansion in intensive farming and rising disposable income that led to increased expenditure on the care of domestic pets. The understanding of disease had developed since 1900 so that it had come to be perceived as caused by a combination of many factors such as pathogens, constitutional susceptibility, nutrition and the environment rather than simply a specific exogenous cause. Consequently:

> as scientific knowledge has increased, the detailed exploration and consolidation of new biological discovery have inevitably become more complex and discursive. They [researchers] pointed out that, in the general process, the aims of individual experimenters have become less ambitious, if more precisely defined, than in the early days of physiology. (Littlewood, 1965: 23)

A further significant contribution to the rise in numbers was the perceived need to use many animals to obtain statistically significant results in tests of new drugs because of individual variability, even between animals of the same species and age (Littlewood, 1965: 24). This was also associated with animal testing requirements for human and veterinary vaccines, sera[16] and drugs that were established by the Therapeutic Substances Acts of 1925 and 1956, and the Diseases of Animals Acts of 1935 and 1950 (Littlewood, 1965: 12–3). Concomitantly, the animal research industry and associated institutional structure had grown significantly, comprising government, academic and commercial sectors, more research workers, research training and professional societies (Littlewood, 1965: 20–1).

Therefore, in the context of professional autonomy over the direction of research and decisions regarding the use of animals, animal experimentation expanded rapidly. In 1920, 70,367 animal experiments were conducted, a figure that rose to 450,822 in 1930, 954,691 in 1939 and approximately 1.5 million in 1948 (Littlewood, 1965: 253, 13). There were no major changes in the formal institutional framework for animal research policy-making during this period, and animal experimentation did not re-emerge as a significant political issue again until after World War Two (Ryder, 1989: 147–8). However, it was the success of animal research interests in protecting their freedom to carry out this growing number of experiments that provoked the new wave of politicisation (Littlewood, 1965: 13–4, 24): this process is explored in more detail in the next chapter.

Conclusion

This chapter has sought to address this study's second research question: Did the animal research policy network evolve into a policy community after 1876? It has been established that during the six years immediately following the assent of the 1876 Cruelty to Animals Act, the network exhibited some features of an issue network. State decision-makers, particularly the Home Secretary, operated at arm's length from both of the opposing lobbies. The pro-vivisection lobby were certainly the more powerful of the two interests, but anti-vivisection interests were not completely excluded from the network. It was through the Home Secretary's operation of his discretionary power that a balancing of interests took place to some degree, partially invoking a 'moral judgment' model of justice. Thus, non-scientific criteria were adopted in the scrutiny of applications, an approach that manifested itself in a 15 per cent rejection rate. It could be argued that the 85:15 ratio represents a reasonable estimate of the political 'horsepower' of each lobby at the time.

However, assumptions of the issue network's long-term persistence rest on a deterministic ontology. By contrast, this present work has adopted a more dynamic, critical realist methodology whereby empirical data are closely interwoven with theoretical insights. This enables the construction of a detailed narrative of the evolution of this policy network, focussing on the dialectical relationship between structure and agency. Eschewing a linear extrapolation from the original network structure, 'path dependency' here means acknowledging that policies emerge from a context that represents the sedimentation of pre-existing decisions, institutions, structures and power relations.

Thus, it can be seen that the original issue network policy process contained the seeds of its own instability. The statute itself already endowed pro-vivisection bodies with significant political resources in the network through their participation in endorsing applications, whereas anti-vivisection and animal protection interests had no formal role in policy-making. This meant that the sole representation for non-scientific interests and values came through the Home Secretary's discretionary power. This was always likely to be a vulnerable toehold in the network for the anti-vivisection lobby because the vagueness of the statute also meant that the Home Secretary had the freedom to change the administration of the Act in ways that served the interests of animal researchers. In other words, there was no constraining macro-structure that might have stabilised the meso-level issue network.

Thus, it has been found that in 1881 and 1882, the pro-vivisectionists acted to remove the unexpected hindrances on their activities that were caused by the Home Secretary's openness to competing values in the network. In particular, they exploited their favourable resource distribution by undertaking strategic action within and outside the network. Externally, physiologists reconstituted the successful coalition of 1876 in order to harness the professional power and prestige of the medical profession. The evolving structural context facilitated the successful deployment of social and technical resources within the network that allowed them to capture decision-making powers from the Home Secretary, and therefore expunge the vestiges of dissenting values from the policy process. This signified a fundamental change in both the network and the pattern of outcomes: the emergence of a policy community where outcomes consistently reflected the beliefs and interests of the pro-vivisection lobby that prioritised the pursuit of knowledge and professional autonomy over animal welfare and public accountability.

In contrast to the embryonic issue network of the late 1870s and early 1880s, the mature policy community was a more cohesive, stable institution. While anti-vivisection managed through exogenous, outsider activity to politicise the issue in the 1900s, they lacked the resources to disturb an institutionalised policy community that was composed of interests and an ideology that arguably enjoyed even greater structural advantages than in the early 1880s. Therefore, the second Royal Commission failed to deliver significant policy changes. From that point until after the Second World War, the structural context offered even less encouragement for reformers, and the policy network remained relatively undisturbed.

However, in 1963 pressure from animal welfare groups finally persuaded the Home Office to establish a departmental committee of inquiry under the chairmanship of Sir Stanley Littlewood. This signified a critical juncture that marked the beginning of a lengthy period of politicisation, which, in turn, resulted in the 1876 Act being replaced in 1986. The next chapter traces these developments to see whether they stimulated any changes in the network and outcomes.

7
The Animals (Scientific Procedures) Act 1986: Emergence of an Issue Network or Policy Community Dynamic Conservatism?

Introduction

The previous chapter has demonstrated that animal research policy had been made by a professional policy community from the 1880s until at least 1950. This chapter continues the narrative by describing and analysing the evolution of animal research policy between 1950 and 1986 and is split into two sections that analyse the two major critical junctures during this period: the Littlewood Report of 1965 and the assent of the Animals (Scientific Procedures) Act 1986.

In this narrative, the ideologies and goals, strategies, interrelationships, resources and power of the various interest groups will be elucidated within the framework of the dialectical network model and critical realist epistemology outlined earlier in the study. Thus, the third of the research questions that structure this work will be addressed: Was the assent of the Animals (Scientific Procedures) Act 1986 a process of dynamic conservatism on the part of a policy community or a genuine response to public concern formulated through an issue network?

The Littlewood Report

The first indication since 1913 that the issue of animal experimentation was a significant political issue appeared in 1962 with the Home Secretary's appointment of a departmental committee of enquiry into the matter, chaired by Sir Stanley Littlewood (Garner, 1998: 177). This section analyses the political processes that led to the appointment of the Littlewood Committee, affected its deliberations, and influenced the Government's response. According to Hampson (1987: 316), the Littlewood Enquiry was instrumental in politicising and setting the agenda in animal research policy in the modern era. Therefore, the following discussion will play an important role in explaining the status of the policy network that formed part of the political and historical context from whence the subsequent Animals (Scientific Procedures) Act 1986 emerged. Most of the empirical data cited in this section are drawn from the report of the Littlewood Committee, published in 1965, which provides a comprehensive and detailed account of contemporary animal experimentation politics.

The origins of the Littlewood Committee: the dialectical relationship between outcomes, network, context and agency

Disturbances in the dialectical interrelationships between policy outcomes, the animal research policy community, its context and political actors led to increases in the politicisation of this policy area and pressure on the network, and hence the subsequent establishment of the Littlewood Enquiry. This section describes and analyses the phenomena involved in this dialectical process.

The policy network in action: the implementation of the Cruelty to Animals Act 1876 prior to the Littlewood Enquiry

This subsection traces the network structure and interactions as expressed in the key areas of the broad institutional framework, the assessment of applications, and infringement action. The aim here is to highlight any salient alterations in the policy community in the years preceding the Littlewood Enquiry.

Initially, it is important to reiterate the observation made at the end of the previous chapter regarding the large, expanding scale of animal experimentation, both in terms of the numbers of animals used, and the size and variety of the interest groups involved. Thus, the number of experiments had risen from 95,731 in 1910 to 1,779,215 in 1950, and they were now conducted by a pharmaceutical industry as well as a growing governmental and academic research sector (Littlewood, 1965: 253, 20–1).

In respect of the broad institutional framework for policy-making, prior to the Littlewood Enquiry the 1876 Act stood in the same form as it had when it was passed, formally endowing the Home Secretary with extensive discretion over its administration (Littlewood, 1965: 30). In spite of the broad nature of the 1876 Act, the Home Office had not issued guidance on its operation, so advice and instruction was at the discretion of inspectors. However, notes were issued by the experimenters' representative group the Research Defence Society (RDS) to assist licence and certificate applications (Littlewood, 1965: 47).

The size of the inspectorate had slightly increased to five full-time posts in 1950, and then to six in 1961. As recommended by the 1912 Royal Commission, inspectors were required to be medically qualified (until 1963) (Littlewood, 1965: 41). Meanwhile, the AC's activity continued to be limited to occasional recommendations on unusual or particularly severe experimental proposals. Until 1952, when one member from the Royal College of Veterinary Surgeons (RCVS) was appointed, all members (except the chairman, who was a senior judge) had been medically qualified (Littlewood, 1965: 43).

However, it should be noted that the entry of the veterinary profession into the network at this time did not represent a significant change as their representative bodies[1] participated in animal experimentation and had close relationships with animal research interests (Littlewood, 1965: 50–1, 64). Conversely, as Ryder (1989: 208–9) notes, the veterinary profession had been conspicuous by its absence from animal protection reform campaigns:

> The president of one august [veterinary authority] explained that such were the vested interests of his members that if he publicly criticized bloodsports or cosmetics experiments on animals or factory-farming he would immediately cease to be president.

Licence and certificate applications still required the signed endorsement of a president of one of the specified professional societies and a professor in the relevant discipline ('statutory signatories'). Individuals with scientific and medical qualifications were then granted licences on the basis of an inspector's judgment: 'that the applicant is a suitable person' (Littlewood, 1965: 32)[2]. The Littlewood Report does not provide any specification of the qualities of a 'suitable person' in this context, and there is no reference to any change of policy since the 1912 Royal Commission indicated that statutory signatories' perceptions of the competence of applicants meant that refusals were extremely rare (French, 1975: 214). This raises the important question of the relative distribution of practical power between statutory signatories and inspectors, a theme that will be revisited below.

In 1932, Home Office practice changed in relation to the duration of most licences, which were now issued for five years rather twelve months (Littlewood, 1965: 33). On its own, the licence gave authority to conduct experiments where the animal was fully anaesthetised throughout and killed before it could recover. Certificates were necessary for licence holders to dispense with anaesthesia or to prolong the life of the animal after anaesthesia (Littlewood, 1965: 34). In fact, certificates were especially important in the implementation of policy because, out of 8,748 licensees on 1 August 1964, only 489 did not hold a certificate (Littlewood, 1965: 52).

In respect of these crucial certificates, the Littlewood Report (1965: 35) observed: 'We were told that the Home Office has always acted on a view that the statutory signatories are free to limit the scope or currency of a certificate or not to do so, in their discretion'. This confirms an arrangement first highlighted by the report of the second Royal Commission, which had allowed the animal research lobby – which included the bodies stipulated as statutory signatories – to control a fundamental aspect of the administration of the 1876 Act.[3]

The influence of scientific bodies over application assessments was consolidated by the approach taken by the Home Office to the assessment of the potential benefits of research proposals stated in certificate applications. Thus, the inspector's task was 'merely to establish that the application is for a class of purpose permitted by the Act: he is not required to evaluate the potential benefit likely to accrue from the proposal' (Littlewood, 1965: 36). Those purposes were defined at Section 3(1) of the Act as 'the advancement by new discovery of physiological knowledge or of knowledge which will be useful for saving or prolonging life or alleviating suffering' (Littlewood, 1965: 77). Meanwhile, the assessment of the utility of research proposals was left to the discretion of scientific bodies: 'If responsibility for assessing the potential merits of research lies anywhere in the present system, it is with the statutory signatories of application for licences and certificates....' (Littlewood, 1965: 77). This confirms that a key policy goal of the broad animal protection lobby – that animal experimentation proposals should be routinely subject to an independent utilitarian assessment (Littlewood, 1965: 77) – continued to be excluded from the network's ideological structure.

The potential for Home Office regulation of animal experimentation was further constrained by the fact that certificates stipulated the particular types of technique that licensees could perform, but they were not:

> restricted in terms to the performance of experiments for particular or specific research purposes. A licence-holder may, if the terms of his licence and certificate are appropriate, use them for an experimental

purpose other than that which he had in mind when applying for it. (Littlewood, 1965: 36)

Therefore, according to Home Office statements, the full disallowance of a certificate was rare as applicants were assumed to be familiar with the legal requirements and worked with inspectors in marginal cases before submitting an application (Littlewood, 1965: 37). This description reinforces the impression of close, cooperative relationships between inspectors and researchers that aimed to facilitate the licensing of experiments. Nevertheless, occasionally the Home Office is said to have disallowed a procedure as 'objectionable in principle' because, for example, it was intended to incapacitate an animal through total blindness (Littlewood, 1965: 37, 125).

The autonomy of researchers appears to have further expanded at the expense of animal protection as a result of another change in Home Office policy with regard to certificates. Thus, it is noted that the Home Office had gradually abandoned attempts to attach conditions on certificates that limited the numbers of animals used in painful experiments. The stated reasons for this indicate that the purpose of the change was to assist licence holders, as such limits:

> greatly increased the complexity of administration and the risk of inadvertent breaches of the conditions of licences... [I]f a responsible licensee found it necessary to work up to and beyond the limit, the Home Office saw no justification to refuse an increase of numbers. (Littlewood, 1965: 35)

'Exceptionally severe' experiments still included a number limitation, although there is no indication regarding whether the Home Office permitted increases at the request of a licensee.

In summary, the open-ended nature of the licensing and certification system indicates that there was virtually no external control of animal research. It appears that in the conflict between animal welfare and researchers' interests, the latter were able to consolidate their domination of the network.

With regard to enforcement and infringement action, the Inspectorate's stated tasks were:

- to scrutinise publications of experiments and ensure they were properly authorised
- to inspect registered premises.

Most detected breaches of the Act or licence conditions were said to have been discovered through inspectors' scrutiny of published articles or licensees' annual returns, while few were observed by inspectors in course of visits (Littlewood, 1965: 44). The majority of these contraventions were deemed by the Home Office to be due to 'misunderstanding or inadvertence' and involved unauthorised or unspecified procedures that would have been permitted if applied for properly. In response: 'All that has been judged necessary is to draw the licence-holder's attention to the irregularities committed, to ask him for an explanation, and to warn him to be more careful in future' (Littlewood, 1965: 44–5). More serious 'technical offences' were those that suggested 'more than ordinary carelessness, e.g. in ignoring a previous warning or neglecting to acquaint himself with the elementary requirements of the Act or licence' (Littlewood, 1965: 45). The detection of these types of breach resulted in a 'severe caution' or a warning of license revocation if repeated. However, revocation was considered by the Home Office to be an 'extreme' response and had not occurred in the years closely preceding the enquiry. Finally, the Home Office stated that they were unaware of any prosecutions under the Act (Littlewood, 1965: 45).

In summary, this description of the administration of the 1876 Act prior to the Littlewood Enquiry confirms that the marginal changes that had occurred since 1912 tended to favour researchers' interests and loosened the control on the number of animals they were able to use. Furthermore, the expansion of the practice of animal experimentation had endowed research interests with greater resources, compared with those committed by the Home Office to inspection. It is also noteworthy that the scientific and medical bodies that had comprised the animal research lobby earlier in the network's life were now joined by the pharmaceutical industry. As the extant animal research policy literature acknowledges, the pharmaceutical industry had, by the 1950s, assumed such importance to the British economy as a whole that the Ministry of Health saw its role as representing the industry's interests in government, which meant preventing the regulation of its activities (Garner, 1998: 48). Indeed, the Board of Trade and the Treasury also had an interest in the commercial success of the UK drug industry, particularly with regard to the revenue generated by its exports. This reflected the Government's general perception of these revenues as essential to the success of a Keynesian economic policy that aimed to fulfil the political imperative of full employment (Abraham, 1995: 58–9). At the same time, the institutionalisation of animal experimentation into a research process that was perceived to deliver health benefits also endowed the practice with considerable legitimacy and hence a tendency towards stability through the process of path dependency.

These additional economic, organisational and political resources were augmented by legal resources, due to the requirements of legislation specifying animal tests for certain medical products. In 1950, these experiments numbered 596,813, or approximately one-third of the total. Furthermore, these tests were devised by other government departments, research institutions, professional bodies and producer groups without Home Office consultation. The stated overriding goal of these regulatory tests was to ensure the quality of the product rather than take animal welfare into consideration (Littlewood, 1965: 48–51). Although policy communities normally contain just one government body, such policy-making structures can exist with more than one such body if there is an accepted hierarchy (Smith, 1993b: 79). In this situation, the Home Office was content to be subordinate to other government departments on the sub-issue of mandatory testing.

In general, it is clear that prior to the Littlewood Enquiry, the animal research policy community remained stable. If anything, the resource distribution between animal research and anti-vivisection groups, which had structural elements, seemed to favour the former more heavily than ever before. However, the Enquiry represented a critical juncture with the potential to disturb this stable policy network.

Responses to policy community outcomes

After the Second World War, anti-vivisection groups undertook political action as a result of their perception of alterations in policy outcomes and their strategically-selective context. Thus, in 1948, the anti-vivisection movement began lobbying the Home Secretary and Parliament for stricter implementation of the 1876 Act and for the appointment of some form of inquiry (Littlewood, 1965: 13–4). Their main arguments in support of this course of action were:

- The substantial annual increases in the numbers of experiments signified the inadequacy of 1876 Act. Between 1939 and 1948, the number of experiments rose from one million to one and a half million. (This concern about the absolute scale of animals sacrificed indicates the belief that each individual animal was of inherent value in addition to the undesirability of increases in aggregate pain. For the anti-vivisection lobby, there were too many experiments, many of which were repetitive.)

- The 1876 Act's implementation permitted 'a great amount of pain and suffering to animals', and the adequacy of anaesthetics was questionable
- The perceived lack of utility of most, if not all, animal experiments
- The inadequacy of inspection and enforcement, due to the discretionary power of licensees and an understaffed inspectorate.

The policy community's apparent success in furthering its interests generated concern not only in anti-vivisection circles but also among the public and Parliament. These concerns were magnified by what the Littlewood Report (1965: 18) termed 'growing public sentiment towards animals'.

Policy community mediation of exogenous pressure and new information

As well as providing a resource for animal protection group activity, the Littlewood Report (1965: 18, 129) argues that these exogenous attitudinal shifts also gave rise to greater interest in animal welfare in research laboratories. Symptoms of this dynamic included the development of laboratory animal science and the formation in 1926 of the Universities Federation for Animal Welfare (UFAW), which produced guides to laboratory animal husbandry and encouraged research into animal pain. Increased social sensitivity to animal welfare had therefore stimulated the production of more information on this topic from within the policy community, which, in turn, recursively affected the climate of public opinion.

As discussed in Chapter 2, the policy network analytical framework conceives that policy communities mediate new information by interpreting it through the lens of their ideological structure (Jordan and Greenaway, 1998: 671–5). Thus, new information may be made consistent with the policy community's goals instead of implying anomalies in its ideological structure that may lead to major policy change. The Littlewood Report (1965: 18) provides some evidence of this when it states that this interest in animal husbandry was also stimulated by the scientific goal of seeking to exclude extraneous animal pain or illness that may have represented an interference in the controlled experiment. Thus, animal researchers were able to internalise the discourse of 'animal welfare', which would be likely to provide some measure of reassurance to a concerned public and therefore enhance the network's legitimacy resources. The appointment of a veterinary scientist to the AC in 1952 appears to be indicative of this growing acknowledgement of the need to consider animal welfare, and also demonstrates the theoretical proposition that although policy learning processes have the potential to destabilise policy communities, actors with technical expertise are more likely than non-experts to be admitted to the policy network (Jordan and Greenaway, 1998: 671).

Thus, the ethical and policy lessons drawn by the research lobby were quite different to the animal protection and anti-vivisectionist lobby. In particular, while groups such as UFAW and the veterinary profession exhibited a 'scientific' approach to animal welfare that sought limited reductions in pain which did not impede research goals, the anti-vivisection lobby wished to highlight what they saw as the moral relevance of animal welfare and its implications for policy on whether painful experiments should be authorised at all. This divergence in the understanding of animal welfare between the lobbies is emphasised by the Littlewood Report's (1965: 61) subsequent observations on the use of analgesia in experiments:

> We have formed the impression that analgesia is not widely used in laboratory practice. We think that there may be more scope for its use than is generally realised. There are some procedures that predictably cause serious pain where some alleviation would not interfere with experimental observations or results.

This provides further evidence of the prioritisation of research goals over animal welfare. Thus, the entry of animal welfare considerations into the network provides an instructive example of how, as discussed in Chapter 2, policy learning and lesson-drawing in technical policy communities tends to be limited by path dependency and other structural and institutional constraints (Dolowitz and Marsh, 1996: 355–6; Rose, 1993: 25–6). Despite the presence of wider public and group concerns, lessons were only drawn from actors with similar professional and ideological affiliations. Consequently, policy change remained secondary and constrained within the existing policy paradigm or dominant appreciative system.

Ongoing politicisation forces policy review

Having previously rejected outright anti-vivisection requests for an enquiry, in 1957 the Government announced that it would keep the matter under review (Littlewood, 1965: 14). This indicates that despite the policy community's apparent internalisation of exogenous pressure, the balance between the network's legitimacy resources and the political resources of reform groups had altered sufficiently to cause a change, if not in actual policy, at least in the actions of policy-makers.

To give an idea of the trajectory of animal experiment numbers in this period, by 1960 the number of experiments had risen to over 3.7 million, compared with 1.5 million in 1948. The RSPCA's perception of these policy outcomes and the changing political context seems to have prompted them to take strategic action between 1959 and 1961. As the most powerful animal protection group, the RSPCA's actions are especially significant (Garner, 1998: 98). The Society counselled the Home Secretary that there was widespread public concern about the 1876 Act, and proposed policy changes. The most significant of these were demands for the introduction of some form of utilitarian assessment of research applications,[4] and for the broadening of the composition of the Inspectorate and AC to include veterinarians and animal welfarists, respectively (Littlewood, 1965: 14). The Littlewood Report (1965: 73–6), published five years later, detected widespread concern for animal welfare in general, but was unsure whether a perceived lack of interest in experiments was due to public acquiescence or ignorance due to secrecy. It concluded that:

> the public generally has accepted in principle the necessity for and value of animal experimentation, but that this tacit acceptance cannot be taken to represent informed assent to all that is done under the Act.

It seems reasonable to infer that this apparent concern about the extent and regulation of animal research may have combined with the RSPCA's actions to enable the case for reform and for an enquiry to have sufficient credibility to maintain interest in Parliament.

This exogenous political pressure finally persuaded the Home Secretary to consider certain reform proposals put forward by MPs during the spring of 1962 (Littlewood, 1965: 14). However, it is noteworthy that these proposals were, in general, less ambitious than those previously advanced by the RSPCA, never mind the more radical anti-vivisection groups. This appears to indicate that the policy community's technical resources meant that its pro-vivisection ideology continued to be a significant constraint on the policy agenda. For example, instead of investigating the idea of veterinarians and animal welfare representatives joining the AC, the Home Secretary considered whether a (scientifically or medically qualified) woman should be appointed. Moreover, the suggested introduction of utilitarian criteria to applications was not pursued. Nevertheless, the demand for an enquiry remained on the

196 *The Politics of Animal Experimentation*

agenda, implying that the legitimacy resources of the animal research community were less formidable than their expertise-related resources.

However, the animal research lobby and the policy community also benefited from a considerable political resource advantage that stemmed from their entrenched domination of the policy network. This network-political resource had structural and institutional aspects, and manifested itself in the way in which the Government responded to exogenous pressure. Thus, in this process of policy learning and formulation, the Home Secretary had little choice but to turn initially to the AC, which, as explained in the last chapter, was structurally pro-vivisectionist.

The Advisory Committee: the reaction of a policy community to external pressure

Although the AC, as a formal institution, was not tightly integrated into the policy network, due to its passive advisory role and the marked infrequency of its meetings,[5] it can still be regarded as, in some respects, connected to, and reflective of, the policy community because of the scientific and medical affiliations and backgrounds of its individual members and their appointment by pro-vivisection institutions. The response of the AC, in the form of a letter in November 1962 from its chairman Lord Morris to the Home Secretary (reproduced at Littlewood, 1965: 208–11), therefore provides an interesting example of the strategic action of policy communities. It may also provide a possible insight into the ideology of the animal research lobby. However, evidence was presented in the last chapter indicating that despite their utilitarian defence of animal experimentation in the public sphere, researchers did not wish for applications to conduct animal research to be subject to a utilitarian assessment (French, 1975: 219). Therefore, it cannot necessarily be assumed that the public discourse of any lobby group is identical to their real beliefs and goals.

In responding to the Home Secretary, the AC began by acknowledging that '...some of the concern that has been expressed arises from the fact that the annual returns to parliament show what at first seem to be very high numbers of "experiments"' (Littlewood, 1965: 208). Interestingly, the tendency of policy communities to try to resist public opinion (Marsh, 1998b: 188–9) can be seen in the attempt by the AC to downplay the validity of this exogenous lay viewpoint. Thus, the Committee suggested that the public might have lacked 'an adequate appreciation of the *necessary* extent of present day research [and a] full understanding of the nature of many of the "experiments"'

(Littlewood, 1965: 208-9 – emphasis added). This is also a clear expression of the long-standing core policy belief of the animal research lobby: that their professional members were the only appropriate judges of the legitimacy of animal experiments. The AC questioned the public's appreciation of the 'nature' of experiments by arguing that many contemporary 'experiments' as counted by the Home Office under the 1876 Act were not, strictly speaking, 'experiments' or 'vivisection', because they involved standardisation testing of medicines by injection. This tactic appears to represent an attempt by the policy community to utilise their technical expertise in order to make authoritative truth claims, thus mediating and neutralising potentially destabilising new information. However, the validity of this argument was undermined by the Littlewood Report (1965: 3), which noted that it was obvious from the beginning of the enquiry that such 'non-experimental' uses of animals 'involve much the same procedures or need for care as in experiments'. In other words, the potential damage or suffering inflicted on animals was similar in both the contemporary tests and the earlier surgical experiments.

Further parameters of the policy community's shared ideological outlook are revealed by the AC's vigorous defence of the increasing scale of animal experimentation in terms of the inevitable demands of expanding scientific inquiry, drug development and consumer product manufacturers:

> The increase in the number of experiments has been inevitable in view of the great advances which have been made in recent years both in science and in medicine, and experiments are necessary in order to increase medical and scientific knowledge. Very many are also necessary in order to safeguard the public from ill effects that might result if untested products were placed on the market. (Littlewood, 1965: 209)

The AC also averred that researchers would not conduct experiments that were 'wanton' or 'lightly undertaken', although these terms are not defined by the AC, which gives the impression that the assertion is mainly related to a perception of their fellow professionals' character.

However, having raised doubts about the validity of concerns underpinning calls for an enquiry, the AC nevertheless suggested that such an enquiry might be useful to maximise public confidence in the regime: 'It may, of course, be that the holding of an inquiry could be the means of allaying any public disquiet that there may be' (Littlewood, 1965: 209). This proposed action would be consistent with Smith's (1997: 45) observation that policy communities can practice dynamic conservatism by internalising changes in their environment though building new resources; in this case, they perceived that the endorsement of an enquiry would introduce valuable legitimacy resources into the network. Yet, in respect of the composition of an inquiry, the ideology of unique professional competence was again prominent in the AC's recommendations: 'If there were to be an inquiry it would probably be most usefully conducted by persons with medical and veterinary and scientific knowledge' (Littlewood, 1965: 209).

This technocratic strand of pro-vivisectionist ideology and the desire of policy communities to control the outward flow of information emerged in response to the proposal that MPs and zoologists should be able to visit animal laboratories. The AC emphasised the need for such visitors to be 'fair-minded and competent to form judgment'. The Committee's overriding concerns focussed on the notions that registered places were private property and that researchers should be entitled to privacy and freedom from disturbance. However, the perceived potential for such visits to be used to legitimise the network meant that they were countenanced on the basis that they might help 'remove the suspicions of genuine seekers after truth' (Littlewood, 1965: 211).

The AC rejected the proposal to reduce the number of registered places (which may have facilitated closer inspection and/or reduced the number of experiments) because they believed that such moves could impede research. The suggestion that benefits of animal experiments should be reported in the official annual returns published by the Home Office was also opposed by the AC on the grounds that it would be impractical to publish such information as it was normally impossible to ascertain benefits 'even some years after their [i.e. the experiments'] completion' (Littlewood, 1965: 211). Here it again appears that the goal of professional autonomy conjoined with strategic action that sought to assert researchers' sole authority to judge whether animal experiments were valid, justified and beneficial.

In a similar vein, the primary administrative change independently proposed by the AC was concerned with serving researchers' interests:

> The Act imposes requirements of obtaining the signature on applications of the Presidents of various medical and scientific societies. Those requirements may under modern conditions be very burdensome

for those concerned. A modern Act might contain some different machinery. (Littlewood, 1965: 210)

However, the AC agreed with the proposal to increase the size of the inspectorate and supported the change whereby veterinary professionals would also be considered for appointment. But the cooperative relationship between inspectors and licensees, and the vague criteria applied by the inspectors, were endorsed by the AC as the appropriate mode of supervision:

> The inspectors have many responsibilities in regard to applications for licenses and they cannot in the nature of things observe very many experiments. It may well be that the inspector's most valuable work is to ensure that only the right kind of people are doing the right kind of work in the right kind of places. The Committee understand from the Chief Inspector that he has not felt that the inspectors have been unable to discharge their responsibilities. The Committee nevertheless feel that public confidence would be fortified if by an increase in the number of inspectors still more frequent contacts between inspectors and licensees and still more visits to licensed premises could be provided. The Committee consider that these contacts and visits provide the best means of ensuring a constant measure of conscientious concern on the part of all persons who are in any way involved. (Littlewood, 1965: 210)

If the policy network dynamics table (2.2) set out in Chapter 2 is used as a heuristic guide, it is clear that the response of the AC to the Home Secretary's consultation corresponds to the postulated dialectical interaction between a policy community and exogenous public opinion. In particular, the policy community exhibited 'dynamic conservatism' in order to maintain network homeostasis. This involved downplaying the main concerns underlying calls for an enquiry, while noting the potential for it to enhance the network's legitimacy. Secondary changes were countenanced, in particular in relation to the size and composition of the inspectorate, which could be considered as the policy community's mediation of exogenous pressure. Such changes were also compatible within the broader ideology of the network as they did not question the cooperative style of regulation.

The AC's position also provides a useful insight into the pro-vivisection discourse and, at least to some extent, the ideology and policy goals that structured the policy community. These beliefs are generally arranged from the most core/fundamental at the top to more specific/secondary at the bottom:

- It is generally necessary and legitimate to experiment on animals to advance scientific knowledge, develop drugs and test consumer products.
- Researchers possess unique professional competence to judge the legitimacy of animal experimentation.
- Animal researchers should not be externally impeded in their activities
- It is impractical to measure tangible benefits of animal experiments, even retrospectively, for several years.
- Researchers can be trusted to decide what animal experiments are necessary and how to perform such experiments.
- Animal experiments are essentially a private matter for researchers rather than necessarily requiring public accountability.
- The appropriate direction of influence between researchers and public is very much from the former to the latter, with an emphasis on policy action by the former which reassures the latter.
- *Cooperative* inspection is helpful in maintaining professional standards and assuaging public concern.
- The present policy process is satisfactory.
- The need for an enquiry (and implicitly, the possibility of policy change) is doubtful, although it may be necessary to streamline the regulatory burden on researchers and reassure the public.

The Home Secretary's statement following his consultation with the AC announced that: 'Having given consideration to the whole position, and in the light of Lord Morris's letter, I have decided to appoint a Departmental Committee to inquire into the working of the Cruelty to Animals Act, 1876' (Littlewood, 1965: 211). The AC's other recommendations were accepted.

The Littlewood Enquiry: scope and participation
Scope of enquiry
The first important aspect of the Littlewood Enquiry is the scope of its terms of reference, including how those terms were interpreted. If this juncture signified a major change towards a more open issue network, one would expect a broad enquiry that examined the concerns raised by all the main interested parties. The Enquiry's terms of reference were: 'To

consider the present control over experiments on living animals, and to consider whether, and if so what, changes are desirable in the law or its administration' (Littlewood, 1965: 1).

Key anti-vivisection proposals to the Littlewood Enquiry focussed on the need for an assessment of the *marginal* utility of animal experiments, which involved three broad issues:

1. whether adequate alternative methods of testing would be stimulated if animal tests were prohibited
2. the relative merits of medicine associated with animal experiments (in particular drug therapy) compared with techniques not associated with animal experiments, such as health education and improvements to diet (Littlewood, 1965: 2, 79–80). In other words, this question implied a broader critique of a dominant, technocratic paradigm in medicine and health policy
3. an examination of the regulatory system in relation to 'the whole field of scientific research with a view to considering the relative merits of biological and other forms of experiment and the case for general limitation of experiments on living animals' (Littlewood, 1965: 2).

However, the Littlewood Committee (1965: 80) decided that such questions were outside their terms of reference and that the 1876 Act did not provide a basis for experiments to be regulated according to broader social judgments about the scientific, medical and public health context of animal research proposals. Thus, the core elements of the anti-vivisection position were excluded from the policy review discourse. The Littlewood Report (1965: 67) confirmed this narrowing of the agenda when discussing its interaction with the anti-vivisection groups: '...we made it clear that we were prevented by our terms of reference from receiving evidence directed at the total prohibition of experiments on living animals'. The broader political significance of this omission is made clear by a memorandum appended to the final report written by one of the Committee members, Mrs Joyce Shore Butler MP:

> I have signed this Report, accepting – with my colleagues – that any attempt to answer the three major questions of which mention is made in paragraph 237[6] [paraphrased above] lies outside our terms of reference. I am convinced, however, that unless or until answers are found to these questions there will remain room for doubt about the need and justification for the use of animals for laboratory purposes. (Littlewood, 1965: 201)

Group and public evidence to the Littlewood Enquiry

The Littlewood Report's (1965: 72) discussion of the sources of evidence it received identifies a basic schism between the groups and the nature of their submissions:

> In the main, the more directly involved [in animal research] organisations...expressed general support for the Act and the way it has been administered...By contrast, animal welfare organisations without direct knowledge of the working of the Act tended to complain that the facts were hard to establish and to criticise what they inferred to be the present position.

In respect of the former group, in addition to extensive responses from university animal researchers, the animal research interest groups provided a unified response under the umbrella of the RDS. The RDS had set up a committee representing several professional groups, which composed a memorandum for the Littlewood Enquiry to which other scientific societies subscribed, as did the Association for the British Pharmaceutical Industry (ABPI) and the BMA with minor reservations (Littlewood, 1965: 64). These were the main participating groups mentioned by the Littlewood Report:

- University Departments and individuals involved in animal research
- RDS
- Royal Society
- Royal College of Surgeons
- Physiological Society
- Pharmacological Society
- Nutrition Society
- Royal College of Physicians
- Pathological Society
- Society for Experimental Biology
- Society for General Microbiology
- British Veterinary Association (BVA)
- ABPI
- BMA
- 'a large number of smaller societies'.

The Littlewood Report (1965: 65) noted that 'the sum total of reaction from the scientific interests was remarkably uniform'. That response was generally supportive of the contemporary mode of regulation, subject to amendments, while primarily enunciating 'a deeply-felt insistence on the principle of freedom in research....' Other stated positions reflected the internalisation of animal welfare concerns within a paradigm that prioritised research goals, such as the prevention of 'needless' pain and suffering, the desirability of a limit on the pain caused by experiment, a minimisation of the number of animals to what was deemed 'essential' and a desire to see optimum levels of husbandry. There was also a general wish for greater official guidance on the administration of the Act to help licensees understand their regulatory responsibilities, but, there was no reference to an ethical assessment of research proposals.

The groups in favour of core policy continuity also included the following organisations that the Littlewood Report (1965: 65–7) categorised as 'animal welfare':

- Animal Health Trust
- UFAW
- Animal Technicians Association
- British Laboratory Animals Veterinary Association (a sub-division of the BVA)
- Laboratory Animals Science Association.

These groups were closely involved in animal research, but were particularly focussed on enhancing the welfare of animals used in experiments, although such goals were generally considered secondary to perceived scientific imperatives.

Also within this 'animal welfare' category was the RSPCA. The differences between the RSPCA and other groups in this set appear to mark the main faultline in this policy area at the time, that between 'scientific' and 'lay' groups. Thus, the Report (1965: 66) stated: '...the matters giving rise to principal disquiet on the part of the [RSPCA] were also the main cause of concern to all groups of witnesses other than those directly engaged in work under the Act'. Those other groups mainly comprised four prominent anti-vivisection bodies: the BUAV, NAVS, the Scottish Society for the Prevention of Vivisection (SSPV) and the Scottish Anti-Vivisection Society (SAVS). In addition to their disquiet about evidence of severe pain and suffering caused by experiment, their shared policy concerns included, according to the Littlewood Report (1965: 66–7):

1. Too many animals were being used in research. The use of animals had become routine when it often seemed avoidable; and the trend was always towards further increases.

2. The apparatus of control appeared to be stringent, but in practice there seemed to be complete freedom. Many experiments were needlessly repetitive, and much research seemed to be duplicated.
3. Too many experiments were allowed for purposes that had no bearing on the relief of human or animal suffering.
4. It was impossible to ascertain how much suffering was being inflicted.
5. There was inadequate veterinary supervision of the working of the Act, experimental procedures and anaesthesia, and of the routine care of animals.
6. There were too few inspectors and too little inspection
7. There had never been a prosecution.[7]

The underlying belief of the anti-vivisection bodies was that animal research should be abolished (Hopley, 1998: 72). However, the structural constraints they faced, as manifested in the limitations on the Littlewood Enquiry's terms of reference, led them to lobby for reforms, similar to those sought by the RSPCA, that they believed would at least tighten the regulation of animal research. The term 'animal protection' is thus used to describe this broad category of groups.

This description of group participation in the Littlewood Enquiry suggests that animal research interests were more numerous, more organised and hence better resourced than the animal protection lobby. Fifty-nine witnesses gave evidence on behalf of eighteen professional and industrial animal research interest groups, whereas only sixteen witnesses represented five animal protection organisations (Littlewood, 1965: 203–5). As was the case at the Second Royal Commission, there was little, if any, explicit professional support for the animal protection position. The participating groups and their policy beliefs are summarised in Table 7.1.

The Littlewood Enquiry (1965: 71–2) also canvassed the churches for their opinion. While the Roman Catholic Church issued a brief summary of general principles, the British Council of Churches undertook extensive enquiries through a specially commissioned working group. The Council submitted memoranda that appear to have echoed the position of the RSPCA, raising concerns for instance about experiments with little connection to the relief of human suffering. Finally, in respect of public opinion, the Enquiry received eighty-five letters which were mostly critical of animal research, but determined that these contributions 'gave us little help because few seemed to be written by uncommitted persons' (1965: 73).

Table 7.1 Littlewood Enquiry: groups and their policy goals

	Pro-animal research lobby	Animal protection/anti-vivisection lobby
Participating groups	University Departments and individuals involved in animal research; RDS; Royal Society; Royal College of Surgeons; Physiological Society; Pharmacological Society; Nutrition Society; Royal College of Physicians; Pathological Society; Society for Experimental Biology; Society for General Microbiology; British Veterinary Association (BVA); ABPI; BMA; 'a large number of smaller societies'; Animal Health Trust; UFAW; Animal Technicians Association; Laboratory Animals Science Association.	RSPCA; BUAV; NAVS; SSPV; SAVS (the latter four had to set aside fundamental objections to animal experiments in submissions to Littlewood Enquiry, so policy beliefs below represent incremental improvements).
Core policy beliefs	• general support for contemporary administration of Act • 'a deeply-felt insistence on the principle of freedom in research…' • animal welfare secondary to research imperatives	• reduction of experiments • preferred government/social control of research, rather than professional autonomy • wrong to sacrifice animals unless for a clear, vital human interest • self-imposed risks of suffering do not warrant animal experiments
Secondary policy aims	• research community can be trusted to mitigate animal pain • severe and enduring pain unacceptable, but to be defined and enforced by researcher • not all duplication is 'unnecessary' and this is a matter for scientific community to determine • applications' sponsorship from professional colleagues • no external scrutiny of scientific validity of proposed research • cooperative, 'consultative' inspection	• pain should be limited by positive Home Office control • research should not cause 'real' or 'obvious' animal suffering • statutory requirements to prevent duplication of experiments • Home Office control over numbers of animals used in tests to prevent wastage through poor design • stopping experiments with no foreseeable contribution to medical advancement • prohibition of tests of cosmetics, smoking, food additives • more rigorous, independent sponsorship of applications. • external review of scientific validity of research proposals • larger, more active, more invigilatory inspectorate • more information published about suffering caused by experiments • greater veterinary involvement in policy formulation and oversight of experiments • more, and stricter infringement action, including prosecutions

The following two sections of this chapter examine the Littlewood Enquiry's deliberations and assess which groups were favoured by the Report's recommendations and the subsequent Government response.

The Littlewood Report: findings and recommendations

In Chapter 3, a paradox was identified in pre-existing work on animal research policy that needed to be explored and, if possible, resolved. Specifically, if the postulation of an issue network model is valid, then one would expect some degree of policy learning, and hence policy change, to have occurred following the Littlewood Report. However, according to Garner (1998: 178–9), the Government's response to the Littlewood Report was generally inert.

The first point to make in respect of tackling this paradox is that the foregoing analysis has already indicated that the Littlewood Enquiry emerged in the context of a policy community. This would appear to lend weight to the first hypothesis of an incorrect issue network attribution. However, at this stage of the narrative, it remains theoretically possible that the Enquiry led to a weakening of the tight policy community and that closer examination will reveal significant subsequent policy changes. In fact, this section will undertake a more detailed examination of the Enquiry and its Report than that found in the current literature, thereby enhancing our understanding of the evolution of network during this period.

The most salient aspects of animal research policy at issue in the Enquiry were those concerning the severity and 'necessity' of inflicted pain, and the question of whether 'wastage' of animals took place in terms of 'unnecessary' experiments. However, the definitions of the pivotal concepts of 'necessary' pain, 'wastage' and 'unnecessary' were highly contentious and dependent on the value judgments at the core of each lobby. This section will examine the Report's deliberations and conclusions on these issues, which are crucial to understanding policy outcomes, as well as its findings on inspection and the AC, which are relevant to the composition and structure of the policy network.

Pain

Animal researchers generally supported the existing Pain Condition (Littlewood, 1965: 58) which, as explained in the previous chapter, left the control of 'severe and enduring' pain to the discretion of the experimenter and in important respects made pain control secondary to the researcher's interest in completing the experiment. The RSPCA proposed a change whereby 'the experimenter should not be the sole judge in the application of the Pain Condition' (Littlewood, 1965: 153). However, the Enquiry (1965: 153) was persuaded by the testimony of senior scientists that 'informal' collective supervision of animal welfare was the norm in research establishments, and so they rejected calls to alter this provision.

The animal protection lobby[8] also proposed an amendment whereby experiments would be terminated immediately if 'real' or 'obvious' suffering were to arise, regardless of the stage of the experiment. While noting that this measure would provide significantly more protection, the Report (1985: 58) rejected it on the grounds that it would rule out many experiments, including statutory animal tests, and lead to a waste of animals killed before useful results were obtained.

However, the Report acceded to a proposal from UFAW, who worked with the scientific community, for a revised Pain Condition (Recommendation #2) that sought to clarify and update rather than substantively tighten the provision. The new proposal consisted of the following clauses (Littlewood, 1965: 58–9):

1. a first clause was proposed that provided a general injunction to licensees to 'take effective precautions' to prevent or minimise animal suffering
2. the second suggested clause was amended to require the killing of animals suffering 'enduring discomfort' (instead of 'severe' or 'enduring' *pain*) when the experiment had finished
3. the third clause appeared to be more preventative than the pre-existing one as it prohibited the licensee from inflicting persistent severe pain rather than requiring the animal to be killed once such a condition occurred.

These were vague principles that were dependent on the judgement of researchers and, in the first two clauses, pain limits were secondary to research goals.

'Unnecessary Wastage'

The Littlewood Report addressed three areas of alleged 'wastage':

1. Duplication or repetition
2. Experimental design
3. Additional unnecessary experiments.

With regard to the first area, the RSPCA and anti-vivisection groups were concerned about:

- The unnecessary repetition of work partly due to ignorance of previous experiments
- Duplication in industry due to commercial confidentiality (Littlewood, 1965: 83).

Scientific witnesses to the enquiry rejected the underlying premise of the complaint about repetition, arguing instead that certain forms of repetition were not scientifically unnecessary, as it could permit confirmation of the validity of earlier studies, and could result in observations of hitherto unobserved phenomena in the experiment. The Report (1965: 83) stated that such repetition would not be irresponsible, indicating once again their prioritisation of the standard operating procedures of scientific research. Furthermore, the Report (1965: 84) accepted the submissions of animal research groups that internal factors in academic and commercial research meant there was no significant 'unnecessary' duplication of experiments.

In contrast, the RSPCA's suggestion of a central register of more detailed applications that would enable the Home Office to detect potential repetition and investigate further was rejected as impractical and superfluous (Littlewood, 1965: 85). Instead, the Report (1965: 86) recommended (#11–3) that applicants, sponsors and inspectors should endeavour to prevent any unnecessary repetition or duplication. This approach typified the Report's preference for control ultimately based on professional discretionary judgment rather than a specific legal requirement (1965: 86).

With regard to the issue of experimental design, the Report (1965: 87) implied that a lack of expertise among researchers in statistical design led to the wastage of animals. Scientific groups also acknowledged that this was a problem to some extent. However, the Report rejected the RSPCA's proposal for Home Office intervention and concluded that this was a matter for the laboratory itself. Once again, an educational approach was adopted with the recommendation (#14) that the management of registered experimental premises should, as a legal requirement, provide for statistics advice to licensees.

UFAW also raised a related 'design' issue, suggesting that the numbers of animals used in tests could be reduced by prioritising the achievement of statistical significance rather than working to a pre-ordained minimum number of animals. The Report (1965: 87) expressed confidence in the commitment of the scientific bodies that formulated prescribed tests to minimise animal numbers, and ruled out control by the Home Office. It did, however, recommend (#15) that formulating bodies be required to consult the Home Office (Littlewood, 1965: 88).

A further source of animal protection concern related to the unnecessary retention of prescribed tests. In response, the Ministry of Health and MAFF stated that prescribed tests needed to conform to international standards, and the Report (1965: 89) endorsed the approach of these bodies:

> We were not able to discover any instance where there had been undue delay in dispensing with a prescribed test, and are satisfied that all concerned in the formulation of prescribed tests are anxious not to waste animals by retaining biological tests longer than necessary.

But perhaps the most serious policy criticism levelled by animal protection groups was articulated by the RSPCA, who made a strong call for the prohibition of tests not 'essential to solve some specific problem of suffering' and for 'self-inflicted' risks such as from the use of cosmetics, smoking, and the consumption of food additives (Littlewood, 1965: 89). The former of these two proposals reiterated the long-standing demand for the amendment of section 3(1) of the 1876 Act to limit the permitted purpose of experiments to those that were deemed to be directly related to, and likely to deliver, a significant medical benefit. However, the Enquiry's response to this central demand revealed its sympathy for research interests and their claims. In respect of the debate over whether an experiment, or class of experiments, was necessary or unnecessary, the burden of proof was clearly placed on the animal protection side. They were faced with having to establish, firstly, that no possible benefit might accrue and, secondly, that the administration of such a restriction should fulfil the preference of the Enquiry that it 'should be clear, simple, flexible, and reasonably rapid, and that control should be seen to be capable of effective enforcement' (Littlewood, 1965: 81).

For the animal protection lobby, the acknowledged uncertainty of benefit led them to define experiments aimed purely for advancing biological knowledge as an 'unnecessary' form of animal discomfort and pain. Conversely, for the scientific lobby and the Littlewood Report (1965: 91), the theoretical possibility of some future benefit meant that such experiments could not be defined as 'unnecessary':

> From our study of the evidence about unnecessary experiments and the complexity of biological knowledge we conclude that it

is impossible to tell what practical applications any new discovery in biological knowledge may have later for the benefit of man or animal. Accordingly, we recommend [#16] that there should be no general barrier to the use of animal experimentation in seeking new biological knowledge even if it cannot be shown to be of immediate of foreseeable value.

Furthermore, with regard to cosmetics testing, the Report (1965: 89) argued that it would be administratively difficult to draw a clear line between the testing of cosmetics and experiments designed to investigate medical problems arising from 'untested' cosmetics. In addition, the Report noted evidence of variable levels of pain in such tests, and so rejected the proposal because it was deemed 'undesirable in principle arbitrarily to determine that particular kinds of purposes should never be served by animal experiments whether or not they involve stress and pain'.

Although unwilling to recommend a general prohibition of any specific purposes of animal experimentation, the Littlewood Report (1965: 120–1) did indicate the need for some form of scrutiny to examine whether proposed experiments were for a permitted purpose, for example, to advance biological knowledge through new discovery:

> [N]otwithstanding that experiments involve exploration of the unknown, research is based upon established knowledge and profound scientific thinking. It follows that the claim that any experiment may throw up unexpected knowledge of incalculable benefit should not be taken as a sufficient pretext for allowing any random exploration to be undertaken without restriction.

In other words, the Report perceived that 'valid' animal research conformed to established academic scientific criteria in terms of the soundness of the underlying hypothesis and the methods proposed to explore it. This raises the issue of how any system of licensing and control of experiments should implement such criteria. The Report was ambiguous about the distribution of practical authority in pre-existing scrutiny mechanisms; on the one hand, it suggested that this has been a duty of inspectors,[9] while on the other hand it stated that the Home Office accepted the endorsement of the sponsoring statutory signatories (1965: 144). There is a similar opacity about its recommendations on this matter, which will now be examined as part of the discussion of the Report's deliberations on licensing, control and inspection.

Licensing, control and inspection

Given the considerable (though imprecise) practical power of statutory signatories over the approval of licensee and research applications, the issue of the sponsorship of applications is germane to the policy process at this time. The sponsorship question comprises two factors: the identity of sponsors and their role. With regard to the first factor, the RSPCA called for the existing statutory signatories – the presidents of various scientific and medical institutions – to be more rigorous in examining the background and suitability of applicants (Littlewood, 1965: 111). However the Report accepted contrary submissions from scientific bodies that the present system had already become anachronistic and impractical due to the volume of applicants relative to the number of potential sponsors, and the sponsors' lack of personal knowledge of applicants. Therefore, it (1965: 112–3) recommended (#34) that the sponsoring signatories should be scientists (including a professor) with personal knowledge of the applicant and/or the laboratory, shared disciplinary expertise, and in a position to vouch for the appropriate training of the applicant and proper conduct of the experiments. The Report (1965: 113) also rebuffed the argument that sponsors should be limited to individuals who had not been licence holders because 'it might appear to foster vested interest in research'. Instead of independent, external scrutiny, the Report preferred the claims of relevant experience and expertise because they were perceived to promote self-regulation and informed sponsorship.

Notwithstanding this preference for licence sponsorship by professional colleagues, the Report (1965: 125–6) seems to have taken on board the recommendations of the RSPCA and the BVA for what appears to have been tighter control of experiments where pain or stress was deemed likely to ensue:[10]

> we recommend [#49] that the licensee should submit a detailed application setting out the purpose in view and the design of the experiment (including procedures, species and numbers of animals) and that approval should be limited to the project as defined in the application...A report on the symptoms of pain observed during the experiment should be required at the end of the project.

However, it would be the sponsors' responsibility (#35) to certify that the research proposal fulfilled the legal requirements concerning the suitability of the applicant, its purposes, and that it was not repetitive or wasteful in its envisaged use of animals (Littlewood, 1965: 113–4).[11]

This discussion of the role of sponsors raises the pivotal question of their relationship with inspectors, an issue which was likely to have had significant implications for the distribution of practical power over policy outcomes. In fact, the Report's recommendations provide useful evidence about the established rules of the game in this relationship, as they are based on a retention of the pre-existing application process (though with more detailed content in certain cases): 'The crux of the application is to provide the Home Office with an informed basis for deciding whether a candidate is suitable for a licence....' (1965: 113).

But confusingly, and contrary to the submissions by UFAW, the Imperial Cancer Research Fund[12] and the RDS,[13] the Report (1965: 146–7) averred (#62) that it was important that inspectors had the power to assess the merits of experimental proposals and provide guidance to licensees, particularly where pain was expected to ensue. However, given that inspectors would already have been provided with an 'informed' endorsement of research proposals, it raises the question of whether inspectors would, in practice, be prepared to challenge such apparently professional expert advice. This question will be addressed below.

The animal protection lobby raised further concerns about the adequacy of the inspection arrangements for animal experiments. The Home Office confirmed that the primary role of inspectors was to cooperate and communicate with licensees to ensure compliance with the law rather than actually inspect experiments in progress (Littlewood, 1965: 143–5). On the other hand, the RSPCA proposed a more pro-active role for inspectors in terms of aiming to reduce the numbers of experiments and the pursuant suffering, as well as conducting more intensive inspection of experiments. The Report (1965: 144) explained the differences between the Home Office and RSPCA stances in terms of their opinions about licensees:

> The Home Office assumes that if he has been judged fit to hold a licence, the licensee can be trusted to comply with the Act unless evidence is found to the contrary. The Society prefers to place its trust in the inspector and assumes that where there is any risk of pain, supervision from outside should be extensive enough to safeguard the animal whether or not the licensee can be trusted to comply with the Act.

When this is taken in conjunction with the responsibility given to sponsors for endorsing applications, it seems clear that the differences between policy-makers and animal protection ran deeper than the issue of trust in licensees. Rather, it concerned a choice between professional self-regulation or external public regulation, and differences in the willingness to devote resources to maximising the mitigation of animal pain and discomfort. While acknowledging the inherent danger in the extant arrangements for licensees to dominate inspectors, the Report (1965: 145) nevertheless concurred with the Home Office stance. Thus, the Report endorsed the 'cooperative', as opposed to 'invigilatory', approach to inspection, and rejected any accusations of pro-researcher bias and the 'rubber-stamping' of applications.

With regard to the numbers of inspectors, the Report noted (1965: 150) that all groups wished to see an increase, though there were differences in the extent and the recommended frequency of visits to research units; the RSPCA proposed a visit every month, whereas research groups suggested quarterly intervals. The Report (1965: 151) recommended (#63) an increase in inspectors to sixteen, plus four superintending inspectors and a chief inspector, in order to allow quarterly visits, with the assumption that between ten and twenty premises would be visited per week. Animal protection proposals regarding the composition of the Inspectorate are not recorded by the Report, which recommended (#63) that the Home Office aim to recruit inspectors in equal number from the medical and veterinary professionals (Littlewood, 1965: 147–9).

The Advisory Committee ('AC')

Due to consistent evidence regarding the increasing complexity of animal experimentation and the desirability of an expert advisory body to coordinate policy and issue guidance, the Littlewood Report (1965: 156–63) seemed prepared to countenance change in this area, with the professed aim of enhancing the practical authority and supervisory status of the AC. It recommended (all AC recommendations are #71) that the AC be given legal standing and, following an RSPCA proposal, that it be empowered to investigate and advise on matters on its own initiative rather than be directed by the Home Secretary. Improved accountability would also be enhanced by the publication of an annual report. Finally, it recommended that the AC be reconstituted to include four lay members out of a membership of twelve.

However, while the remaining AC members would be appointed following consultation with various professional and industry groups, the Report (1965: 162) rejected animal protection requests for a similar role and decided to exclude such involvement for animal welfare societies: 'Most witnesses, however, felt that it would be undesirable for lay members to be appointed as formal representatives of animal welfare societies'. It is unfortunate that the Report does not explain the reasons given for excluding these groups.[14] Nevertheless, as a potential consequence, animal protection groups would be largely excluded not only from the implementation, but also from the formulation network.

Summary of recommendations

In all, the Littlewood Report made eighty-three recommendations and called for new legislation to implement most of the reforms. As this section has demonstrated, it did not advise that fundamental policy or institutional changes were required[15] but instead highlighted a number of technical administrative weaknesses that had arisen due to the rapid increase in the scale of animal experimentation. Thus, it called for the following secondary changes:

- more detailed guidance on the operation of the Act
- more instruction for licensees in legal requirements
- a greater role for 'advisory machinery'
- a larger inspectorate
- accelerated granting of licences and certificates
- more oversight of painful experiments (1965: 80–1).

Insights into the evolution of the animal research policy network

The Littlewood Report is significant because it affords some degree of insight into the particularly impenetrable 'black box'[16] that is the animal research policy process. If the data provided by the Report are related to the insights of policy network analysis as outlined in Chapter 2, it is possible to reconstruct an account of the network in terms of the relations between actors, dominant values, the rules of the game and strategies, and the resource interdependencies that affect power distribution and hence structure interactions.

One of the most pivotal issues to have arisen in the foregoing analysis of the Report concerned the network interactions between inspectors, researchers and sponsors in the scrutiny of applications for painful research. This question is germane because it focuses on a policy implementation process that involves the discretionary application of vague statutory criteria to technically-complex cases. Those vague criteria include whether the research is for a statutory permitted purpose, such as the advancement of knowledge through new discovery or medically useful knowledge. In policy domains of this type, implementation has a particularly significant impact on the pattern of policy outcomes

(Smith, 1997: 21, 28). By the same token, this makes it more likely that the advisory, formulatory body would be peripheral.

It was noted above that the practice had developed whereby the expert endorsement of applications for animal experimentation by researchers and their sponsors was seen to provide an 'informed basis' for inspectors' licensing decisions. This raised the question of the extent of inspectors' autonomy to challenge such expert endorsement. There are two related elements of the implementation process that are relevant to this question: resource distributions and ideological structure.

The character and implications of the network's resource distribution are illuminated by the Littlewood Report's (1965: 166) citation of Home Office evidence to the second Royal Commission. It reveals that since at least the 1906–12 period, the Home Office had deemed it unfeasible for inspectors to be expert in all areas of biological research. Consequently, the Home Office felt that it was not in a position to be able to make an informed judgement about whether a research application was sufficiently scientifically valid to fulfil the statutory criteria:

> The Secretary of State would be placed in a position of grave responsibility and much difficulty if he were to undertake to formulate and act on a departmental opinion as to whether the scientific results of a research – possibly of an obscure character and conducted by an expert of unique qualifications – were sufficiently established or promising to justify the pursuit of the investigation.

In other words, despite the Inspectorate's formal political resource of authority to licence, it was perceived to lack the necessary organisational, knowledge and information resources to scrutinise applications, and so its authority was partially dependent upon the information supplied by researchers. Therefore, when examined using policy network analysis (Smith, 1997: 35), it seems reasonable to infer that the inspector would have been reluctant to refuse applications because the scientific rationale of such a decision would be open to challenge. The dominance of professional and economic interests over policy implementation is further confirmation of the presence of a policy community, and also helps to explain its stability up until that point, as the monopolisation of technical knowledge facilitates insider groups' resistance to change that may be contrary to their interests (Smith, 1993a: 98).

The pro-research lobby's domination of the Home Office, which was achieved through an asymmetry in information and knowledge resources, would have encouraged the development of the network structure whereby policy decisions were co-produced at the laboratory level, and militated against attempts to formulate more detailed general guidance (Smith, 1997: 211). Indeed, the Littlewood Report's repeated identification and legitimisation of cooperative relations between these actors demonstrates the persistence of positive-sum, exchange relationships that, once again, are indicative of a policy community.

The shared professional background of inspectors and researchers and the ideological consensus that bound the policy community helped to reinforce this exchange relationship. That ideological consensus included two beliefs. Firstly, animal experiments that were proposed by research scientists were assumed to be justified, as they were considered instrumental to the advancement of knowledge or to the development and testing of medical products. Secondly, the welfare of animals, though not entirely irrelevant, was a secondary consideration to the pursuit of research goals. Therefore, only expert resources that could either facilitate the perceived compliance of research proposals to statutory criteria, or mitigate animal suffering without interfering with the experiment, were perceived as valuable. Broader ethical and public health considerations and associated actors were excluded from the network.[17] Within this ideological context and resource structure, inspection was facilitative rather than the neutral or preventative process desired by animal protection groups. The 'rules of the game' involved co-operative relations, trust and, especially in respect of achieving compliance with any limits such as the Pain Condition, a reliance on the regulatee to provide the information rather than the inspector seeking it out.[18]

Implications for future network evolution

The animal research policy network as revealed by the Littlewood Report demonstrates an entrenchment of the policy community that had first emerged in the 1880s. The establishment of the Enquiry represented a potential perturbation to this policy community, but the ability of professional and industrial communities to influence agenda-setting[19] could already be seen in the narrow terms of reference that constrained the enquiry. The Report (1965: 79–80, 190) implied that significant changes in animal research policy would have required a broader examination of the direction and methods of biological research: 'There is no basis for regulating the objectives or scope of animal experimentation without reference to the general organisation and development of biological science'. But this task was explicitly ruled out by the Enquiry and considered beyond the legal and, given the Home Office's perception of scientific autonomy and unique expertise dating back fifty years, knowledge and organisational resources of the government. This strongly suggests that, in practice, the resources most influential over animal research policy appear to have rested in the hands of the research profession and related industries. This resource advantage was consolidated by the strategic action of the RDS, which brought together major animal research interest groups to submit a joint memorandum to the Enquiry. Therefore, the consistent unanimity of researchers in support of animal experimentation and professional autonomy, and the corresponding lack of overt expert support for the animal protection lobby, appear to have placed a major constraint on the possibility of policy change.

The limited potential for change can be seen by analysing the likely impact of arguably the most significant recommendation. This entailed the submission of more specific detail about the purpose, procedures and numbers of animals in applications to conduct painful experiments on animals, which would then form the limiting conditions on the licence. But in the context of researchers' domination of their positive-sum exchange relationships with inspectors, it can be averred that these specifications would largely be a reflection of the wishes of researchers. In other words, such a requirement may have had little potential impact on policy outcomes. The degree of freedom of researchers to exceed the initial specifications is uncertain, but once again it is likely that, given the previous willingness of the Home Office to revise number limitations upwards and the institutionalised obstacles to more police-style inspection, even such self-imposed limits may have been ineffective.

These conclusions help to answer the secondary research question posed at the beginning of this section: If the issue network postulation has any validity, then the Littlewood Report would have led to some significant network and policy changes. However, the foregoing analysis suggests that the Report promoted a continuation of a policy community and outcomes that favoured animal research interests. Given Garner's observation that the government neglected to implement even the minor, secondary changes proposed by the Report, the issue network hypothesis appears somewhat implausible. However, it is necessary to reconstruct subsequent events in order to draw firmer conclusions on this issue.

This study deploys a critical realist methodology that incorporates insights from historical institutionalism and path dependency. Therefore, this assessment of the routinised standard operating procedures and ideas that shaped the policy community at this time provides a basis for understanding the subsequent impact of any governmental implementation of the Littlewood Report. Furthermore, depending on the extent of any changes introduced by the government, these structures may comprise a relevant context for an analysis of the emergence of the Animals (Scientific Procedures) Act 1986 and its subsequent impact. It is now time to address the government's response to the Littlewood Report.

The government's response to the Littlewood Report

The consensus in studies (Hampson, 1987: 316; Hopley, 1998: 74; Ryder, 1989: 243; Brooman and Legge, 1997: 129; Balls, 1986: 7) with some relevance to the evolution of animal research policy concurs with Garner's (1998: 177–9) description of government inaction in response to the Littlewood Report.[20] Despite the 'very moderate' (Ryder, 1996: 169–70) nature of the Report, anti-vivisection groups such as the BUAV perceived that its recommendations had the potential to tighten the regulation of animal research to some degree and undertook strategic action to stimulate debate in Parliament in an effort to keep the momentum for change on the political agenda (Hopley, 1998: 74). The government's reluctance to acknowledge, never mind act on, the Report (Ryder, 1989: 243) indicates a shared position with pro-research interests, who also argued against the need for any reforms (Garner, 1998: 180).

One of the few administrative changes made by the Home Office was a gradual increase in the number of inspectors. But even eight years after the Report's publication, there were still only fourteen inspectors, some way below the minimum of twenty-one that the Report – in line with the stated position of research groups – had recommended (Straughan, 1995: 45). The Home Office also rejected the recommendations regarding the AC on the grounds that they were too radical (Balls, 1986: 7). The Home Office's unwillingness to make even minor changes may reflect Ryder's (1996: 171) perception of their 'high respect for science and medicine'. Conversely, the Home Office was highly suspicious of even moderate animal protection lobbyists (Ryder, 1996: 171). The strong cohesion between state and insider research groups represented an endogenous institutionalised force for stability that enabled the animal research policy community to remain virtually intact until at least 1975.

Nevertheless, the exogenous perception that the 1876 Act was out of date and in need of reform did not dissipate. Indeed, when viewed in terms of the dialectical network model, it could be argued that the perceptions and strategic actions of policy community actors, as constrained by the tight ideological and social structure of the network, failed to take sufficient account of some shortcomings in legitimacy resources of the network. For example, Garner (1998: 179) recounts how the Home Secretary in 1971 continued to justify the Government's inaction by reference to a perceived lack of public interest and the theme in the Littlewood Report that there was nothing fundamentally flawed about the 1876 Act. In the context of a rise in the number of procedures from 4.75 million in 1965 to 5.61 million in 1971 (Straughan, 1995: 45), insufficient adaptive secondary change on the part of the policy community meant that, according to Hampson (1987: 316):

> The failure of the government to implement the eighty-three legislative and administrative changes recommended by this Committee led to a spate of private members' bills in both Houses of Parliament throughout the late 1960s and early 1970s.

These observations make it possible to review the first two ambiguities identified in the issue network model proposition back in Chapter 3:

1. If the policy network resembled an issue network model, then one would expect some degree of policy learning and hence policy change to have occurred following the Littlewood Report. However, the Government's response appears to have been inert and consonant with the position of the animal research community
2. The growing politicisation of animal research may be partly the result of policy outcomes favouring animal research interests (e.g. an exponential rise in animal experiments), indicating the possibility of a pre-existing policy community.

The foregoing analysis strongly suggests that on both counts, a postulation of a policy community is far more plausible than an issue network. Even if, for the sake of this analysis, it is accepted that public interest was relatively weak, an issue network policy-making arena would have facilitated secondary policy learning in the network in line with the moderate recommendations of the Littlewood Report.

The path to the reform the 1876 Act: the Animals (Scientific Procedures) Act 1986

The case for the persistence of an animal research issue network, as opposed to a policy community, rests significantly on the claim that 'the more moderate section of the animal protection movement played a central role in the formulation and passage of the [Animals (Scientific Procedures)] Act' (Garner, 1998: 182).[21] However, Chapter 3's analysis identified empirical and theoretical ambiguities in this interpretation.

Even before these ambiguities are investigated further, the foregoing analysis has indicated that the basis for this conception of the passage of the 1986 legislation – the persistence of an animal research issue network since the assent of the 1876 Act – is unsound. Nevertheless, the question remains whether the alleged participation of the animal protection lobby in the formulation of the new legislation was related to a contemporaneous network change towards the issue network end of the typology. It is this question that must be addressed in order to tackle the third research question: Did the passage of the Animals (Scientific Procedures) Act 1986 signify a core change in policy or an instance of dynamic conservatism?

Specifying the group and network ideologies

One key task for network analysts is the adequate specification of relevant group ideologies and policy beliefs. Previously, it was found that the 'pro-animal research' ideology has been defined too loosely as encompassing all non-abolitionist positions, when in fact it contains two distinct core policy positions, as demonstrated by the foregoing analysis of the evolution of the policy network. Those positions are:

1. (i) animal welfare is secondary to research goals
    (ii) animal experiments are considered 'necessary' and hence permissible in the pursuit of knowledge without immediate or foreseeable human benefit
    (iii) resistance to utilitarian scrutiny of experimentation proposals
    (iv) support for professional self-regulation and the avoidance of lay interference
2. (i) animal welfare may take priority over research goals
    (ii) animal experimentation only considered 'necessary' and hence permissible when essential to satisfy urgent and pressing human needs
    (iii) support for independent utilitarian analysis of experimentation proposals
    (iv) lay control is required to ensure consideration of the wider public – and animals' – interests

Interestingly, the validity of this distinction is supported by its consistency with Orlans' (1993: 22) classification of animal-related belief systems in modern 'western societies'; it corresponds, respectively, to her 'animal use' and 'animal welfare' ideological categories. Thus 'animal use' represents the position of the animal research interest groups, while 'animal welfare' represents the position of groups who seek policy reforms that increase the limits on animal experimentation, such as the RSPCA (Orlans, 1993: 22). The distinction also helps to express the differences between the way the two ideologies view the moral implications of the welfare of animals, as discussed above in regard to the animal research policy community's learning in response to growing awareness of, and sensitivity to, animal welfare.

This, of course, leaves the 'animal rights' category, which includes the major anti-vivisection lobby groups. As Orlans (1993: 22) notes, their basic belief is that the underlying justification for inalienable human rights to protection from exploitation and killing applies to at least most animals, which therefore entails the abolition of animal experimentation. However, their activities include regulatory as well as abolitionist goals. As discussed in Chapter 3, the major groups in this category did, indeed, undertake strategic action in relation to the replacement of the 1876 Act whereby they made regulatory demands in line with the animal welfare position rather than lobby for the abolition of animal research.

Structure and agency: the dialectic between perceived policy outcomes, exogenous pressure and strategic action

As noted above, during the late 1960s and early 1970s, the Home Office's failure to act on the Littlewood Report provoked a gradual increase in pressure for reform from exogenous areas, particularly public opinion and the ideological context. An autobiographical article by the psychologist and lobbyist Richard Ryder (1996) provides an insight into these developments, as well as highlighting the interrelationships between agency and structure, particularly the potential for agents to interpret and alter their structural context. Ryder's previous experiences of animal research as an experimental psychologist led him to believe that the public's general acquiescence to animal experimentation in the late 1960s was based on a false view of the validity, benefits and level of suffering involved (1996: 171–2). This resulted in strategic action on his part, beginning in 1969 in the form of campaigning to raise public awareness through the media, leafletting and protests.

Ryder's activities also complemented and reinforced a philosophical upsurge of interest in the ethics of human-animal interactions in this period, the first product of which was the 1971 publication of a collection of essays entitled 'Animals, Men and Morals',[22] to which Ryder contributed (Ryder, 1996: 168). Singer's (1996)[23] review of the book made an ideological link between this emerging intellectual critique and previous ones against racial and sexual discrimination, describing it as: 'a manifesto for an Animal Liberation movement'. Singer's subsequent publication of his book 'Animal Liberation' in 1975 is widely credited with providing a groundbreaking intellectual springboard for a revitalised animal protection movement that questioned the orthodox idea of 'speciesism', which refers to a perceived assumption of human moral supremacy and the notion that the moral status and treatment of individual beings should depend on their species membership (Radford, 2001: 168; Ryder, 1989: 247; Sperling, 1988: 82; Garner, 1993: 2; Hopley, 1998: 80; Nuffield, 2005: 23). Instead, Singer argues that just as it is wrong to exploit or cause pain to all human beings regardless of their intelligence or ability make moral choices, it is wrong to exploit or cause pain to other sentient animals that also have an interest in not suffering (Garner, 1993: 12–1). As Orlans (1993: 24–5) observes:

> The rise of the animal rights movement dates back to Peter Singer's publication in 1975 of his revolutionary book *Animal Liberation*. This important book changed the way many of us look at animals; it inspired a worldwide movement... Other philosophers soon followed with significant challenges to long-held views about the moral relationship between people and animals.[24]

The potential impact of these ideological perturbations on the animal research policy community would be partly dependent upon the media coverage gained by this 'new social movement' and public opinion (Richards and Smith, 2002: 183–4). If the policy community's 'animal use' paradigm becomes discredited, then state actors may be persuaded to introduce new actors and ideas into the policy network in order to maintain the network's legitimacy. However, such movements tend to focus on wider social change rather than the policy process, as they are wary of strategic compromises that are perceived to be inconsistent with their underlying moral principles (Richards and Smith, 2002: 184–5). There appears to be a potential for this to occur in the animal liberation movement because in absolute terms, it implies the pursuit of the abolition of animal experimentation rather than incremental policy change (Garner, 1993: 21, 128–31). Nevertheless, the dialectical interaction between this changing exogenous context and the animal research policy network is examined below.

1975: the intensification of animal research politicisation

One of the ways in which policy communities may attempt to contain potentially destabilising exogenous forces is through their control of the outward flow of information concerning policy outcomes.[25] The renowned secrecy of animal research policy (Littlewood, 1965: 164) may have assisted the policy community in this manner. However, whether such secrecy had any real impact would depend on whether the emergence of unusual, direct evidence of animal research was anomalous in terms of the public's previous perceptions of the practice: '...that all experiments were for valid and important medical research and involved little suffering' (Ryder, 1996: 171). In other words, if new information conformed to the policy community's public discourse, then secrecy would be immaterial to the policy process, apart from possibly engendering suspicion in itself.

In fact, in early 1975, Ryder (1996: 172) was involved in initiating an 'exposé' in the *Sunday People* of tests of a tobacco substitute performed by ICI that required beagles to inhale smoke. According to Hopley (1998: 81) the furore was unprecedented and '...caused outrage among people who hadn't previously protested against vivisection and there was a public awakening about just how widespread vivisection had become'. The public response appears to have demonstrated significant support for the animal protection lobby's belief that self-imposed risks of suffering do not warrant animal experiments.[26] A fortnight after the smoking beagles story emerged came the publication of Ryder's book *Victims of Science*, which aimed to provide new insights into animal research practices in the context of an 'anti-speciesist' ideological framework. This work complemented the tabloid media coverage for the exposé by stimulating extensive discussion throughout the rest of the media (Ryder, 1989: 245).

There were a number of exogenous consequences of this novel information and the ideological challenge to the animal use belief system that structured policy. Firstly, the extensive public response to the story is said to have changed the attitude of the media to animal welfare in general from one of derision and apathy to interest and sympathy (Ryder, 1989: 245–6). Secondly, with the assistance of activists and reinvigorated campaign groups, intensive public protest and parliamentary activism was stimulated. These combined to have an immediate, though minor, policy effect: the experiments were stopped, and the Home Office announced that dogs would no longer be used in tobacco experiments (Hopley, 1998: 81–2; Ryder, 1989: 245–6). In addition, four lay members were appointed to the AC, ten years after the Littlewood Report had first recommended this measure, 'although the Government stressed that it did not intend to alter the functions of the Committee' (Balls, 1986: 7).

In the light of the strategic actions of animal protection lobbyists, public concern remained high. This, in turn, forced the issue up the political agenda and is said to have been one of the instrumental factors in forcing the eventual replacement of the 1876 Act with the Animals (Scientific Procedures) Act 1986 (Ryder, 1996: 171). However, this intensifying exogenous pressure continued to be largely resisted by an obdurate policy network (Ryder, 1989: 247).

The evolution of proposals for new legislation

New legislation on the political agenda

Perceiving the need for strategic action that might affect the network's resistance to new legislation, Ryder and a small number of sympathetic parliamentarians and scientists associated with the RSPCA formed an ad hoc reformist lobby group later in 1975 (Ryder, 1989: 247; Hollands, 1995: 33). This strategic action was based on a perceived need to argue for incremental reform rather than abolition, avoid accusations of emotionalism and ignorance, and deploy scientific expertise to carry the message to government (Ryder, 1996: 172–3). One key figure in this group was Lord Douglas Houghton, former chairman of the Parliamentary Labour Party. He brought two important resources to the group: connections with members of the Labour Government and political expertise in the form of advice to try to make contact with central government rather than expending resources on backbench MPs (Ryder, 1996: 170–2). Only one anti-vivisection group was represented, the SSPV, under the leadership of Clive Hollands.

This strategy was assisted by support from a small number of scientists who had, to varying degrees, begun to question existing animal research practices, and thus lent greater credibility to the reform group's cause (Ryder, 1996: 172–3). However, Langley's (1989: 200–3) account of the defensive and vituperative reactions of research institutions to occasional internal criticism of animal welfare provisions suggests that the scientific profession's:

> hierarchical structures and the conformity of its practitioners, has prevented advances in laboratory animal welfare which could otherwise have been implemented much sooner. The insular and reactive nature of the scientific community, on this subject in particular, has also suppressed free debate on the morality of using animals as tools in research, testing and education.

Thus, the extent of scientific support for reform, and hence the 'expertise' resources of the animal protection movement, remained relatively limited.

Initial attempts to propose animal welfare reforms and build cordial relationships with the Home Office in 1976 were rebuffed with some hostility from the Minister and Chief Inspector (see Chapter 3). This response appears to indicate the presence of a strong 'animal use' ideological consensus in a highly cohesive policy community.

The official reaction may also be related to a change in the economic and political context of animal research policy that occurred at the same time. Abraham's (1995: 74–7) study of pharmaceutical industry regulation notes the abrupt switch from a Keynesian to a monetarist economic policy in 1976 that required the government to prioritise industrial competitiveness at the expense of other social goals. Abraham (1995: 76) states:

> Of particular significance, the National Economic Development Council's "sector working party" on the pharmaceutical industry concluded in 1976 that, in order for the pharmaceutical sector to maximise its contribution to a positive UK balance of payments through expansion of direct exports and import substitution, the DHSS [Department for Health and Social Security] should seek to have the minimal impact on the industry's research.

This made the government more responsive to complaints regarding alleged over-regulation from the pharmaceutical industry, who threatened to relocate their research abroad. Thus, state moves in the early 1970s to increase and develop a more adversarial approach to the regulation of the pharmaceutical industry were reversed from 1976. Though Abraham's analysis is focussed on DHSS policy networks, it should also be noted that throughout this time, the industry was in 'complete control' of its animal research practices (Abraham, 1995: 251). Furthermore, the DHSS was a sponsor of the pharmaceutical industry's interests throughout government. It is therefore probable that Home Office regulators of animal research also felt similar deregulatory pressures.

Nevertheless, continued public and political interest and a change in Minister led to dialogue and an indication from the then-Home Secretary Merlyn Rees that he would try to slowly reform the regulatory system in the direction advocated by the ad hoc group (Ryder, 1989: 248). With a reform process apparently initiated, in 1977 the ad hoc group formalised itself under the acronym CRAE, the Committee for the Reform of Animal Experimentation (Hollands, 1995: 33). However, the Home Office did not take any discernible action towards legislative reform (Hollands, 1995: 34; Ryder, 1989: 200).

This absence of tangible action on the part of government led CRAE to perceive the need to undertake what amounted to further strategic action to alter the network's structural context and hence apply pressure to the policy network. A campaign was initiated that successfully persuaded the major political parties to include pledges to update the 1876 Act in their 1979 General Election manifestoes (Hollands, 1995: 34; Ryder, 1989: 200–1). Clearly, exogenous public and political opinion was considerably more supportive of reform than the animal research policy community.

The development of legislative proposals: the 1983 and 1985 white papers

Following their election victory, the new Conservative administration reiterated their intention to introduce new legislation on animal research (Hollands, 1995: 34). In an attempt to pre-empt legislative change, the RDS sponsored a private members' bill introduced by Lord Halsbury in 1979, which Garner's (1998: 180–1) states would have weakened animal research regulation. Garner speculates that the Government blocked the Bill because of concerns about opposition from animal welfare groups, although other evidence suggests lack of parliamentary time (Hollands, 1995: 34) and the need to comply with impending EU legislation (Hopley, 1998: 89–90).

In fact, the RDS Bill exerted considerable influence over the subsequent evolution of legislative proposals contained in two White Papers published in 1983 and 1985 (Ryder, 1989: 248–9). The House of Lords Select Committee that scrutinised the Bill produced a report in 1980 that outlined an 'enabling' statute which would have granted the Home Office broad discretion rather than seeking to prohibit specific categories of research. It also suggested a more powerful advisory committee, but one that was dominated by professional and industrial interests with very little animal welfare representation.

At the same time, many animal welfare individuals (including Ryder, who was chairman of the RSPCA's relevant committee) and all bar one of the animal welfare groups (the SSPV led by Hollands) split from CRAE. This was a strategic response to their perception of an unfavourable political context in the shape of an unsympathetic Thatcher administration; a belief that the tendency towards a high degree of discretionary power for the Home Office would result in decision-making favouring professional and industrial insider groups; and a judgment that in general the degree of compromise envisaged with pro-research interests was too high (Ryder, 1989: 248–9).

In fact, this perception of structural constraints on policy change may have been reinforced by the new Conservative Government's deepening of the state's commitment to the pharmaceutical industry's commercial interests and desire for self-regulation (Abraham, 1995: 76–7). This intensified a trend that had begun with the introduction of a monetarist economic policy in 1976 that entailed the belief that industrial competitiveness was essential to Britain's economic well-being. The discussion of policy community dynamics in Chapter 2 noted that state actors' perceptions of the extent of insider groups' contribution to the national economy was proportionate to those groups' autonomy, and promoted positive-sum exchange relationships where the group actors had most resources. Therefore, it appears that economic and political trends were likely to represent powerful macro-structural constraints on change in the animal research policy community.

Nevertheless, in order to acquire sufficient organisational resources to be able to lobby for incremental reforms, the remaining CRAE figures, most notably Hollands and Houghton, formed an alliance in early 1983 with two scientific groups, FRAME and the BVA ('the CRAE Alliance') (Hampson, 1987: 318; Hollands, 1995: 34). However, according to Hollands (1995: 34), the Alliance submitted a policy statement to the Home Secretary in March 1983 which included CRAE's four main objectives as originally outlined in 1979:

1. the restriction of pain through a ban on any 'severe' pain (first proposed by UFAW back in 1965) and, more innovatively, through a cost-benefit assessment where the degree of permissible pain was linked to the purpose of the research – for example, experiments aimed at life-saving research would be permitted to have higher pain thresholds (Hampson, 1987: 318)
2. 'a very substantial reduction in the number of animals used'
3. promotion and adoption of alternative methods
4. increased public accountability.

Later in 1983, the Government issued its first White Paper, but its close affinity with the RDS Bill of 1980 resulted in criticism from all the animal welfare groups, and even CRAE (Hampson, 1987: 317; Hollands, 1995: 34). The perceived failure of the Government to stipulate adequate restrictions on the infliction of pain was a particular concern. However, it is interesting to note that the CRAE Alliance did not appear to be concerned about the broad discretionary powers granted to the Home Secretary (Hollands, 1995: 34–5), indicating that they did not share the interpretation of the animal protection groups who were perturbed about the potential insider influence of the pro-animal research lobby.[27]

The 1983 White Paper led the animal protection lobby to split further into three main factions due to their different strategic responses. The RSPCA condemned the White Paper for failing to satisfy its four main demands, which corresponded to CRAE's 1979 objectives except with a stricter goal whereby all painful research would be prohibited (Hampson, 1987: 317; Ryder, 1989: 249). A report on pain in experiments had been submitted by the RSPCA, but the Home Office cited alleged uncertainties about the definition of pain in order to justify inaction on this question in the 1983 White Paper. Nevertheless, the RSPCA continued to focus on what they saw as the central issue of pain, and worked with a leading expert[28] to persuade the Home Office to shift its position (Ryder, 1989: 249–50; Hampson, 1987: 318). Thus, the RSPCA tended to adopt an equivocal stance towards the impending legislation. They did not overtly welcome the prospect of a new Act because the proposals were inconsistent with their own policies, but they were reluctant to completely reject the process for fear of losing any influence over its final form (Hampson, 1987: 318–9).

Meanwhile, the anti-vivisection groups fundamentally disagreed with the 'enabling' form envisaged in the White Paper and regarded the proposed legislation as essentially the same as its Victorian forebear (Hopley, 1998: 89–90). Therefore, the four largest groups – the BUAV, NAVS, Animal Aid and SAVS – established a coalition called the 'Mobilisation Against the Government White Paper' (Hampson, 1987: 317; Hopley, 1998: 90). The Mobilisation group rejected the White Paper and sought a fresh approach that would ensure the prohibition of areas of research they had long campaigned against and that they perceived to be widely opposed by the public. Those areas included two particularly severe and scientifically controversial toxicity tests – the LD50 (Stephens, 1989: 165) and Draize eye irritancy assays (Langley, 1989: 194) – military experiments, cosmetics, alcohol and tobacco testing and invasive psychological research (Ryder, 1989: 254). The Mobilisation group also sought an 'independent' advisory committee which was not dominated by pro-animal research interests (Hopley, 1998: 90; Ryder, 1989: 254). Despite the fact that these demands represented a major compromise relative to the groups' underlying abolitionist position, and that the categories of research they wished to ban accounting for a mere 11 per cent of animal experimentation, the demands were seen as too radical (Hampson, 1987: 317–8). Their ability to influence Government was further eroded by the perceived ambiguity in the BUAV's policy on illegal action: 'The organisation had been excluded from any negotiations about the Bill on the grounds of its alleged support for illegitimate and illegal activities' (Hopley, 1998: 91).

In contrast to the outsider status of the anti-vivisection groups, the CRAE Alliance sought access to both pro-research groups and the Government in order to try to influence the direction of the legislative proposals. According to Hollands (1995: 34–5), a leading figure in CRAE, the Alliance achieved 'insider status' with the Home Office, which resulted in a 1985 supplementary White Paper that is said to have incorporated many of the CRAE Alliance's suggestions (Hampson, 1987: 318). One of the most salient alterations in the Home Office's position was its rejection of pain restrictions identical to the 1876 Act in favour of the adoption of a cost-benefit assessment of research proposals, submitted in the form of 'project licence applications', which linked the degree of permissible pain to the purpose of the research (Hampson, 1987: 318). Furthermore, the previous Pain Condition was tightened so that any severe pain, regardless of its expected duration, was to be immediately stopped by the animal's destruction. A more equivocal development in the 1985 White Paper concerned the role of a new AC: this did not explicitly allow for an AC to act on its own initiative, but it could 'suggest topics which it thinks need to be considered' (Balls, 1986: 9). In other words, the Home Secretary would still retain ultimate control over a new AC's agenda. In terms of the AC's composition, the new White Paper introduced a 50 per cent limit on the number of current licence holders, though at least two thirds of the Committee were required to be scientists.

However, the 'Mobilisation' coalition rejected the 1985 White Paper because of its failure to incorporate any of its demands (Hopley, 1998: 90; Ryder, 1989: 254). The Government's acceptance of a cost-benefit assessment of project licences in that Paper made its refusal to countenance a ban on cosmetics testing on animals appear highly paradoxical, given that widespread opposition to such tests was based upon the belief that the utility of the product did not warrant the infliction of animal suffering. It could thus raise suspicion about the way the Government intended to implement the cost-benefit assessment. Indeed, subsequent comments from a Home Office Inspector, Straughan (1995: 41), suggest that the project licence system was introduced as:

> an ideal way of obtaining sufficient information and applying sufficient controls to properly meet the fundamental objectives of the EC Directive [86/609/EEC] to "ensure that the number of animals used for experimental or other scientific purposes is reduced to a

minimum" and that, where pain, suffering, distress and lasting harm are unavoidable, "these shall be kept to the minimum"!

In other words, contrary to CRAE's understanding (Hollands, 1995: 35–6), project licences would not be subject to a recognisable form of cost-benefit assessment that would countenance the refusal of animal experiments in the pursuit of any scientific purposes as defined by researchers. Instead, the official interpretation was that, in keeping with the animal use ideology, the information in project licence applications would merely allow animal welfare to be taken into account in a way that did not interfere with over-riding scientific aims. For the RSPCA's part, while the new White Paper partially reflected the group's desire to limit pain, the remaining gaps between the draft legislation's provisions and their policies meant that the Society maintained its ambivalent stance (Hampson, 1987: 318–9).

The passage of the Animals (Scientific Procedures) Bill

The Government threatened not to include the measure in the 1985 Queen's Speech because of its contentious nature and limited Parliamentary time (Hollands, 1995: 35). The validity of this rhetoric is brought into question by Straughan's (1995: 40) revealing observation concerning the important changes in the European context:

> In 1980, the Council of Europe's Convention for the Protection of Animals used for Experimental and Other Scientific Purposes was in preparation... Subsequently, though this is rarely acknowledged, it became clear that the EC was likely to embody the Council of Europe Convention in a Directive and that, to comply with this likely Directive, the UK would need new legislation. Indeed Directive 86/609/EEC and the Animals (Scientific Procedures) Act were both finalised in the same year, 1986.

It is not unreasonable to speculate that the Government's professed hesitancy may have been a tactic to ensure that the new law reflected the 1985 White Paper. For, perceiving the possibility of no legislation at all, the CRAE Alliance decided it was necessary to seize the present opportunity and thus acted to bolster support for the new law from both main parties (Ryder, 1989: 257). Ryder (1989: 254) suggests that Houghton was responsible for Labour's official support for the measure, and it could be argued that this went some way towards neutralising concerns among some Labour MPs that the Bill was not sufficiently restrictive.

Although the RSPCA continued to lobby the Government over the proposed new law: '... it fell to none of the leading animal welfare societies to become official advisers to the government on the Bill's passage. This role was adopted by the alliance between CRAE, the BVA and FRAME' (Hampson, 1987: 319). The one set of interesting amendments won by the CRAE Alliance during the passage of the Bill concerned the proposed new advisory committee, to be known as the Animal Procedures Committee (APC). One amendment gave the APC the power to act on its own initiative, while another consisted of a broad requirement for the Home Secretary, when appointing members, to 'have regard to the desirability of ensuring that the interests of animal welfare are adequately represented' (Section 19 (3)(b); cited by Balls, 1986: 9).

While the CRAE Alliance together with pro-research interests generally supported the Bill, the animal protection movement opposed it on the following grounds that relate closely to the goals first set by CRAE in 1979, when its membership was wider and more animal protection-oriented:

> [it] fails to prevent a single experiment that takes place at present, does little to encourage, in the long term, the replacement of animals in experimentation, does little to increase the protection for laboratory animals and does not allay public concern for the need to have greater public accountability in the field of animal experimentation. (Ryder, 1989: 255)

Worries about the lack of accountability were exacerbated by the confidentiality clause (Section 24) that had been inserted at the request of research interests, and was deemed to be likely to increase the secrecy enveloping animal research. Additional concerns surrounded the Bill's failure to specifically ban non-medical categories of experiment, and the lack of mandatory training for researchers in pain relief (Ryder, 1989: 256–7). But the general objection to the Bill, which reflected the complaint about the 'enabling' style of the White Papers, was that it 'contained no firm guarantees of improvement' (Ryder, 1989: 257).

The Mobilisation groups continued their efforts with large lobbies of parliament and meetings with sympathetic backbenchers. However, without access to policy-makers these moves made no difference to the final legislation, passed on 20 May 1986 (Hopley, 1998: 90–1). Meanwhile, the RSPCA's equivocation led it to be reticent in both pushing for the amendments it sought as well as opposing disagreeable clauses (Ryder, 1989: 256). Combined with Labour support for the Bill, these factors meant that there was little effective opposition to its passage (Ryder, 1989: 257).

The policy network implications of the assent of the Animals (Scientific Procedures) Act 1986

In order to interrogate the key claim that sections of the animal protection movement were integral participants in the 1986 Act, it is necessary to assess whether the CRAE Alliance was representative of the animal protection movement and whether its perceived insider status was genuine or peripheral.

In the first instance, it is clear that from 1983 onwards, the CRAE Alliance was quite *unrepresentative* of the animal protection movement, both in terms of its composition and its goals for the 1986 Act. While there was some common ground with the RSPCA in terms of the aim of introducing pain limits to experiments related to the level of predicted benefit, relations with the RSPCA were fairly loose (Hampson, 1987: 318–9) and there was outright antagonism between CRAE and the Mobilisation grouping (Hollands, 1995: 35). Furthermore, the position of FRAME and the BVA was, as noted in Chapter 3, equivocal in terms of straddling the border between the animal welfare and animal use ideological positions.

This issue is also connected with the question regarding the nature of the pro-animal research consensus that the CRAE Alliance needed to conform to in order to gain access to policy makers. It has already been established that an animal use ideology structured the animal research policy community in the mid-1970s. The question is: Did the formulation and assent of the Animals (Scientific Procedures) Act 1986 correspond with the institutionalisation of a new animal welfare network ideology? Answering this question will depend on the implications of the key innovations in the new law: the cost-benefit assessment and the APC.

The implications of the new cost-benefit assessment

To explore whether the cost-benefit assessment introduced this animal welfare requirement, it is necessary to examine the mechanism for its operation envisaged by the new Act. The 'enabling' and discretionary nature of the legislation meant that licensing decisions were still to be made on a project-by-project basis. Furthermore, the statutory criteria, involving an assessment of costs and benefits and a subsequent 'weighing' of the two concepts, remained a subjective, expert-dependent process that required the discretionary application of vague statutory criteria to complex multi-faceted cases. Therefore, this remained a policy area where the implementation 'stage' was pivotal to outcomes.

The analysis earlier in this chapter of the network implications of the implementation regime as described by the Littlewood Report indicated that the Inspectorate lacked the resources to effectively question the expert judgments expressed in licence applications. Furthermore, inspectors and researchers shared professional backgrounds and an animal use ideology. This represented a policy community with a structure of resource interdependency that allowed animal researchers to dominate policy implementation through positive-sum power games. As discussed in Chapter 2, this type of implementation structures allows regulatees to circumvent the goals of formal policy formulation networks and promotes regulatory co-production with implementers from a position of negotiating strength (Hill, 1997: 144). Therefore, any changes in the policy community would require changes in the membership of the implementation network and its ideological, resource distribution and power structures.

Firstly, it would appear that substantial increases in Inspectorate resources would have been required to potentiate a change in the power relationship between inspectors and researchers. Writing shortly after the passage of the Act, Hampson (1987: 320) asserts that the Inspectorate was to be 'expanded and become more specialised, [and] assisted by a panel of independent expert assessors established by the Home Office for consultation as considered necessary'. To what extent any such changes affected the resource interdependency would remain to be seen, though any core change in policy outcomes would also require a concomitant ideological change in the structure of the network.

However, the pre-existing medical and veterinary scientific background of inspectors was enshrined by the 1986 Act (Straughan, 1995: 41). Therefore, even if the Inspectorate was given much greater resources, there does not appear to have been any change in the composition of the network. Even commentators sympathetic to animal research, such as Monamy (2000: 64), question the Inspectorate's role:

> This has been criticized because, although they have scientific training, inspectors have been given no formal instruction in ethics. This may lead to them having a biased view in favour of experimentation. It is also unlikely that 25 or so inspectors adequately represent or reflect an evolving public attitude to experimentation.

Inspectors' and researchers' shared membership of professional groups is likely to affect their interrelationships and thus may promote major discrepancies between intended formal rules and actual policy outcomes that reflect the goals of research interests (Hill, 1997: 191). The secrecy surrounding the policy process, the inability of the 'victims' of the regulated activity to complain, and the fact that only researchers had access to formal grievance procedures in the form of a statutory right to appeal may, together, have had the effect of reinforcing the network's isolation from public accountability. Thus, the implementation structure for animal research policy envisaged in the Animals (Scientific Procedures) Act 1986 appears to have facilitated an élitist 'professional treatment' model of decision-making that serves the interests of 'regulated' professions and associated industries, rather than a pluralistic 'moral judgement' model that involves independent regulation and incorporates broader social values.

In respect of any ideological structuration of the implementation process, it is noteworthy that pro-research interests came, as explained in Chapter 3, to accept the cost-benefit assessment, and Hollands (1995: 35) records that such actors 'considered [the 1986 Act] nothing more than expensive window dressing'. This contentment with the 1986 Act may be due to the fact that the 'advancement of knowledge' continued to be a 'permissible purpose' (Section 5(3)(d); cited by Smith and Boyd, 1991: 255) and is thus defined as a potential benefit under the Act. Furthermore, the 1986 Act (Section 5(3)(a)) also included the 'testing of any product' as a permissible purpose. This translated in practice to a failure to specify bans on animal toxicity tests for non-medical items such as cosmetics and toiletries, despite contrary cost-benefit arguments. Thus, the new law was interpreted by animal protection groups as allowing the continuation of the infliction of severe suffering for trivial ends (Ryder, 1989: 258).

Formally, the cost-benefit assessment appeared to represent change through lesson-drawing: both the CRAE Alliance and the RSPCA were able to enlist support from scientific experts that allowed policy learning to take place in regard to the theoretical feasibility of specifying different degrees of pain. However, in the absence of specific bans, the question of how the different degrees of pain would be defined, the adequacy of their predictions and enforcement, and how they would be weighed against the postulated benefits of a research programme, continued to be effectively left to the discretion of an implementation network comprising researchers and Home Office inspectors and structured by an 'animal use' ideology. Consequently, it is far from certain that CRAE's goal of achieving additional pain restrictions and a substantial reduction in the numbers of animals used through the cost-benefit assessment of project licence applications (Hollands, 1995: 35–6) would be realised. The legislative outcome of lesson-drawing suggests that it was constrained within pre-existing policies and an institutionalised paradigm. It also tends to indicate broader structural asymmetries favouring professional and economic interests.

By formally incorporating policy changes such as the cost-benefit assessment and the related concept of pain limitations, the policy community appears to have maintained legitimacy resources and thus mitigated the potentially destabilising impact of the dialectical interaction of exogenous public concern and the implementation network on the resource distribution in the network. This process may have been aided by the technical expertise in the network that would have privileged members' truth claims and allowed them to mediate the public's understanding of the effect of formal policy change.

Therefore, when the available empirical information regarding the evolution of the Animals (Scientific Procedures) Act 1986 is evaluated through the policy network analytical framework, the most plausible conclusion is that the new cost-benefit assessment did not represent a core network and policy change. However, given the wide scope of discretion in its operation and insufficient available data regarding the interactions between research interests and state actors during the formulation of the new Act, the most reliable test for ideological change would be the subsequent policy outcomes: the subject of the next chapter. These policy outcome implications will also serve as a test for the theoretical interpretations offered in this section.

The implications of the new Animal Procedures Committee (APC)

The establishment of the APC was also mooted as a source of increased public accountability in the network (Hampson, 1987: 320). The Committee was now permitted to investigate the operation of the Act and advise the Home Secretary on its own initiative rather than be directed by the Home Secretary, and was required to present an annual report to Parliament (Hollands, 1995: 37). However, scientific domination of the APC was enshrined in the 1986 Act. Ryder (1989: 252) comments:

> British governments have rarely had the courage to invite into their very closest counsels leading and genuine reformers, and this has been especially true in the field of animal research...Indeed, two-thirds of the old Home Office advisory committee, for instance, were scientists sympathetic to research, many of whom had well-known public records as defenders of nonhuman experimentation. The

Animal Procedures Committee appointed after the passage of the new legislation was composed similarly, with twelve of its twenty members having clear connections with science or industry and only five having any sort of record as reformers.

Ryder goes on to note that although the majority of the public was opposed to certain types of animal experiments, such as cosmetics testing, this position was not reflected by the APC.

Conclusion: the Animals (Scientific Procedures) Act 1986 and network change

This chapter has addressed this study's third research question concerning the degree of policy change related to the assent of the Animals (Scientific Procedures) Act 1986. In order to answer this question, the trajectory of animal research policy from the events leading to the establishment of the Littlewood Enquiry has been traced.

The analysis of the Littlewood Enquiry and subsequent Report has revealed that animal experimentation had developed into an institutionalised, industrial-scale practice that was integral to biomedical research, in both academia and the pharmaceutical industry, and the marketing of products that contained new chemical compounds. The animal research policy community internalised growing exogenous concern for animal welfare, but defined the concept scientifically as a technical, secondary consideration that was subordinate to research goals, rather than a utilitarian ethical problem that called into question any particular animal experiments. Therefore, the composition and structure of resource dependency in the network continued to be dominated by research interests with the cooperation of the Home Office. The implementation of animal welfare considerations remained at the discretion of researchers rather than subject to pluralistic influence. With the assistance of a favourable economic, political and technical structural context, the research community was responding to exogenous pressure through dynamic conservatism rather than embracing significant policy change. In summary, the Littlewood Enquiry and its aftermath strongly indicated the persistence of a stable and cohesive animal research policy community.

Furthermore, the degree of change implied by the main innovations in the Animals (Scientific Procedures) Act 1986 – the cost-benefit assessment and the APC – was, at best, equivocal rather than a core change in policy which would have entailed the adoption of an 'animal welfare' policy paradigm. The unrepresentative nature of the CRAE Alliance, and the significant differences between the goals of the animal protection movement and the form of the eventual Animals (Scientific Procedures) Act 1986, indicate that the animal protection movement did not have a significant role in the formulation of the Act. The new law's frameworks for the operation of the cost-benefit assessment and the APC did not represent an unambiguous change in the membership, ideology, resource distributions and power relationships in the network.

This analysis of the evolution of the 1986 Act also provides some insights into the relationship between structure and agency and how an adequate understanding of this relationship can help provide a better understanding of the network's evolution. The case for an ongoing issue network appears to be based largely on the observation that there was some degree of animal protection involvement in the formulation of the Act. But, as discussed in Chapter 4, there is an intentionalist assumption underlying this interpretation that tends to overemphasise individuals as the main driver of political events, and the foregoing analysis suggests that the issue network thesis does not take sufficient account of the structural constraints facing actors.

Thus, the observation that animal protection actors had to accept a 'pro-animal research' ideology to gain access to policy makers requires greater precision. This chapter has provided evidence that the ideological structuration – the 'animal use' position – was narrower. This formed the strategically selective context that constrained reformist actors' goals through its effect on their perceptions and learning regarding what policy change was feasible: CRAE Alliance actors' access was conditional upon a greater modification of their goals than previously realised. The narrowness of these constraints can be seen by the semi-exclusion of the RSPCA and the complete exclusion of the anti-vivisection groups. Thus, instead of a pluralistic issue network, this analysis of the evolution of the animal research policy network reveals an élitist policy community environment that is strategically selective in favour of 'animal use' groups to the disadvantage of animal protection actors. Within those constraints, it can be seen that the CRAE Alliance applied their skills and resources to achieve certain formal secondary changes, such as the introduction of pain assessments and a more powerful APC.

But these acts of agency were constrained within a tight 'action setting' or level of structure. While the strategic agency of animal protection actors such as Ryder impacted on elements of the broader structural context, the economic, technological, knowledge and social resources of the policy community actors were such that they were able to exclude most of the animal protection movement. Thus, Ryder (1989: 257) observes: 'Our campaigns stretching over some fifteen years had undoubtedly caused the legislation to happen, but they had not fashioned its content'.

In fact, there is some evidence to suggest that one major factor behind the replacement of the 1876 Act was the need to implement a new EC Directive, and that fulfilling these requirements did not require a cost-benefit assessment in the way understood by CRAE. This is unlikely to be an example of EU policy that required British policy network change because the EU requirements could be made compatible with the extant policy community. Indeed, the EU policy and its apparent intended interpretation by the Home Office strongly resembles the generally conservative Littlewood Report's recommendation [#49] made twenty years earlier for a secondary change in the form of closer monitoring of experiments through the provision of more detailed information about purposes, procedures, expected severity and numbers of animals to be used. This mediation of the cost-benefit assessment would also explain animal research groups' strategic acceptance of the cost-benefit assessment. In summary, these observations imply that the animal research policy community may have successfully resisted core changes demanded by animal protection groups and instead maintained network homeostasis through adaptive, secondary change.

As a consequence, it is unsafe to infer that the evidence of animal protection actors' involvement in the formulation of the 1986 Act signals a change in the network's ideology or policy outcomes. The participation of the CRAE Alliance appears to have been as 'peripheral' insiders and thus an instance of symbolic, rather than substantive, interaction whose main effect was to endow the formulation process with legitimacy. This is an example of a policy process that exhibits attenuated (due to the unequivocal exclusion of the majority of the animal protection movement) plurality as opposed to a genuinely pluralist power structure. The structure of the animal research policy network continued to be dominated by a combination of economic, professional and state interests. This situation could be said to be a manifestation of the asymmetric power model (Marsh et al., 2003); in other words, the evolution of this network appears to reflect broader social patterns of structured inequality that endow such privileged groups with resource advantages that constrain the policy agenda and outcomes (Marsh, 2002). Thus, in answer to the third research question, it appears most plausible to interpret the assent of the Animals (Scientific Procedures) Act 1986 as an instance of dynamic conservatism rather than representing a core change in policy.

However, these conclusions are necessarily provisional. While it may be possible to ascertain the relative likelihood of policy outcomes from a legislative and administrative framework and from the historical trajectory of the policy area, the evolution of its implementation is an inherently uncertain question. Therefore, the final empirical section of this book examines a pivotal case study of animal research policy that consists of unique evidence of policy outcomes. By relating such empirical observations to the necessarily more theoretical narrative offered in this conclusion, it is hoped that a more developed understanding of this policy process can be developed.

8
Imutran Xenotransplantation Research Case Study

Introduction

The previous three chapters have employed a dialectical policy network model within a critical realist methodology to construct a narrative of the historical background to British animal research policy up until the assent of the Animals (Scientific Procedures) Act 1986. That narrative strongly indicates that the postulation of a stable issue network dating back to 1876 is inaccurate. Instead, while the formative network displayed certain issue network characteristics, the lack of a constraining macro-structure promoted instability in the meso-level issue network. This facilitated a network transformation in 1882, whereby the better-resourced animal research interests established close, exclusive relations with Home Office actors and instantiated a policy community structured by an 'animal use' ideology. Furthermore, this policy community appears to have persisted to at least 1986.

The previous chapter has argued that the assent of the Animals (Scientific Procedures) Act 1986 signified limited secondary changes at most. Therefore, it appeared probable that the animal research policy network would continue to fit a policy community model where animal research interests are dominant. However, because the extensive scope of Home Office discretion that characterised the 1876 Act was retained in the new law, there is a degree of uncertainty regarding its implications for the network and outcomes. Therefore, this chapter aims to bring the study of this policy area up to date by examining the fourth and final research question: Does the implementation of the Animals (Scientific Procedures) Act 1986 reflect an issue network or a policy community? To answer this, this chapter analyses a crucial case study of animal research under the Animals (Scientific Procedures) Act 1986 based on primary documentation relating to implementation and outcomes.

The relevance and utility of this case study – Imutran's[1] primate xenotransplantation research – to the understanding of this policy issue has been discussed in detail in Chapter 4. However, based on the salient aspects of animal research policy that were identified in the previous chapter, the particular questions that will be addressed in this chapter are:

1. The operation of the statutory cost/benefit assessment of animal research projects. Specifically, the extent to which the interests of affected parties are promoted or damaged by policy outcomes. This is done by looking at how costs and benefits of research are assessed and weighed in project licence applications and monitored during the subsequent research programme
2. The impact of the Animal Procedures Committee in terms of public accountability and policy outcomes
3. The relative quality of access of interest groups to policy-makers.

The uniquely detailed and reliable description of these empirical issues provided by this case study facilitates a re-appraisal of: the network relationships; whether the network's ideology is 'animal use' or 'animal welfare'; the extent to which the various interest groups achieved their goals in this policy area; the balance of power in the policy network; and hence its position on the Marsh/Rhodes typology. Thus, it plays a vital role in addressing this study's hypothesis: **the interests of animals are given scant consideration in an élitist policy process characterised by research interests' domination and the effective exclusion of animal protection groups.**

The Animals (Scientific Procedures) Act 1986 and policy network analysis

This section begins by setting out the contrasting empirical criteria of the policy process that would indicate either the issue network or policy community models. It then provides a summary of the formal rule structure introduced by the Animals (Scientific Procedures) Act 1986, which will assist subsequent analysis of how the Act is implemented.

The issue network scenario

An animal research issue network would be indicated by the occurrence of major, core policy change due to the assent of the Animals (Scientific

Procedures) Act 1986. Those changes would have involved the operation of the cost-benefit assessment in a manner that reflected an animal welfare, rather than an animal use, ideology. The animal welfare position implies:

1. Independent and balanced assessments of the likely costs and benefits of a research project. This would involve the conscientious interrogation of project licence applications by the Inspectorate and, possibly, the APC
2. The possibility that the perceived costs to animals may be considered to outweigh the predicted benefits of the project, and thus that a project licence application may be refused, or the licence subsequently revoked, if the actual costs and benefits of the research are different to the original predictions (Hampson, 1989: 240–1). This is an essential mechanism to achieve a reduction in the number of animal experiments, which was a key stated expectation of the CRAE Alliance, the responsible Minister[2] and the Opposition Home Affairs Spokesperson[3] during the passage of the 1986 Act (Hollands, 1995: 35, 37)
3. The cost-benefit assessment and related controls such as severity limits on procedures would be operated in a manner consistent with the published Guidance (Home Office, 1990) laid before Parliament. This would indicate that the 1986 Act had introduced refinements to the severity of animal experiments. This would also signify that policy implementation was responsive to broader social concerns and some influence from animal protection groups
4. The APC would contribute to public accountability by setting and enforcing broad rules governing research regulation, and/or participating in the scrutiny of particular applications to ensure adequate implementation. Public statements in annual reports would also accurately reflect the policy process and outcomes
5. Tight controls would be imposed on research. Procedures would be in place to assist the detection of regulatory breaches. Detected breaches of regulations would be dealt with by effective infringement action consistent with stated policy
6. Grievance procedures would be neutral, open and responsive to allegations of Home Office maladministration and/or regulatory infringements
7. Broader policy changes through policy learning would be considered.

In general, an issue network model of animal research policy would reflect a pluralistic 'moral judgment' model of justice.

The policy community scenario

In the case of a policy community, the regulation of animal research would have undergone, at most, minor secondary change due to the 1986 Act. The cost-benefit assessment would be implemented in a manner consonant with the pre-existing 'animal use' ideological policy community structure:

1. Researchers' assessments of costs and benefits would be adopted as the basis for licensing decisions. Thus, economic and professional interests will tend to dominate the interests of animals and the goals of animal protection groups. Researchers and inspectors would cooperate, though researchers would be the dominant party, but in a positive-sum relationship
2. Home Office assessments of the costs and benefits of project licences would not envisage refusal or revocation, even if actual costs and benefits did not reflect those predicted in the original application
3. Discrepancies would exist between, on the one hand, public accounts of the operation of cost-benefit assessment and its components such as severity limits, and, on the other hand, actual implementation. Implementation would favour animal research interests
4. The APC would fail to exercise effective oversight of the research or offer significantly improved public accountability
5. Conditions on research would be broadly defined. Regulatory infringements would be difficult to detect and researchers would be given the benefit of the doubt by the Home Office's discretionary interpretation of rules. There would be a prevalent ideology of self-regulation
6. Grievance procedures, including via the APC, would lack independence and thoroughness. The Home Office would tend to defend researchers' interests and obstruct animal protection groups' goals. The Home Office and researchers would exploit confidentiality to control information about controversial events. Infringement action would be weak and tokenistic
7. Policy learning and change would be significantly constrained.

In general, a policy community model of animal research policy would reflect an élitist 'professional treatment' model of justice.

In order to facilitate understanding of the possible network implications of the Imutran xenotransplantation research case study, it is necessary to provide a detailed account of how the rule framework for the implementation of the new law developed, particularly in relation to the pivotal cost-benefit assessment.

The regulatory framework for animal experimentation after 1986

As explained in previous chapters, the controls imposed by the 1876 Act relied on a single licence that signified that an individual was considered to be suitably qualified to perform specified experimental 'techniques'. But there was minimal legislative control over the experiments in which licensed techniques could be applied (Smith and Boyd, 1991: 252). The Animals (Scientific Procedures) Act 1986 appears to overcome this lacuna by introducing a dual licensing system. Firstly, there are *personal* licences that are, in many ways, similar to the licenses issued under the 1876 Act's regime. They are issued to individuals and specify the categories of procedure that the licensee may perform and the species of animal they may experiment upon. The granting of a personal license requires the sponsorship of a senior scientist who vouches for the applicant's adequate training in the relevant procedures and their knowledge of husbandry and health requirements for the relevant species (Hampson, 1989: 222; Home Office, 1990: 16). However, under the 1986 Act, personal licences do not represent permission to carry out any experiment, which are supposed to be controlled by a *project* licence valid for a maximum of five years.

Thus, the primary innovation introduced by the new law was the cost-benefit assessment of *project* licence applications. Section 5(1) of the 1986 Act states:

> A project licence is a licence granted by the Secretary of State specifying a programme of work and authorising the application, as part of that programme, of specified regulated procedures to animals of specified descriptions at a specified place or specified places.

The cost-benefit assessment of project licences is widely acknowledged to be the cornerstone of the regulatory system (APC, 1998: 43; Nuffield, 2005: 226; Hampson, 1989: 240–1; Hollands, 1995: 36). This core requirement is set out at Section 5(4) of the 1986 Act:

> In determining whether and on what terms to grant a project licence the Secretary of State shall weigh the likely adverse effects on the animals concerned against the benefit likely to accrue as a result of the programme to be specified in the licence.

Additional controls include the requirement that experimental facilities must, according to Section 6 of the 1986 Act, be: 'a place designated by a certificate issued by the Secretary of State...as a scientific procedure establishment'. Section 10 of the 1986 Act sets out a basic set of conditions that the Home Secretary must attach to licenses or certificates, and empowers the minister to set further conditions.

However, the wording of the 1986 Act is minimalist regarding its implementation (Hampson, 1989: 222). Therefore, Section 21 requires the Home Secretary to lay guidance before Parliament (Home Office, 1990), subject to consultation with the APC, that describes the administrative machinery in more detail. The Guidance on the Operation of the Animals (Scientific Procedures) Act 1986 ('the Guidance') is therefore an essential source of information about the regulatory framework, the assessment of severity and benefit, and sets out many additional conditions. Further useful sources of policy information include:

- A paper on the cost-benefit assessment written by the Chief Inspector and published as part of the APC's review of the operation of the Animals (Scientific Procedures) Act 1986 (APC, 1998: 43–62). The timing of the publication of this paper – 1998 – makes it particularly relevant to this case study that covers animal experiments between 1995 and 2000
- The APC's (2003) 'Review of Cost-Benefit Assessment in the Use of Animals in Research'. This report followed the APC's conclusion in its review of the operation of the Act (APC, 1998: 28–110) that there was widespread uncertainty about the operation of the cost-benefit assessment (APC, 2003: 5).

Factors in the cost-benefit assessment of project licence applications

The project licence application form requires the submission of the information necessary for the performance of the cost-benefit assessment (Home Office, 1990: 9). Thus, it is should include detailed descriptions of:

- the objectives of the proposed research programme
- the scientific justification
- the procedures to be undertaken in pursuit of those objectives
- the numbers of animals that will be used and their species
- the likely severity of those procedures
- the personal licensees and facilities involved.

Particularly salient parts of the project licence are sections 17 ('Background, objectives and potential benefits'), 18 ('Description of plan of work') and 19 (involving a description of the different protocols, including the likely adverse effects on the animals and the steps taken to minimise those effects).

The Chief Inspector's paper (Home Office, 1998a) indicates that this information is then used to derive the main factors of the cost-benefit assessment:

- animal welfare implications of the sourcing and transportation of animals to be used in the proposed project
- judgements on the likely severity of the adverse effects on animals in terms of pain, suffering, distress or lasting harm
- standards of care and accommodation
- the technical competence of the people and establishments to be involved in the project
- if appropriate, the relevance of the animal model to the human condition under investigation
- the likelihood of success in terms of meeting the project's stated objectives
- the soundness of experimental design in relation to meeting the stated objectives
- how the data generated will be used
- the utility of the product or substance being developed.

The paper also illuminates a number of additional important features of the implementation of the cost-benefit assessment:

- The discretionary, subjective nature of this crucial decision-making mechanism: 'The assessment must...be performed by balancing "costs" and "benefits" which cannot be quantified in terms which allow any formal mathematical calculation to be performed' (Home Office, 1998a: 51)
- It is supposed to be an iterative process throughout the life of a project licence (Home Office, 1998a: 59)
- Not only is a project licence application meant to pass the cost-benefit assessment, but the assessment is also supposed to require: 'that the cost is minimised and the benefits maximised' (Home Office, 1998a: 58).

The discretionary nature of the Home Office's operation of the cost-benefit assessment is emphasised by the RSPCA when it notes that the Chief Inspector's paper 'describes the factors that are taken into account by the Home Office Inspectorate when assessing costs and benefits, but does not really explain how this is done in practice' (Jennings et al., 2002: 27).

The cost-benefit assessment process

According to the Home Office (1998a: 56–7), the Inspectorate's dialogue with researchers regarding their project licence applications may involve up to four stages:

1. an 'idea' stage prior to the award of funding or the specification of the project
2. a 'draft application'
3. a 'second draft'
4. a 'formal application'.

The Home Office (1998a: 52–7) claims that the preliminary work in the first three stages consumes 'a considerable proportion of resources' and is 'essential' to eliminate inadequate applications and to refine potential projects so that costs are minimised, and benefits maximised. The Chief Inspector's paper (Home Office, 1998a: 57) also gives a vague impression that a large proportion of proposals are rejected prior to formal application: 'Most proposals do not progress beyond [the idea] stage.... [E]ven at [the second draft] stage, the "dropout" rate is high'. Straughan (1995: 42) hints that such early rejections may be exceptional: '[Preliminary advice] allows proposals to be amended or *even* declined outside the formal mechanisms of the 1986 Act' (emphasis added).

A recent Home Office Inspectorate report reveals new data about licence applications and decisions. It states that in 2010, 515 project licence applications were granted, and 1645 project licence amendment requests were advised upon, while initial advice was provided on forty-four preliminary applications that were not proceeded with (Home Office, 2011: 6–7). The report does not clarify how many of these forty-four applicants were advised not to formally apply or how many were abandoned for other reasons. The 515 formal applications enjoyed a 100% success rate (Featherstone, 2011), while the fate of the 1645 amendment applications is not stated.

Interestingly, despite the apparently vital role and resource-intensive nature of these preliminary interactions between inspectors and applicants, 'such preliminary considerations and actions are not recorded routinely for statistical use, or sufficiently quantified at present to allow their use in the calculation of work-loads' (Straughan, 1995: 42). It is only at the final stage that the application is officially recognised and counted; in 1998, the Home Office stated that at this point refusal is rare (1998a: 57), a situation which, as noted above, has continues until at least 2010. The approved licence represents the authority and conditions under which the specified animal experiments may be legally undertaken (Home Office, 1998a: 52).

These observations can be related to the dialectical policy network framework, particularly the comparative table of policy community and issue network interactions with their exogenous context. In particular, it invites the proposition that official assertions of rigorous preliminary scrutiny may represent a policy community's strategic, symbolic action to legitimise the high formal approval rate of applications, rather than a transparent and accurate account of the regulatory process. This case study will test this proposition to some degree. However, there are a number of other observations that can be made that are relevant to the network's structure of resource interdependency and hence the cost-benefit assessment of project licences.

It was noted in the previous chapter that any policy changes instigated by the Animals (Scientific Procedures) Act 1986 would be related to changes in the implementation network, i.e. in the interrelationships between Home Office inspectors and animal researchers. While the new law did not change the ideological composition of the network, substantial increases in Inspectorate resources were identified as a necessary, if not sufficient, condition of a change in the structure of resource interdependency that would empower Inspectors to question and refuse applications on utilitarian or animal welfare grounds. The number of inspectors is also a crucial factor in maintaining public confidence in the adequate administration of the 1986 Act (Straughan, 1995: 47).

However, an analysis of the Inspectorate's workload[4] demonstrates that in 1997, the Inspectorate still lacked sufficient new resources to facilitate the meaningful questioning of the professional adjudications implicit in project licence applications. Furthermore, the absence of rules regarding how costs and benefits should be weighed against each other increases the scope of discretionary, contestable judgement and the hence the impact of bargaining within the structure of resource interdependency (Smith, 1997: 28). Indeed, the APC's (2003: 71, 77) review of the cost-benefit assessment concluded that project licence applicants and holders were *primarily* responsible for the conduct of the cost-benefit assessment. Similarly, the APC (1998: 16) has also noted that: 'The successful operation of the 1986 Act depends upon self-regulation by the scientific community, assisted by the Home Office'.

However, although this resource and institutional structure would tend to *constrain* or *enable* certain modes of policy implementation and favour certain interests, it would not *determine* the administration of the Act because of the ineluctable role of agency in interpreting that structure and undertaking strategic action. This is why the empirical evidence of network interactions provided by the forthcoming case study is important to understanding this policy process.

The case study: Imutran's primate xenotransplantation research

A summary of Imutran's primate xenotransplantation research and related data

The case study focuses on thirty-two separate xenotransplantation studies conducted by Imutran at Huntingdon Life Sciences (HLS) between 1995 and 2000. The studies mainly involved the transplantation of pig hearts and kidneys into two species of higher primates – cynomolgous macaque monkeys and baboons – and the administration of various combinations of immunosuppressive drugs.[5]

The confidential final draft reports for these studies form a central part of the primary data for this case study, supplemented by confidential meeting minutes, feasibility studies and internal progress reports, correspondence between Imutran and their collaborators, particularly HLS, and communications with the Home Office. In addition, confidential documentation leaked from the Home Office includes copies of project licence authorities and communication among Imutran, the Home Office and the APC. These documents emerged into the public domain in 2003 following the settlement of legal proceedings brought by Imutran against the author and his lobby group, 'Uncaged'. The disclosure occurred after the Defendants had argued that there was an overriding public interest in the publication of information (Uncaged Campaigns, 2003) demonstrating misconduct on the part of the Home Office and Imutran, and included:

- The study reports for the Imutran studies, each of which is prefixed with either 'ITN' or 'IAN' and followed by a number
- Documents prefixed 'I', 'WCB', 'CY' and 'hlsapp', which were those photocopies included in the original leak from Imutran Ltd in spring 2000

- A redacted form of the original Diaries of Despair report (Lyons, 2003)
- Documents prefixed 'ND' are those leaked from the Home Office in October 2002.

This primary documentation provides an unparalleled insight into the policy process and outcomes, particularly the costs incurred by animals used in research, thus enabling a more comprehensive evaluation of the operation of the cost-benefit assessment than hitherto possible (Jennings et al., 2002: 9, 35). These confidential primary documents are related to secondary and tertiary sources in the public domain in order to investigate the implementation of the Animals (Scientific Procedures) Act 1986. Those secondary and tertiary sources include reports of the research in scientific journals and Home Office statements regarding animal research policy in general and this research programme in particular.

One particularly useful source is a report on the matter written by the RSPCA (Jennings et al., 2002). This followed the RSPCA's successful application to the High Court in October 2000 for an exception to the then-injunction to permit scientists in the Society's Research Animals Department (RAD) to examine the confidential documentation and produce a report (Jennings et al., 2002: 1–2). The RSPCA (Jennings et al., 2002: Abstract) explain that in addition to the confidential Imutran papers, they had access to some (but not all) further relevant information in the form of Imutran's confidential response to the Uncaged report[6] and video footage of a baboon who had transplant surgery. Furthermore, they state:

> The Head of the RAD is a member of the Animal Procedures Committee (APC) and the United Kingdom Xenotransplantation Interim Authority (UKXIRA) so she was also aware of essential background information which was not publicly available. In addition, we had discussions with the Home Office Inspectorate and with HLS staff.

Utilising these sources, this case study examines the regulation of Imutran's primate xenotransplantation research programme, with a particular emphasis on the implementation of the cost-benefit assessment. This, in turn, will help to address the seven variables – identified at the beginning of this chapter – that indicate whether the animal research policy network resembles an issue network or policy community model. In extrapolating from this example to animal research policy in general, it is important to appreciate the status of this case study. Thus, as discussed in Chapter 4, because these experiments appear to have involved the closest regulatory scrutiny that is afforded animal research projects, it represents a critical case most likely to demonstrate an issue network model of policy-making that would contradict the hypothesis addressed in this study. Therefore, if, despite this *relatively* stringent regulation, a policy community type of policy area is indicated, then this case study will be particularly useful for drawing wider generalizations about animal research policy.

Determining the 'cost': severity assessments

Severity classifications

The 'cost' element of the cost-benefit assessment was primarily based on the severity assessments of animal experiments set out in the project licence application. In relation to 'cost', the Home Office (2003: 3) stated that:

> assessments are undertaken using the detailed narrative descriptions of the procedures, the likely adverse effects, and the endpoints to be applied, provided at sections 18 and 19 of the project licence application form – and these, in turn, determine the endpoints that must be applied.

There are four classifications of the severity of adverse effects: 'unclassified', 'mild', 'moderate' and 'substantial' (Home Office, 1990: 10). In addition, there was a formally prohibited 'severe' category, as indicated by standard condition 14 of personal licences (Home Office, 1990: 55):

> In all circumstances where an animal which is being or has been subjected to a regulated procedure is in severe pain or severe distress which cannot be alleviated the personal licensee must ensure that the animal is painlessly killed forthwith.

This was supposed to enforce Article 8 of European Directive 86/609, which stipulated:

> If anaesthesia is not possible, analgesics or other appropriate methods should be used in order to ensure as far as possible that pain, suffering, distress or harm are limited *and that in any event the animal is not subject to severe pain, distress or suffering.*[7] (emphasis added)

This case study primarily focuses on the distinctions between 'moderate' and 'substantial' adverse effects, although this distinction is complicated by the Home Office's (1990: 10) perception that: 'It is not possible to lay down hard and fast rules about how potential severity should be assessed'. Nevertheless, the following broad guidance was provided (Home Office, 1990: 10):

- 'Moderate' severity corresponds to: 'toxicity tests avoiding lethal endpoints; and most surgical procedures, provided that suffering can be controlled by reliable post-operative analgesia and care'
- 'Substantial' severity: 'acute toxicity procedures where significant morbidity or death is an endpoint...; some models of disease and major surgery where significant post-operative suffering may result'
- 'Severe' pain, however, was not defined by the Home Office, and the RSPCA (Jennings et al., 2002: 19–20, 40) questioned whether 'substantial' severity did not in fact represent 'severe' pain and distress.

It can be seen that one apparently clear distinction between 'moderate' and 'substantial' severity is that death is not envisaged as an endpoint in the former category. These severity categorisations were applied to project licences through two separate but related mechanisms:

- severity *limits* that correspond to each of the protocols that comprise the project licence
- a severity *band* that represents the overall severity of the project.

Severity limits and bands

The severity limit was supposed to set the crucial 'severity condition' on the project licence to *control animal pain and suffering* (Home Office, 1990: 11). In practice, both the assessment of 'costs' and the derivation and implementation of the severity limit/condition required the specification in the project licence application of the 'endpoints' to be applied in the protocol – i.e. the point at which the animal is either withdrawn from the experiment or euthanased to prevent further suffering (Home Office, 2003: 3).

Furthermore, the Home Office (1990: 10) stipulated that the severity limit:

> should reflect the maximum severity expected to be experienced by any animal. It should not take into account the numbers of animals which might experience the maximum severity or the proportion of the animal's lifetime for which it might experience severe effects.

In other words, even if it were possible that only one animal could experience 'substantial' pain or distress for a short period of time, the severity limit should reflect that worst-case scenario. The Guidance (1990: 11) appears to have encouraged further restrictions on severity by stating:

> Licence holders are required by conditions in both project and personal licences to minimise any pain, suffering or distress and they should approach the limit of severity which has been authorised only when absolutely necessary.

However, this appears to be a vague, discretionary principle, rather than an easily enforceable rule, as it is unclear how 'absolutely necessary' was defined or who defined it. But together with the previous statement concerning the derivation of the severity limit, it gives the impression that the intensity of adverse effects suffered by animals would not normally reach the severity limit.

Furthermore, the effectiveness of the severity limit condition as a control on animal pain and suffering appeared to be attenuated by the administrative definition of a *breach of the licence condition*. Thus, even if the severity limit was exceeded, the licence condition was only considered to have been breached if the animal had suffered 'significantly more' than the severity limit and the Inspector was not notified (Home Office, 1990: 11). Furthermore, the condition was not considered as breached if the severity limit is exceeded: 'either unexpectedly or for extraneous reasons' (Home Office, 1990: 11). The penalties for such infringements should also be taken into account: breaches of the severity limit were excluded from the list of possible criminal offences under the Act (Home Office, 1990: 49, 54).

If these rules are related to theories of bureaucratic and implementation structures discussed in Chapter 2, then it seems possible that rule enforcement may be compromised. For in such a highly complex and technical area of policy, the question of whether adverse effects are unexpected or extraneous is potentially contestable. Therefore, the ability to enforce compliance in such circumstances is likely to be affected by the relationships in the implementation network and the structure of resource interdependency.

The whole project's severity *band* was used to represent: 'the "cost" to be taken into consideration when applying the cost/benefit assessment' (Home Office, 1998: 54). Categorising the severity band is an even more complex, multi-faceted task than severity limit classification as it requires an assessment of the degree of suffering expected to be experienced by the average animal used during the project. This is complicated by the fact that a project may contain several protocols with different severity limits. Furthermore, allocating the severity band requires estimating the proportion of animals that are likely to experience the maximum severity indicated by the limits and the length of such exposure. Another way of expressing this calculation is that it is supposed to reflect the 'actual suffering likely to be caused as a result' of each protocol that makes up the proposed project (Home Office, 2003: 4).

However, it can be seen that severity limits and bands are related because, while the allocation of an overall severity band contains a greater element of discretion than severity limit classifications, if the potential suffering caused by a protocol, represented by the severity limit, is underestimated, then logically this will distort the determination of overall severity and hence the cost-benefit assessment. This case study's discussion of severity assessments will therefore focus on severity limits, with the understanding that those discussions will help to elucidate the operation of the wider cost-benefit assessment.

Severity limits also played additional, crucial roles in the operation of the regulatory system. Thus, in the case of primate experiments, the severity limits allocated to procedures *affected the level of scrutiny* received by the application. Proposals to conduct procedures of 'substantial' severity on primates were routinely considered by the APC (APC, 1998: 2), whereas the Committee did not usually examine 'moderate' severity procedures unless they involve wild-caught primates.

In addition to the regulatory role played by both severity limits and bands, they also played a role in informing the public about animal research and the levels of suffering involved. Apart from leaks and undercover investigations, the only information available to the public that gives any indication of the levels of suffering experienced by animals was to be found in the breakdown of project licences into severity bands (e.g. APC, 1996: 11). Thus, severity bands performed a relatively important, if highly limited and unverifiable, role in informing the public debate about animal research. Any underestimations in the severity band assessment were thus liable to mislead the public about the extent of suffering caused by animal research.

The severity assessment of Imutran's protocols and projects

The Imutran primate xenotransplantation research was licensed under three different project licences. The vast majority of the experiments were licensed under project licence PPL 80/0848 (ND1), which was first issued in 1994 and covered all of the primate xenotransplantation protocols until 1998 (Home Office, 2003: 1). In 1997, the APC (1998: 13) stated:

> We felt that the current project licence was too large to monitor properly and recommended that the licence holder be asked to consider splitting the work into more manageable projects, each with its own separate licence.

Thus, in 1998, a separate licence – PPL 80/1223 – was issued for heart xenotransplantation research (ND21). However, both the leaked documentation and a ministerial Written Answer indicate that no procedures were conducted under this licence (O'Brien, 2000a). In 1999, when the original project licence expired, a new licence – PPL 80/1366 – was issued for kidney xenotransplantation (Home Office, 2003: 1). Only one of the internal study reports, IAN022, includes descriptions of research conducted under this licence.

By analysing the study reports, the project licence PPL 80/0848, and Home Office statements, it is possible to calculate that approximately 445 monkeys and baboons were used in various types of heterotopic[8] heart and kidney xenotransplantation procedures, while sixteen baboons were used in orthotopic[9] heart xenograft experiments. (The various protocols are discussed in detail below.) It can also be deduced that the heterotopic protocols were allocated a 'moderate' severity limit, while the orthotopic procedures were classified as 'substantial' (Home Office, 2003: 5).

The Home Office (2003: 4–6) has provided an account of the allocation of the severity limits for Imutran's research in response to a letter from the House of Commons Home Affairs Select Committee. According to the Home Office, the severity limits were based on estimations of the adverse effects caused by:

1. 'the surgical procedures and the likely course of the post-operative recovery period'
2. 'the likely effects of the immuno-suppressive treatments'
3. 'the known consequences of uncontrolled or uncontrollable organ rejection'.

Interestingly, it is revealed that Imutran's own severity assessments were accepted by the Home Office. The stated reasoning behind the allocation of the severity limits for the different types of protocols forms a crucial aspect of the scrutiny of this regulatory process, so it is worth quoting it in full:

1. a moderate severity limit would be appropriate in the case of heterotopic organ transplants, where the recipient animal's own organ would remain in situ and continue to function. (The Inspectorate's assessment was that the surgical procedures and post-operative care are similar to those used in human clinical practice and other established research contexts. Failure or rejection of the rejected organ should not (in those cases where the animal's own organ remained in place) seriously impair the welfare of the animals, rather it would cause local problems and not interfere with the normal working of the animals' own organ.)
2. a moderate severity limit might, depending on the treatments and endpoints, suffice in the case of heterotopic kidney transplants, with removal of the animals' own kidneys. (In this case failure or rejection of the transplanted organ resulting in renal failure was recognised to be the outcome, though it was not the outcome intended. However with proper clinical management and appropriate clinical and biochemical endpoints the welfare costs were still potentially within the broad moderate category. Untreated non-transient renal failure would result in gradual deterioration of the general health of the animal over several days – sufficient time for the problem to be identified by the routine blood tests and remedied or for the animal to be killed before the level of suffering merited a 'substantial' limit.)
3. A substantial severity limit would be appropriate in the case of orthotopic heart transplants (where the transplanted heart replaced the animals' own heart). (This was consistent with the Inspectorate analysis that failure or rejection of the transplanted heart would impose an immediate high welfare cost on the animals, and it was recognised that in the case of acute failure of the transplanted organ some animals might die before appropriate clinical investigation and management, or euthanasia, could be applied). (Home Office, 2003: 5)

Relating this reasoning to the three aspects of adverse effects listed just above, it can be seen that there is no reference to the severity of the immunosuppressive treatments, indicating that this was not formally considered to be a likely source of significant adverse effects. This issue is explored further below. It is also important to note that this Home Office account of their severity assessment reiterates a fundamental distinction between moderate and substantial severity: the latter countenances death as an endpoint. These severity estimations provide a useful benchmark for the forthcoming analysis of the operation of the cost-benefit assessment of Imutran's research. Because severity limits are supposed to represent the worst adverse effects that any animal might endure for even a short period of time during the protocol, assessing the adequacy of severity limit categorisations requires particular attention to the most severe cases.

The benefit assessment

The formal requirements for benefit assessments

Section 3 of the Animals (Scientific Procedures) Act 1986 listed the permissible purposes for the granting of a project licence:

(a) The prevention (whether by the testing of a product or otherwise) or the diagnosis or treatment of disease, ill-health or abnormality, or their effects, in man, animals or plants
(b) The assessment, detection, regulation or modification of physiological conditions in man, animals or plants
(c) The protection of the natural environment in the interests of the health and welfare of man or animals
(d) The advancement of knowledge in biological or behavioural sciences
(e) Education or training otherwise than in primary or secondary schools
(f) Forensic enquiries
(g) The breeding of animals for experimental or other scientific use.

This list provides a preliminary insight into what the Home Office considers to be legitimate potential benefits. In addition, the Home Office (1998: 56) takes into account certain economic benefits that may also accrue from the pursuit of the above purposes, although the Home Office view of justified economic benefits does appear to have evolved. Thus, in 1993, profitability, employment and the conservation of natural resources were considered as benefits. However, by 1997, the Home Office states that the profitability of the company applying for the licence and researchers' career prospects were no longer counted as benefits, although broader economic benefits from cheaper healthcare – thus allowing animal research for 'me-too' products – were considered as justifying animal experimentation (Home Office, 1998a: 56).

While costs are formally categorised according to the degree of expected severity, the administrative framework for the assessment of benefits has been considerably vaguer. The Guidance on the operation of the 1986 Act merely states:

> The benefit of work may sometimes be difficult to assess in advance and fundamental research where no immediate benefit is sought other than the increase of knowledge is valid and permissible... [I]n all cases, applicants should set out the potential benefit of their specific project. It is recognised that research into life-threatening disease may necessitate a degree of severity which might be difficult to justify in other research. (Home Office, 1990: 9)

Subsequently, the Home Office (1998: 56) has said that benefit assessments of applications for the development of new products are affected by the utility of the proposed new material: 'In the case of new medicines, this may be deemed to be high; in the case of cosmetics, low'. However, the Home Office states that it is the particular benefits that can be expected to arise directly from the research plan specified in the project licence that are considered in the assessment of the application. In other words, although an applicant may state that their ultimate goal is to cure cancer, the benefit assessment is supposed to rely on the scientific merit of the proposed research. The hypothesis must be scientifically justified, and the experimental design must be adequate to achieve the stated objectives (O'Brien, 2000a). Thus:

> The essential determinants of benefit remain the likelihood of success, and how the data (or other product) generated by the programme of work will be used, rather than the importance of the field to which the research relates. (Home Office, 1998a: 50)

This benefit assessment cannot be expressed in quantitative terms (APC, 2003: 42). Furthermore, the Home Office provides no indication of how an assessment of these determinants is applied to arrive at an 'informed judgement' (APC, 2003: 42) on the weighting to be attributed to the perceived benefits from either the putative 'likelihood of success' of a research programme or the utility of the data or product (APC, 2003: 73; Jennings et al., 2002: 27). Such assessments are further complicated by the intrinsic uncertainty of the consequences accruing from scientific research and the subjective, contestable nature of determining and valuing potential benefits (APC, 2003: 35). However, in order to mitigate the inherent difficulties in predicting the outcomes of research, the Home Office states that the assessment of benefits is an ongoing process through the period of the licence that aims to establish that the predicted benefits are being realised in practice, rather than a one-off event at the initial applications stage (Home Office, 1998a: 56).

The initial benefit assessment of Imutran's research

The Home Office (1998b: 60) states that the judgement of benefit 'is made on the basis of information set out in an application for a project licence (PPL)'. This information is co-produced by the Home Office Inspectorate and the applicants, and forms the justification – in cost/benefit terms – for the proposed programme of work. Therefore, the assertions in the project licence forms are particularly indicative of the Home Office's benefit assessment.

Section 17 of Imutran's first xenotransplantation project licence (PPL 80/0848), dated February 1994, sets out the 'Background, Objectives and Potential Benefits' of their research (ND1.6–1.7). The licence argued that there were insufficient human organs available to meet the demand for transplants, and that it would not be possible to satisfy that demand by improved donor organ retrieval systems. It stated that Imutran's research was aimed at trying to develop animal organs as a potential solution to the inadequate organ supply. Thus, Section 14 of the application (ND1.5) signified that the primary purpose of the project was 'control of disease, ill-health, abnormality', with 'advancement of biological or behavioural science' as a secondary purpose.

Prior to applying for the project licence to conduct pig-to-primate xenotransplantation, Imutran had performed experiments that indicated that their hDAF transgenic pig hearts, unlike genetically-normal pig organs, were not subject to hyperacute rejection (HAR) when perfused with human blood (ND1.6–1.7; Concar, 1994).[10] On the basis of this evidence, Imutran sought permission to test hDAF organs in cynomolgus monkeys. This species was chosen because their 'complement' system – the aspect of the immune system that the hDAF pig organs were supposed to deactivate – was believed to be most similar to human complement.

The objective of Imutran's project was to work their way to a position where they could apply for permission to perform clinical trials of hDAF pig organs (ND1.7). Publicly, Imutran were stating that permission for this from regulators such as the US FDA or the Department of Health in Britain would require at least 70 per cent of transplanted monkeys to survive for more than one year: they expected clinical trials to start in 1996 (Concar, 1994: 24, 28). In order to achieve these results, Imutran planned to perform pig-to-primate transplant experiments to address three sequential questions, set out at section 17(B) of the licence (ND1.7):

1. Will the hDAF organ prevent HAR, and what are the subsequent rejection mechanisms?
2. Will available immunosuppressive techniques prevent post-HAR rejection processes in the long-term?
3. Will the organ function sufficiently to support life in the recipient for a prolonged period?

Imutran's 1994 licence application emphasised clinical trials of pig organs as the project's objective and potential benefits, which would require positive answers to the three questions above (ND1.7).[11] Home Office policy statements on the assessment of benefits imply that the licence was granted on the basis that the Home Office believed that these objectives were likely to be realised (O'Brien, 2000a).

However, in respect of question (ii) above, scientists already recognized the potential for post-HAR rejection processes to be much stronger with pig organs than conventional allotransplantation[12] (Concar, 1994: 29; Sachs, 1994). This raises the question of whether such considerations were taken into account in the Home Office assessment of the likely benefits of the research. Given the lack of consideration for the potential adverse effects of immunosuppressive drug toxicity, it seems probable that the Home Office were confident – or accepted Imutran's confidence – that it would be possible to prevent rejection of the xenograft in the long-term without causing serious adverse effects to the recipient.

In addition to the immunological obstacles to xenotransplantation, the initial benefit assessment should also have included consideration of the physiological compatibility of pig organs. This is relevant to the third Imutran objective listed above and, as one xenotransplantation researcher put it: 'clearly this question is of greatest clinical importance' (Platt, 1998).

One highly salient factor regarding the compatibility of pig organs in the human body is the evolutionary distance between pigs and humans:

> A chain is only as strong as its weakest link; this is also true for xenogeneic immunology, physiology, and pharmacology. The phylogenetic distance between man and pig comprises 180 million years. This tremendous distance has to be bridged by new and still unknown methods to outwit evolution. (Hammer, 2003)

For instance, pre-existing empirical data described systemic species disparities between pigs and humans relating to blood circulation and coagulation that could adversely affect xenograft viability (Hammer, 1991; cited in Langley and D'Silva, 1998: 20). Similarly, kidneys perform crucial complex metabolic and hormonal activities, and there are numerous species differences in the specialist functions of this organ (Poulsen, 1976; cited in Langley and D'Silva, 1998: 36). Meanwhile, although the heart is commonly viewed as a mere pump for the circulation, in fact it is constantly responsive to small changes in the demands of the body via several feedback control systems. It also responds to hormonal control and to pressure, flow and resistance factors. Direct anatomical comparison of pig and human cardiac structure has highlighted several potential significant differences (Crick, 1998).[13] Additionally, there appears to be a number of incompatibilities related to the nerve supply necessary for a pig heart to function in the human body (Bharati, 1991). However, the initial licence application does not discuss the question of physiological compatibility, barring the brief comment that once Imutran have overcome the immunological hurdles to xenotransplantation, they intend 'to confirm the functional integrity in maintaining life of these xenografts by performing orthotopic transplants' (ND1.6–1.10). It therefore appears that this important factor that may have constrained the likelihood of the success of these experiments was also overlooked in the original cost-benefit assessment.

Regulation and policy outcomes in the Imutran case

Because the cost-benefit assessment is supposed to be an iterative process throughout the duration of the project licence, it is appropriate to analyse its operation by discussing the xenotransplantation research programme chronologically.

The starting point for Imutran's primate research

By early 1995, Imutran believed that the experiments performed prior to those recorded in the disclosed internal documentation had shown that the insertion of the hDAF gene into pigs had prevented HAR in nonhuman primate recipients of transgenic pig organs. Thus, an update to the project licence dated February 1995 (ND1.11) states:

> Experiments undertaken as outlined in the plan of work as submitted on 16th February 1994 and as modified on 5th October 1994 have demonstrated that the possession of a transgene for human decay accelerating factor [hDAF] does protect a discordant

pig heart xenograft from hyperacute rejection by a primate. This data therefore, provide the answer to the question addressed in B(i) section 17.

However this statement appears to exaggerate the degree of progress achieved. For the prevention of HAR only addressed the first half of the objective stated at Section 17 B(i); it did not elucidate the 'subsequent rejection mechanisms'. Therefore, at this stage Imutran and the Home Office appear to have assumed that the post-HAR immune response to the hDAF organs would be similar to known phenomena seen in clinical allotransplantation.[14] Thus, the February 1995 licence (ND1.12) goes on to authorise experiments to address objective B(ii): 'to establish if existing immunosuppressive protocols will produce prolonged survival of the xenograft'.

Heterotopic abdominal cardiac xenografting in cynomolgus monkeys

Those experiments, entitled 'Heterotopic abdominal cardiac xenografting (discordant)' (ND1.32–1.36), were performed first as study ITN3 – the earliest procedures covered by the documents, and then study ITN7, between March 1995 and January 1996. Internal Imutran documents demonstrate that the monkeys in ITN3 were expected to survive an average of a hundred days (Lyons, 2003: 77).

In these studies, transgenic pig hearts were grafted into the monkeys' abdominal cavities and attached to the major blood vessels below the renal (kidney) arteries. The object of these experiments was to investigate the rejection of the organ and test immunosuppressive regimes rather than to discover whether the pig heart could sustain life in the monkeys, whose own hearts were left in place. This protocol therefore belongs to severity assessment group #1 (moderate) as described on page 256. Because this study raises a number of key issues that recur throughout the case study, it will be examined in detail.

Technical failure rate

Out of sixty-one animals transplanted in ITN3, thirty-three died within twenty-four hours as a result of 'technical failures': i.e. failures in the surgical procedure. This meant that no useful data were gained from these thrity-three procedures regarding the main objective of the study: long-term immunosuppression and xenograft rejection. The high technical failure rate, which was a theme of the whole research programme, is particularly relevant to the 'technical competence' aspect of cost-benefit assessment (Jennings et al., 2002: 44).

The twenty-eight remaining monkeys were regularly administered a cocktail of immunosuppressive drugs. They survived for an average of twenty-five days before being killed (or dying) due to either (or a combination of) xenograft failure, opportunistic infection arising from the immunosuppression, or immunosuppressive drug toxicity.

In ITN7, all four monkeys transplanted in January 1996 died due to technical failures within a day of the operation, leading to the study being aborted.

Immunosuppressive drug toxicity

In an internal report on study ITN3 dated 7 August 1995, an Imutran surgeon observed that one major cause of the primate deaths was: 'general debility and non-specific diarrhoea' (hlsapp5b.2). The surgeon deduced that the monkeys had lost weight and become weak: 'due to nausea secondary to the immunosuppressive agents'. Similarly, Imutran scientists have stated that as a result of drug toxicity in this study, 'five animals...had to be euthanased due to gastrointestinal toxicity, resulting in severe diarrhoea' (Van den Bogaerde and White, 1997: 915). These adverse effects were not consistent with Imutran and the Home Office's aforementioned regulatory severity assessment that such procedures 'should not seriously impair the welfare of the animals, rather it would cause local problems....' (Home Office, 2003: 5). These outcomes raise the issue of the assessment of the adverse effects caused by immunosuppressive drugs.

In trying to assess the adequacy of this aspect of the severity assessment, it is interesting to note that the confidential study report, *which would not normally be available to the Home Office* (Imutran, 2000),[15] lists immunosuppressive drug toxicity as an event that would require the euthanasia of the animal (ITN3: 23). However, the 'Description of the procedure' section of the project licence states: 'The blood levels of some [immunosuppressive] agents will be measured to *ensure* that therapeutic and not toxic levels are being achieved' (ND1.33, emphasis added). In other words, the project licence authorities asserted that drug toxicity would *not* cause significant adverse effects. The same approach was adopted in the assessment of immunosuppressive regimes for all the heterotopic procedures performed under PPL 80/0848.

This is borne out by the information in Section 19b(vi) of the licence form, which requires that applicants set out a 'Description of the possible adverse effects, their likely incidence and proposed methods of prevention and control' (ND1.35–1.36). In order to do this and therefore fulfil his/her responsibilities under the 1986 Act, the project licence applicant must be aware of the effects the research is likely to have on the animals involved. The internal, confidential study report shows that Imutran *were* aware of the potential for drug toxicity. Yet, in this section of the licence under 'Possible adverse effects' due to 'b) Administration of substances', the licence refers only to possible infection due to suppression of the animals' immune response. There is no reference to drug toxicity (ND1.35–1.36). In any case, once drug toxicity was perceived to be a serious welfare problem, then the severity assessment and cost-benefit assessment should have been revised.

Animals 'Found Dead'

The study report for ITN3 records that three animals – W741m, W264f and W747m – were 'found dead'. According to the Home Office (2003: 4), a moderate severity limit for this type of procedure was advanced by Imutran and accepted by the Home Office on the basis that problems with the organ: 'should not...seriously impair the welfare of the animals, rather it would cause local problems and not interfere with the normal working of the animals' own organ'. Furthermore, the Guidance's (Home Office, 1990) account of the implementation of severity limits for this project (as related above) indicates that allowing an animal's condition to deteriorate until it dies, rather than euthanasing the animal before it suffers serious adverse effects, is an outcome that corresponds with the administrative definition of 'substantial' rather than 'moderate' severity. The fact that animals were found dead indicates that the adverse effects went well beyond 'local problems', indicating a breach of the moderate severity limit. Despite W741m being found dead, the severity limit was not revised.

In respect of the circumstances of these deaths, it is significant that the narrative in the project licence authority (ND1.32–1.36) specified the 'possible adverse effects', 'methods of prevention or control' and the 'application of specified human endpoints' as follows:

> Possible adverse effects:... Particular attention will be paid to any sites of intervention and any more general signs of discomfort/distress/ illness. The appropriate measures will then be taken to specifically identify any problems and treat them. Any scenario felt to be untreatable or causing undue distress will lead to the animal being withdrawn from the experiment and being humanely killed...
>
> Analgesia will be given regularly post operatively to reduce any discomfort...The animals will be regularly examined post operatively... If the xenograft ceases to function or complications ensue the animal will be withdrawn from the experiment and humanely killed.

As the animals were 'found dead', there appears to have been a failure to comply with the above requirements, as well as a contradiction with the assessment that such protocols would only cause 'local problems'; these formed, respectively, the basic conditions for compliance with, and the justification for the allocation of, the moderate severity limit. In fact, the Home Office (Burnham, 2006) has since claimed that two of these animals were among others found dead who: 'had welfare problems that had been diagnosed and were receiving appropriate symptomatic, supportive and specific treatment'. If this is true, it raises two questions about the implementation of the cost-benefit assessment and the Animals (Scientific Procedures) Act 1986:

1. Why were the animals' conditions allowed to deteriorate until they died instead of them being 'withdrawn from the experiment and humanely killed'?
2. Why was the severity limit and hence the cost-benefit assessment not revised in the light of these outcomes?

The Home Office, whose response will be examined later in this case study, denied that these incidents were inconsistent with the moderate severity limit (Burnham, 2006).

With regard to the third animal found dead, W747m, the Home Office (Burnham, 2006) has asserted that this animal:

> had been in previously good health and...the evidence suggests that [this animal's death] resulted from a combination of the immunosuppressive regimens and organ rejection....The evidence available at the time also suggested that death occurred quickly with little suffering being experienced.

In fact, this animal was noted to be consistently quiet for several days before he died; experts in primate husbandry state that this is not normal for a healthy animal (Jennings et al., 2002: 33). It is also significant that one of the Imutran surgeons involved in these procedures observed that a major cause of the primate deaths was 'general debility and non-specific diarrhoea...the monkeys eat well for the first week or so and then seem to lose their appetites and thus progressively lose weight...their poor appetites may be due to nausea secondary to the immunosuppressive agents' (hlsapp5b.2).[16]

Once again, the circumstances surrounding the death of this animal contradict the Home Office's justifications for a moderate severity limit.

On their account, the illness caused by these procedures should not have resulted in the animal deteriorating until he was found dead or, indeed, suddenly dying. In fact, what happened to this animal, due apparently to the foreseeable circumstances of immunosuppression and organ rejection, would have, according to the Home Office (2003: 5), corresponded more closely to a substantial severity limit where: '...some animals might die before clinical investigation or management, or euthanasia, could be applied'.

Imutran-Home Office interactions: 'Ensuring Smooth and Rapid Passage' of applications

Two notes of internal Imutran meetings held shortly after ITN3 commenced in April 1995 reveal the strategic actions undertaken by the company in pursuit of their goal of conducting clinical trials of hDAF organs as quickly as possible. This involved a 'ramping up of the schedule resulting from the success to date' (I3.2). That perceived success had led Imutran to set 'early Q4 as a goal for 1st human clinicals (over 1 year ahead of original schedule)' (I3.2). This schedule implies that Imutran assumed that the heterotopic procedures they were performing under ITN3 had – or would – confirm the long-term prevention of post-HAR rejection. In order to meet the clinical trials target, Imutran urgently needed to gain permission for life-supporting pig-to-primate xenotransplantation.

The first meeting note (I3) indicates Imutran's belief that they could persuade their Home Office Inspector to facilitate approval of their research. They planned for one of their researchers to discuss an application to perform orthotopic heart xenografts in baboons with the Inspector in order to influence him to 'ensure smooth and rapid passage' of the application. They perceived him as an advocate for their aims:

> Important that [the Inspector] understands the issues (technical difficulty, imminence, etc.) and will give us upward support of the application for orthotopic work. We have to work to make him look like a jolly good bureaucrat and yet achieve our goals as well!'. (I3.1)

This note also reveals Imutran's desire to obtain results rapidly from kidney xenotransplantation experiments, possibly within five weeks. Interestingly, Imutran were planning to try to arrange surgeons to conduct these procedures without a licence being in place at the time. The acquisition of such a licence was a key topic at a meeting held at HLS (then named 'Huntingdon Research Centre' (HRC)) nine days later on 28th April 1995 (CY14.1). At the time, the pharmaceutical company Sandoz were supporting Imutran's research with both finance and advice: Sandoz subsequently bought Imutran in April 1996. This document provides an intriguing insight into the interaction between Imutran/Sandoz and the Home Office in relation to the regulation of the research: 'The Home Office regard heart transplants in primates as severe procedures. Sandoz have suggested kidney transplants, the Home Office will attempt to get these classified as moderate procedures'.

At first sight, the meaning of this comment in relation to the classification of the kidney grafts is somewhat ambiguous. There are two possible interpretations:

1. It could represent a genuine attempt by Imutran and the Home Office to have protocols and endpoints laid down in the project licence that would ensure that the procedures complied with a moderate severity limit
2. It could signify an intention to set down a detailed project licence narrative that would merely give the appearance of moderate severity protocols. This would streamline and accelerate the regulatory process because it would avoid the convention whereby substantial severity procedures on primates were examined by the APC. It would also affect the cost-benefit assessment by reducing the weighting given to costs, thereby making it easier to justify the licensing of the research and facilitating a 'moderate' rather than 'substantial' severity banding for the entire project.

Given Imutran's aim to conduct kidney procedures within a matter of weeks and their attempts to arrange surgeons in advance, the second interpretation appears more plausible. Nevertheless, it will be useful to explore these two alternative scenarios as this will highlight other important factors in the licensing and performance of the experiments.

The RSPCA (Jennings et al., 2002: 25) make observations relevant to the former interpretation concerning the setting of endpoints when they note that Imutran desired to keep animals alive as long as possible for two reasons. Firstly, attaining maximum survival times would facilitate that approval of clinical trials by regulatory authorities. In fact, the same meeting note cites Sandoz as recommending that the 'top concern' should be 'the survival of the animals' (CY14.1). The second reason would be to gain as much information as possible about organ rejection and immunosuppression from each experiment. However, as the RSPCA (Jennings et al., 2002: 25) point out, these scientific aims were in direct conflict with the goal of setting early endpoints.[17] Therefore, it would not have been in Imutran's self-interest to apply a moderate, rather than a substantial, severity limit.

In relation to the second possible scenario, a request made by Imutran to the Home Office five months later is pertinent: 'We therefore urgently request authority to use five of the forty-four baboons, *without further reference to the APC. ...*' (ND7.1, emphasis added). Furthermore, the 28th April meeting note also records that Imutran planned to obtain a kidney protocol licence by June or July, as part of a general effort to urgently produce data for an application for clinical trials (CY14.1–14.2). In the event, the licence is dated 28 June 1995 (ND1.17, 1.40–1.42) and the study reports reveal that the procedures commenced the following month (Lyons, 2003: 85). The speed of the application approval and analysis of the APC's annual reports and relevant Home Office statements seems to confirm that, in line with APC normal practice, it did not consider this 'moderate' severity protocol application under the initial licence PPL 80/0848. If the intention of Imutran and the Home Office's moderate severity limit classification was to avoid APC scrutiny, then it appears to have succeeded. Later events, discussed below, concerning the progress of the heart procedures in baboons would seem to confirm that the APC had the potential to have some effect on the research programme, which would explain Imutran's desire to evade the Committee's involvement.

This primary data suggest that Imutran undertook strategic action in the form of interaction with the Home Office which was likely to minimise regulatory oversight of their research activities. Given that the Home Office did, in fact, accept Imutran's argument for a 'moderate' severity limit for these procedures, and Imutran's timeframe was satisfied, then their interaction may represent a phenomenon identified by Hill (1997: 144), whereby regulatory 'co-production' takes place between implementers and regulatees in order to circumvent formal policy requirements. The subsequent policy outcomes, examined below, will help to further illuminate this question.

However, at this juncture, the primary focus of Imutran's research remained the heart xenotransplantation programme. Therefore, before the results of the kidney procedures are analysed, this narrative will continue to trace the progress of the heart xenografts.

Cardiac xenotransplantation in baboons: the final step to clinical trials?

In addition to indicating Imutran's desire to manipulate their Home Office inspector, the note of the 20 April 1995 internal meeting also showed that the company had planned to prepare their application for orthotopic cardiac procedures so that it was 'in its envelope ready and waiting for the day we completed all appropriate cyno hetero. work [i.e. heterotopic cardiac xenotransplants into cynomolgus macaques]. We did not want a delay while we got the application together' (I3.2). However, the best immunosuppressive regime in study ITN3 achieved a median survival of forty days and had been accompanied by lethal drug toxicity (Van den Bogaerde and White, 1997: 915). Nevertheless, in August 1995 a licence application was drawn up which claimed that these experiments had:

> demonstrated the inhibition of hyperacute rejection of heterotopic discordant cardiac xenografts using pig hearts which express human DAF in cynomolgus monkeys. We have established an effective immunosuppressive regime which ensures long-term survival in this model. Thus we have demonstrated that our genetic manipulation has created organs which remain free from immunological destruction when implanted into animals as close phyllogenetically [sic] to man as possible. The one question remains before clinical trials is whether or not these organs can maintain functional integrity and support life in a non-human primate for a prolonged period when implanted as orthotopic transplants....
>
> As discussed in detail in the accompanying document we have now reached the zenith of our studies with this final step remaining before consideration of implantation of human DAF transgenic pig organs into man. (ND1.52–1.53)

The 'accompanying document' includes a comment regarding their plans to biopsy the xenografts: 'Until the first anniversary biopsies will be done two monthly and thereafter annually!!!' (ND5.22). This indicates the survival times that Imutran were advancing as the likely outcome of the next stage of their research and that would pave the way for an application for permission to conduct the first human trials of xenotransplantation. Thus, these licence authorities were drawn up on the understanding that Imutran had provided positive answers to both the second half of objective (i) and objective (ii) that related to the immunological obstacles. In other words, Imutran and the Home Office were interpreting a range of survival times from six to sixty-two days with toxic levels of immunosuppression as 'ensuring' long-term prevention of xenograft rejection.

At the same time, in September 1995, Imutran were making parallel public statements that received widespread media coverage. Although

Imutran were about to miss their target for human trials in the fourth quarter of 1995, they now claimed that hDAF pig hearts would be ready to be tested in humans the following year. Interestingly, other scientists' were sceptical about Imutran's predictions of imminent success. But an article in Nature (Dickson, 1995) quotes an Imutran spokesperson:

> 'A lot of people have said that [tackling the problem of HAR] was only the first of many hurdles,' he says. 'But as far as we can see, the other hurdles have not raised their head in the timeframe of our experiments'.

The perceived 'final step' in animal trials commenced in January 1996 with heterotopic cervical cardiac xenotransplantation in baboons.[18] This procedure, performed by Imutran under study numbers ITN6 and ITN11 from January to July in 1996, involved the implantation of a transgenic pig heart into the neck of a baboon (ND9.2; ND10.1–10.3). As with all the non-life–supporting heterotopic transplants, this procedure was given a moderate severity limit through severity assessment (a) above. Three baboons were used in each study.

However, in the first study, W205m suffered a stroke after two days with consequent collapse, limb spasms and paralysis. For W201m, a combination of drug toxicity, graft rejection, and infection complications appear to have resulted in his deterioration over a number of days. For the last ten days of his life, the transplant and the primate's neck were observed swollen, with the animal consistently noted to be 'quiet and huddled'. The wound was frequently 'seeping yellow fluid', and the primate was also noted to be 'unsteady' on several occasions. Eventually, the afternoon *after* it had been observed that the baboon was 'showing obvious discomfort' and 'reluctant to move', he was 'sacrificed'. The adverse effects suffered by these animals exceeded the 'local problems' envisaged in the moderate severity assessment/limit. This would be due not only to the immunosuppressive drugs, but also to circulatory problems caused by blood clotting and tissue damage caused by rejection of the vascularised heart. The xenograft in the third baboon, V683m, stopped beating after thirteen days.

Imutran continued to present these outcomes as successful on the basis that HAR had been overcome, and were thus given permission to conduct five orthotopic heart xenotransplants on baboons (ND9.2). This was the only protocol in this project to be categorised as of 'substantial' severity (ND1.58). The procedure involved more damaging surgery than the heterotopic procedures as it required the splitting of the breastbone in order to access the chest cavity: this may have been a further reason for the Home Office's belief that a substantial severity limit was unavoidable.[19] The baboons' own hearts were then replaced with those of the pigs, with the aim of investigating how effectively the grafted organ could support life in the baboon. If the graft started to fail because of rejection or some other complication, deterioration and death was expected to be swifter and harder to manage than in the heterotopic model.

The five orthotopic procedures were conducted in spring 1996.[20] Two of the five were technical failures. The first animal used was W213m, who was killed on day nine due to bone marrow damage, which was a 'serious side effect' of the toxic doses of the immunosuppressive regime (Schmoekel et al., 1998: 1574–5). As a result, the animal suffered haemorrhages and a wide range of acute welfare problems. Imutran claim that they reduced the drug regime in the subsequent baboons to overcome toxicity (ND8.1). It seems that this led to acute vascular rejection (AVR) of the xenografts that killed the two remaining baboons after five days (ND8.1). V687m was destroyed when he became weak and had difficulty breathing. Interestingly, in a report to their Home Office inspector, Imutran claim that W211m was quickly killed once he became breathless, contrary to the internal study report that states he was found dead (ND7)[21]. One cannot be sure which version is the true one, but the euthanasia scenario would give the more favourable impression of Imutran's adherence to severity limits and concern for animal welfare.

Although Imutran acknowledged that these results would not permit clinical trials, they nonetheless deduced in their applications to the Home Office that: 'a pig heart can support life in non-human primates' (ND9.2; ND8). According to Imutran, all that was now necessary before clinical trials could begin was the optimisation or 'fine-tuning' of the immunosuppression (ND8, ND9). Thus, in May and June 1996, they asked the Home Office for permission to conduct five more orthotopic grafts to achieve increased survival times (ND8.2) that would 'justify clinical trials' (ND9.2). Interestingly, Imutran stressed the urgency of this request and asked for it to be granted: 'without further reference to the APC' (ND7.1). This would seem to support the observations made above regarding Imutran's desire to minimise any potential disruption to their research caused by APC scrutiny. Indeed, the APC's report for 1996 (1997: 10) appears to confirm such a potential:

> The speed of the development of this work and its sensitivity makes it essential that the [Primate] Sub-committee and, indeed, the full

Committee keeps fully appraised of the progress of this work and its direction. It is also essential that the work is carefully and closely controlled. We accept that this will place extra regulatory burdens on those undertaking such work and that work may be delayed as a result. We do not apologise for this.

This comment emphasises the 'critical' nature of this case study as it indicates that this research programme is particularly likely to support the opposite hypothesis to the one explored in this study (Burnham et al., 2002: 54). In other words, it would seem to be the case most likely to provide evidence of a pluralistic issue network type policy network involving conscientious policy implementation that reflected broader social concerns for animal welfare, rather than an élitist policy community model involving professional self-regulation, a disregard for animal welfare, non-compliance and a lack of accountability.

The request to perform five more orthotopic grafts was authorised, and they took place in September 1996. However, the survival times were lower than in the earlier series. Three more experiments were technical failures. The strength of the immunosuppression caused X215m to die of a pneumonia infection on day four (Schmoekel et al., 1998: 1574–5), having suffered breathing difficulties, abdominal swelling, weakness and diarrhoea. X221m, having been observed suffering from diarrhoea and vomiting, died of organ rejection on day five while he was being examined.

Further immunosuppression studies in baboons

In the meantime, in June 1996 Imutran had asked the Home Office and the APC for permission to conduct a further series of heterotopic cervical cardiac experiments in up to thirty-nine wild-caught baboons in order to test new immunosuppressive drug combinations (ND9.2). Imutran's stated hope was that these might be more effective than the existing drug regimes and thus help to establish the optimum approach for use in humans. Once again, Imutran laid significant emphasis on the desirability of continuing primate experiments to achieve clinical trials (ND9.3).

This request was granted, but the first three experiments saw technical complications in the surgery (ITN11). Therefore, the cervical model was abandoned, and permission given instead for the pig hearts to be transplanted into the baboons' abdomens (ND10). In making this request to the Home Office, Imutran reiterated that they were confident that this would enable them to progress to clinical trials (ND10.3). These heterotopic abdominal heart xenograft procedures were, like all the heterotopic experiments, allotted a moderate severity limit in line with severity assessment (a) above (ND1.55–ND1.57).[22] Imutran performed these procedures under study numbers ITN19 and ITN25, between October 1996 and May 1997.

Once again, primates were found dead or close to death during these procedures. In ITN19, X227m was observed: 'Lying on front across perch, no movement, eyes closed' before being 'sacrificed'. This indicates that he was enduring 'significant morbidity' and was very close to death, and therefore his trauma had exceeded a moderate severity limit. Similarly, in the same study, X214f was observed on her last day: 'Collapsed on cage floor. Abdomen swollen and appears fluid filled. Salivating. Very laboured breathing. Extreme difficulty trying to walk', before she was also 'sacrificed'. X222m 'Died suddenly'. In study ITN25, X218f was 'found collapsed in cage with no detectable respiration or pulse'.

Five baboons lived a relatively long time: twenty-six, thirty-two, thirty-seven, forty-four and ninety-nine days. For the most part, these animals alternated between quiet and alert periods, with bouts of vocalising, vomiting and diarrhoea, until they were killed. In four cases, the animals were euthanased when the grafts stopped beating due to rejection, while in the longest surviving animal, X240f, the graft was still beating when she was killed. For the last three weeks, however, her condition appears to have been particularly poor, as she grew weaker and developed body and head tremors. She was killed because of a persistent 'pyrexia' or fever (ND19.2; Bhatti et al., 1999), which normally would be associated with infection (Martin, 1996: 246) and thus likely to be caused by immunosuppression.

For both these studies, despite extensive previous experience of drug toxicity, there was no reference to adverse effects due to drug toxicity under the relevant section of the project licence (ND1.56). This is despite the fact that an Imutran application to the Home Office for subsequent licences states that the procedures involved 'oral doses which are about eight times higher than in a human being' (ND18.7).

A 'Cavalier Attitude' and 'Violation of Trust'

An interesting chain of correspondence reveals that these experiments were taking place in the midst of what appears to have been a deterioration in the relationship between Imutran, the Home Office and the APC. This initially arose because of delays in Imutran's submission of reports and their use of sixteen more baboons than had been stated in their earlier application paper to the Home Office and the APC (ND13.1).

Imutran failed to provide information requested by the Home Office concerning the procedures each of the baboons was used in. The company did, however, report that the results in the heterotopic abdominal heart xenografts:

> ... encourage us to believe that [redacted name of drug], in combination with [redacted name of drug], will form the basis of an immunosuppressive regime which would be approved by the Department of Health Xenotransplantation Interim Regulatory Authority for clinical use. (ND14.6)

The Home Office's reply (ND15.1) reveals that the APC was sceptical about Imutran's claims: 'Members expressed concern about the continuing high loss rates in the immunosuppressive programme. It was suggested that you may have moved too quickly to procedures which were too complex...'. Moreover, the APC was 'incensed' by the company's false claims of authority for the additional usage of baboons. Interestingly, this letter indicates that Imutran's claims regarding the significant potential benefits and rapid progress of their work had shaped the perceptions of the Home Office and APC, thereby encouraging them to take a flexible approach to Imutran' compliance with the project licence authorities. However, it was now felt that their implied trust in Imutran's willingness to abide by the 'spirit and letter' of the 1986 Act had been 'violated' (ND15).

Imutran's response (ND16) asserted their 'good faith' and defended their actions by claiming that they had achieved their main objectives:

> The purpose of these studies in baboons to which you refer in your letter was to develop a safe, clinically applicable immunosuppressive regime to provide prolonged survival of a transgenic pig heart in a non human primate.... From the studies, we believe that we have achieved our primary objective – that is a clinically acceptable immunosuppressive regime.... I hope that... we would be allowed to submit formal research plans to allow us to progress a research programme of enormous potential scientific and clinical benefit. (ND16)

Those plans involved testing this immunosuppressive regime in orthotopic transplants.

However, Imutran's report also disclosed that they had experimented on two baboons on 6 May 1997 despite being told at the end of March and 1st May that the Home Office had accepted the APC advice to halt

Imutran's procedures pending the required explanations and justifications for future procedures (ND17).[23] Furthermore, Imutran had still failed to provide the requested information, even after they had been chased up by telephone. The Home Office told Imutran that at the next APC meeting:

> Members strongly expressed the view that your response was still not satisfactory and that the failings listed above indicate a cavalier attitude to the controls of the Act. They are also extremely concerned that this attitude may extend to the care and welfare of animals (ND17.2).

Nine baboons had not yet been experimented upon but, despite the Home Office and the APC's apparent concern about Imutran's conduct, the APC believed that these animals 'should not be wasted' (ND17.2). In other words, they would be experimented upon rather than released from the controls of the Act by return to Kenya for return into the wild or given to private care.[24] Imutran supported their subsequent application to conduct orthotopic procedures on these nine animals with the following report on previous progress:

> These animals suffered no side effects from the immunosuppression.... Given the heterotopic nature of the transplant, this does not compromise the animal in any way and all remained healthy throughout the study. (ND19.1)

Once again, these claims – which relate to studies ITN19 and ITN25 described above – substantially exaggerated the success of the experiments while downplaying their adverse effects.

Heart xenografting abandoned

Nonetheless, further orthotopic procedures were licensed and conducted under study IAN007, which accounted for six baboons between November 1997 and April 1998. Two died quickly because of technical failures, with the remaining animals surviving eleven, twelve, twenty-two and thirty-nine days. One interesting point to emerge from a study of the documentation is a further discrepancy between the internal data and those submitted to the Home Office. Thus, according to the internal study report, X206f died before she could be euthanased. However a surgery report for March 1998 submitted by Imutran to their Home Office Inspector claims she was *euthanased* (WCB31).

X201m's survival time of thirty-nine days meant that his single case was reported in the scientific literature. Imutran were keen to highlight his case because he had earned the distinction of achieving 'the longest survival recorded to date of a discordant orthotopic cardiac xenograft' (Vial et al., 2000).

In this paper, Imutran claim: 'Throughout the first 38 post-transplant days the baboon was active and energetic, moving freely about his enclosure, displaying interest in food and interacting socially with a pairing partner' (Vial et al., 2000: 226). However, his clinical signs record casts doubt on the accuracy of this description. Although this record declares him to have been 'alert and active' on several mornings and afternoons, this was certainly not the case 'throughout' his last thirty-nine days, particularly during the last week. The fact that this was the only experiment from this study that was publicly reported, and that the animal's welfare was not as good as the impression given in the published paper, highlights the difficulties faced in ascertaining the true costs to animals in the absence of the type of primary data presented in this study. It also appears to signify one of the features of policy community and professional judgment models of policy-making: the use of secrecy to constrain the public debate and the operation of grievance procedures.

Analysis of the costs incurred by the baboons in these orthotopic procedures demonstrates that some animals did, indeed, deteriorate rapidly and die before they could be treated or euthanase, which is consistent with the broad definition of substantial severity in the Guidance (1990: 10). However, as with the heterotopic protocols, neither the project licence authorities nor the retrospective Home Office account (2003: 4) fully reflect the adverse effects caused by immunosuppression. Thus, the project licence notes the potential for increased risk of infection and that 'Animals will be examined daily for any signs of infection and any infection that can not be humanely and swiftly treated will lead to withdrawal from the experiment and subsequent humane killing' (ND1.62). But there is no acknowledgement of immunosuppressive toxicity, manifest as pancytopenia,[25] despite the experience of previous experiments. Furthermore, questions could be raised regarding the length of time it took Imutran to euthanase X215m and X202f who appeared to suffer significant adverse effects due to infections for at least twenty-four hours and three days, respectively. The assessment of 'costs' depends ultimately on the: 'the detailed narrative descriptions of the procedures, the likely adverse effects, and the endpoints to be applied, provided at Sections 18 and 19 of the project licence application form' (Home Office, 2003:

3). Therefore, despite the 'substantial' severity limit categorisation, the project licence still did not represent the full extent of the costs experienced by animals.

Following the APC's recommendation to split project licence PPL 80/848 (APC, 1998: 13), in 1998 Imutran were granted a separate licence for heart xenotransplantation. However, no further heart xenotransplantation procedures were conducted by Imutran, despite the company going to considerable lengths to import twenty-eight wild-caught baboons for this purpose in May 1999 (WCB1; WCB3; WCB14; Lyons, 2003: 31–36; O'Brien, 2000a). The available evidence reveals that the baboons remained at HLS until at least the end of 1999 (Lyons, 2003: 44). This appears to have breached a commitment to the Home Office to use the animals as quickly as possible because of the particularly serious welfare implications of keeping wild-caught primates in laboratory cages for several months (WCB26). The documentation suggests that one reason why the baboons were not experimented upon involved dilemmas arising due to resource limitations, either in terms of Imutran's budget or the availability of facilities at HLS (WCB26; Lyons, 2003: 130). Imutran's breach of its commitment to the Home Office also raises another question about the effectiveness of the regulatory system in minimising animal suffering. This will be discussed below.

Kidney xenografting

Earlier, Imutran and the Home Office's co-production of a 'moderate', rather than a 'substantial', severity limit for kidney xenotransplantation procedures was discussed. This raised the question of whether this severity assessment was intended to limit animal suffering by setting early endpoints, or whether it was an inaccurate label intended to circumvent formal regulatory requirements such as APC scrutiny. The answer to this question depends on whether the adverse effects were, in practice, kept within the moderate severity limit through the clinical management and euthanasia of the animals before any of them suffered the 'significant morbidity or death' associated with 'substantial' severity (Home Office, 1990: 10).

In fact, the majority of primate xenograft experiments conducted by Imutran involved the transplantation of porcine kidneys into the abdomens of cynomolgus monkeys. These procedures were licensed as Protocol 19b6 of Project Licence PPL 80/848: 'heterotopic renal xenografting' (ND1.40-ND1.42).[26] These renal xenograft procedures were different to the other heterotopic transplants in that, although the pig kidney was not transplanted into the normal anatomical position, the 'native' kidneys were removed in the vast majority of surgeries, so the xenotransplant was usually life-supporting. Thus, these procedures correspond to severity assessment (b) above: the Home Office (2003: 5) has stated that it accepted Imutran's moderate severity limit assessment, which appears to have identified renal failure (due to rejection, for example) as the main adverse effect that would lead to:

> gradual deterioration of the general health of the animal over several days – sufficient time for the problem to be identified by the routine blood tests and remedied or for the animal to be killed before the level of suffering merited a 'substantial' severity limit'.

Adverse effects: moderate or substantial?

An examination of the adverse effects caused by heterotopic renal transplantation (see Table 8.1) suggests that a significant proportion, if not most, of the monkeys suffered quite serious problems caused by a range of complications. The RSPCA report interprets many of the animals' clinical signs – shaking, grinding of teeth, haemorrhaging, vomiting, weakness, diarrhoea – as 'severe, ... serious and very unpleasant' (Jennings et al., 2002: 32–4). In particular, the failure to euthanase animals before they collapsed and/or were found dead indicates adverse effects that extend beyond the 'controlled' harms associated with 'moderate' severity. It is hard to reconcile the fate of these animals with the severity assessment agreed by Imutran and the Home Office whereby the main welfare problem would be gradual renal failure that would allow the problem to be remedied or the euthanasia of the animal before it suffered significant morbidity or death.

Throughout the renal xenograft programme, immunosuppression caused a range of serious adverse effects, including gastro-intestinal toxicity, haemorrhaging and various infections. Yet, as with the other protocols, the first authorities under PPL80/0848 contained assurances that toxicity caused by immunosuppressive treatments would be avoided (ND1.41) and thus did not acknowledge the potential for adverse effects caused by drug toxicity (ND1.42). The second licence, PPL 80/1366, which covered a latter portion of the final published study report IAN022, admitted: 'as a result of immunosuppression, animals may become nauseous, vomit and have diarrhoea' (ND24.19). However, the project licence goes on to state that action will be taken to ensure that the animal recovers:

> In most cases these symptoms are transitory but should symptoms persist the animals may become dehydrated, this will be treated by antiemetics, administration of fluids and electrolyte therapy and

reduction or withdrawal of the immunosuppressive agent until the animal recovers. (ND24.19)

The problem with this approach is that reducing immunosuppression would tend to lead to adverse effects – including nausea, vomiting, lethargy, swelling and eventual death – caused by rejection-induced renal failure. This illustrates the difficulty of controlling the severity of these procedures. Nevertheless, these procedures were still classified as of moderate severity. In the RSPCA's review, the moderate rating is criticised, and the Society argues that a substantial rating was 'without doubt' necessary 'to alert the scientists and technicians involved to the need for greater vigilance, and in order to ensure a meaningful, realistic and honest cost-benefit analysis' (Jennings et al., 2002: 24).

This second kidney xenograft project licence was considered by the APC, despite the procedures being classified as 'moderate', presumably because of the Committee's established interest in the programme and its acknowledged sensitivity. Documents relating to the APC's deliberations on the PPL 80/1366 licence application provide further insights into two important aspects of the policy process: the relationships between inspectors and researchers and the regulatory role of the APC.

APC scrutiny: a 'rubber-stamping' exercise?
In a fax between Imutran and HLS dated 19 November 1998 (CY7.2), an Imutran official relates:

> For your information he [the Inspector] also told me that our application for a kidney transplant licence has been reviewed by the inspectorate and that we should expect to have some 'I's to dot and some T's to cross' before it goes to the APC.

A subsequent report suggests that the inspectorate's review of this licence was an example of cooperation with applicants designed to maximise their chances of successfully negotiating APC scrutiny. Thus, an Imutran document, 'Progress Review Meeting Minutes', dated 30 March 1999 (CY24), reports under the heading 'Home Office update':

> The new kidney Project License goes before the APC next Thursday...The existing kidney Project License expires on the 21st April 1999. [The local Home Office Inspector] has on several occasions expressed his view that the new License will be approved before the existing license is revoked and that Thursday will be merely a 'rubber stamping' exercise.

Once again, this evidence suggests a highly cooperative relationship between the Home Office and the licensees, combined with a lack of respect for the APC. According to the APC report for 1999 (2000: 4–5), the recommendation to approve the new license was actually a close decision:

> In the main Committee, we were concerned about why some of the individual procedures themselves did not merit a substantial rating (though we accepted that the Home Office Inspectorate had interpreted the rules properly). After much discussion we agreed, on balance, to advise the Home Secretary that the license be granted.

Given the actual severity of the previous procedures as revealed in the clinical signs, and the evidence of a close relationship between Imutran and the Home Office Inspectors, it raises the question of whether the 'moderate' rating was approved by the Inspectorate and – with some hesitation – granted formal confirmation by the APC through a partial and manipulated process, and without the APC being in full possession of all relevant information. The RSPCA report discusses this process and the lead author, Dr Maggy Jennings, was a member of the APC at the time. The report's comments were, however, constrained by confidentiality requirements. Consequently, this was the approach taken by the RSPCA in discussing important aspects of the regulation of Imutran's research: '[T]hroughout the report we have had to reiterate questions to which we know the answers because we believe these questions are important and require answering in the public domain' (Jennings et al., 2002: ii). Bearing this approach in mind, it seems reasonable to interpret the following comments as implying that the RSPCA believe that the second renal xenograft project licence application did not accurately reflect the known adverse effects of these procedures:

> [T]his project licence was written with a great deal of experience of seeing and dealing with the adverse effects. Thus, the application to renew the licence *should* have been informed by the results of the previous studies, and therefore would have been expected to accurately reflect the adverse effects previously seen in practice.
>
> The details of this licence cannot be discussed here without breaching Section 24 of the [Animals (Scientific Procedures)] Act [1986]. We therefore ask the APC and the Home Office to compare the effects recorded in the observation sheets in the study reports from work done under the original licence, with the *predicted* adverse

effects outlined in the 19b reference number 2 (vi) of the new licence to see if these correlate and if any useful information can be learned to inform future decisions.

We also ask what the mechanism is in general, and in this specific case, by which all those involved in drawing up and assessing licences (including the Home Office and the APC), and those carrying out and reporting on the work, can ensure the adverse effects are described as honestly and accurately as possible and with real empathy and understanding for what they mean for individual animals. We would seriously question whether this is adequately done. (Jennings et al., 2002: 36; emphases in original)

Implications for Home Office–Imutran relationships

This evidence of policy outcomes and Home Office–Imutran interactions in relation to kidney xenotransplantation enables a response to the questions that arose at the beginning of this section concerning the meaning of the Home Office's attempts in 1995 to classify these procedures as 'moderate'. That evidence strongly suggests that the severity of these procedures throughout the lifetime of the project exceeded the moderate severity limit and the project licence narrative. Therefore, the most plausible explanation of the Home Office's intention to classify these procedures as of 'moderate' severity is that, rather than it being aimed at limiting suffering, it was designed to facilitate the approval of the application. This would be consistent with Imutran's interests in keeping animals alive as long as possible, while minimising regulatory scrutiny.

Thus, the moderate severity categorisation, which was proposed by Imutran and accepted by the Home Office, was an example of regulatory cooperation leading to policy requirements not being fulfilled. Furthermore, the Inspectorate's interactions with Imutran over the replacement licence indicate that Imutran achieved significant success in their strategy to work with inspectors to minimise the impact of regulation on their activities.

An overview of the cost-benefit assessment of Imutran's primate xenotransplantation research

The adequacy of the cost assessment

The consequence of this industry-dominated regulatory co-production was an underestimation of the costs or severity of the procedures, particularly in relation to the adverse effects of immunosuppressive drugs. This is summarised in Table 8.1 below.

Table 8.1 Imutran xenotransplantation protocols: severity assessments and actual severity

Protocol	Studies	Severity limit	Predicted upper-limit adverse effects and humane endpoints[27]	Worst observed adverse effects and endpoints
Heterotopic abdominal cardiac xenotransplantation in cynomolgus monkeys	ITN3, ITN7	Moderate	Local problems caused by rejection or failure of xenograft; assurance of non-toxic drug doses; endpoints involving euthanasia to prevent any serious welfare impairment.	Drug toxicity caused systemic adverse effects – gastro-intestinal toxicity, nausea, vomiting, severe diarrhoea, anorexia; three animals 'found dead'; organ rejection cause of death in at least one case.
Xenotransplantation of foetal pig islets	ITN5	Moderate	If graft fails or 'other complications' occur causing distress, animal will be destroyed. Researchers will 'ensure' that drugs are therapeutic and not toxic.	Immunosuppressive drug toxicity: Huddling, diarrhoea, vomiting, unsteadiness, congested breathing, sneezing and nasal discharge, slow laboured movement, grating teeth, piloerection, swelling, passing blood; two animals found dead.
Heterotopic cervical cardiac xenotransplantation in baboons	ITN6, ITN11	Moderate	Local problems caused by rejection or failure of xenograft; assurance of non-toxic drug doses; endpoints involving euthanasia to prevent any serious welfare impairment.	Stroke, limb spasms, partial paralysis, wound infections, weakness, 'obvious discomfort', immobility.

Heterotopic abdominal cardiac xenotransplantation in baboons	ITN19, ITN25	Moderate	Local problems caused by rejection or failure of xenograft; assurance of non-toxic drug doses; endpoints to prevent any serious welfare impairment.	Collapse, salivation, breathing difficulties, abdominal swelling and abscess, 'cerebral incident', diarrhoea, shivering, weeping, wounds, passing blood, vomiting, immobility, tremors, fever, haemorrhage; found dead.
Orthotopic cardiac xenotransplantation in baboons	ITN9, IAN007	Substantial	Animals might suddenly become ill and die due to heart failure. Infection problems due to immunosuppression: if untreatable, animal will be swiftly destroyed.	Heart failure leading to rapid deterioration. Some animals found dead. Serious side-effects caused by immunosuppression: infections *and drug toxicity*, leading to bone marrow damage. Oedematous swelling, diarrhoea, breathing difficulties, collapse, haemorrhage, tremors, vomiting, infected wounds, lethargy.
Heterotopic kidney xenotransplantation in cynomolgus monkeys	ITN4, ITN12, ITN13, ITN16, ITN18, ITN21, ITN26, IAN001, IAN002, IAN004, IAN005, IAN008, IAN009, IAN010, IAN013, IAN017, IAN018, IAN020, IAN022	Moderate	Gradual renal failure will be detected through blood sampling or observations of listlessness or anorexia; it will either be controlled or animal put down before significant suffering, morbidity or death occurs; assurance of non-toxic drug doses.	Oliguria, fluid retention and uraemia caused by renal failure leading to listlessness, vomiting, weakness, swelling. Immunosuppressive toxicity causing internal and petechial haemorrhage. Infections affecting various organs such a spleen, kidney, lungs, pancreas and gut. Signs include: huddling, recumbency, collapse, trembling, immobility, teeth grinding, acute diarrhoea, nasal discharges, nystagmus, breathing difficulties, salivating, retching, piloerection, unusual posture, low body temperature, discolouration, rolling eyes, open and seeping wounds, 'very distressed', 'pained facial expression'.

The overall technical failure rate of 25 per cent (Lyons, 2003: 8) also raises questions regarding whether the level of technical competence – a component of the cost-benefit assessment – was adequately considered. In certain cases, such as that of the cervical heart transplants in baboons, problems with technical failure led to a change in procedure to try to overcome this. However, on other occasions, Imutran admit that a stuttering surgery schedule exacerbated the technical failure rate (ND19.2). Thus, the costs were not consistently being 'minimised' because the research was allowed to continue through these periods.

The underestimation of severity has two further implications. First, it raises the question of how the Home Office dealt with possible regulatory breaches during the project and subsequent allegations of non-compliance. At the same time, the Home Office's response to accusations of regulatory failure or bias comes into focus. This is the subject of the last part of this case study chapter, which provides further data and analysis relevant to the type of network in this policy area and the dynamics of the network.

Second, the underestimation of severity is connected with the assessment of benefits insofar as Imutran appeared to believe that their transgenic pig organs would be able to keep primates alive for extended periods and in reasonable health, and thus would be likely to provide a source of viable transplant organs for patients on the waiting lists who might otherwise die. It is to this question of benefit assessment that this case study now turns.

The adequacy of the benefit assessment

The above discussion has explained that during 1995 and 1996, Imutran had obtained approval for primate xenotransplantation procedures on the basis of claims that they had:

1. overcome HAR
2. established an effective immunosuppressive regime that ensured long-term survival
3. demonstrated that a pig heart could support life in a non-human primate.

By May 1996, Imutran were claiming that the last step before clinical trials was the fine-tuning of the immunosuppressive regime. A year later, Imutran submitted that they had achieved their principal goal of a clinically applicable immunosuppressive regime. However, in contrast to Imutran's regulatory submissions, at the 4th International

Xenotransplantation Conference in Autumn 1997 there was an acknowledgement of the lack of knowledge regarding later rejection mechanisms that were considered to be just as challenging as HAR: 'The immune system is unbelievably complicated and poorly understood, making xenotransplantation one of the most speculative of all areas of biotechnology' (Butler, 1998: 323).

Yet, the Home Office stated in spring 1998 that the Imutran research was licensed on the basis that 'the main and ultimate benefits of this research can only accrue if xenotransplantation can be used in clinical practice' (Wilkes, 1998a). Then, in responding to the submission of a critical cost-benefit analysis of the Imutran research which concluded that the experiments should not be licensed, the Home Office defended its position by explaining that 'Xenotransplantation is a potential solution to this shortage [of donor organs for transplant]' in its discussion of the 'potential benefits to humans' that, as required by the Animals (Scientific Procedures) Act 1986, were weighed against the 'welfare of the animals involved in the development of xenotransplantation' (Wilkes, 1998b).

However, in early 2000, Imutran informed the Home Office:

> The basic problem facing the clinical application of xenotransplantation is illustrated in the figure below... This shows essentially normal renal function is provided by the pig kidney until approximately day 48 when graft rejection starts to occur. At this time there is no therapy of which we are aware that will reverse this process. (WCB24.1)

Subsequently, a confidential document entitled 'Primate Development Plan' produced by Novartis (Imutran's parent company) in April 2000 reveals that it had decided to terminate the contract with HLS and instead conduct xenotransplantation research through a network of seven research centres across Europe and North America (Lyons, 2003: 135). This planning document also reveals that the lack of progress in overcoming the immunological obstacles to xenotransplantation had led Novartis to set an eighteen-month time limit from April 2000 to decide whether xenotransplantation with hDAF pigs was at all feasible. Through the coordination of the international collaborators, Imutran were charged by Novartis with achieving the effective management of rejection and substantial increases in survival times towards criteria laid down by the United States Government's regulatory body, the Food and Drug Administration (FDA), as a prerequisite for clinical trials (Lyons, 2003: 17). Thus, the target was a median survival time of three months (ND24.7), whereas the existing results were twelve days for orthotopic heart transplants and approximately a month for kidney transplants – with immunosuppressive regimes that were too toxic for clinical use (FDA, 1999: 19–20).

The Department of Health's xenotransplantation advisory committee, the United Kingdom Xenotransplantation Interim Regulatory Authority (UKXIRA), having considered the leaked confidential Imutran documents, averred in their annual report for 2000 that the likelihood of clinically-viable pig organ transplants was 'receding' (UKXIRA, 2001: 18). *New Scientist* magazine interpreted this as a polite way of saying that the technology was 'dead in the water' (Anon., 2002). Meanwhile, a transplant surgeon sitting on the UKXIRA told the open meeting to launch the 2000 annual report that Imutran's research had turned out to be a 'blind alley' (Dark, 2001).

In summary, Imutran's research had made little tangible headway beyond the first half of the first of their three objectives that had been stated in their initial project licence in 1994. In other words, Imutran had largely confirmed the avoidance of HAR by early 1995, but despite another five years of experimentation, subsequent immune responses were far from understood, never mind controlled. The fact that Imutran's research was permitted or allowed to continue, in spite of the lack of progress in achieving the objectives that had formed the basis of its authorisation, raises questions about the Home Office's regulatory performance. The Home Office's response to these concerns and similar issues regarding severity assessments is the subject of the next section.

The Home Office response
The form of the Home Office's response
Uncaged first attempted to publish their report based on the confidential Imutran documentation on 21 September 2000. The report argued that the Imutran papers demonstrated regulatory breaches and raised doubts about: 'the commitment of the Home Office to regulate animal experimentation effectively' (Lyons, 2003: 152). On the basis of this perception, the Home Office was urged to change its policy to implement the Animals (Scientific Procedures) Act 1986 in a more independent and rigorous fashion. The report's central recommendation was for the establishment of an independent judicial inquiry with the following terms of reference:

> To investigate the circumstances surrounding xenotransplantation research conducted on animals by Imutran Ltd in order to identify the lessons to be learnt regarding the ethics of animal experimentation and legislative and executive approaches to the matter. (Lyons, 2003: 152)

However, the Home Office's initial response within a week of receiving the report demonstrates its reluctance to submit to external scrutiny: '[The allegations] all relate to administrative or regulatory issues and my immediate thoughts are that it would be entirely proper for the Home Office to investigate them subject to certain conditions' (O'Brien, 2000b). In the meantime, Imutran had gained a temporary injunction banning public disclosure of the report and the confidential documentation, leaving three reports in the national press as the only public record of the case at that time (Johnston and Calvert, 2000a, 2000b, 2000c).

These events were unfolding in the context of controversy surrounding the Home Office's response to earlier charges of regulatory non-compliance. In April 2000, the APC had expressed serious concerns about the adequacy and balance of a report by the Home Office Inspectorate into allegations against a breeding establishment:

> it was felt by a majority of members that the Inspectorate's report left a number of outstanding questions. Many members felt that the report sought to exonerate Harlan-Hillcrest, with the risk of creating the impression that the conditions which prevailed there were deemed acceptable by the Inspectorate.
>
> Looking to the future, a majority of the Committee were in favour of encouraging the Home Office to consider incorporating an independent element into any enquiries that might be initiated into allegations which suggested not merely particular breaches of the Act, but the possibility of a more generally significant failure of the system of compliance, monitoring and enforcement. (APC, 2000b: paragraphs 5.6–5.7)

The apparent result of this consideration came through a Written Answer the day before a meeting between the Home Office and Uncaged to discuss the Imutran case. In answer to a Parliamentary Question regarding 'incorporating an independent element in future investigations by the Animals (Scientific Procedures) Inspectorate of allegations against establishments and individuals licensed under the Animals (Scientific Procedures) Act 1986', Mike O'Brien (2000c), Parliamentary Under Secretary of State for the Home Office, told Parliament:

> I have considered the introduction of an independent element into future investigations under the Animals (Scientific Procedures) Act 1986.

I have concluded that the appointment of a small independent scrutiny team, drawn from the Animal Procedures Committee, and reporting directly to the Secretary of State would be the best means of providing assurance that any future Inspectorate investigations have been carried out with the necessary objectivity and thoroughness. I am grateful to the Committee for agreeing to undertake this role following an approach to them in June 2000.

In fact, Uncaged's allegations went beyond the conduct of licensees as Home Office regulators were also critiqued. Nevertheless, this statement appeared to indicate the format of an inquiry into the Imutran case. While this appeared to signal a minor policy shift, this failed to satisfy Uncaged's request for an inquiry independent of the Home Office. However, no such minor policy change took place. On the 29th November 2000, the Home Secretary confirmed that not only would he not set up an independent judicial investigation of the Imutran case, but the APC would not be given the role envisaged less than a month earlier:

I have asked the Chief Inspector of the Animals (Scientific Procedures) Inspectorate to examine, as part of the Inspectorate's normal statutory inspection and reporting function, the available evidence relating to compliance with the authorities granted to Imutran for its xenotransplantation work between 1995 and 2000. (Straw, 2000)

Interestingly, this response confounded the expectations of Imutran. Prior to this announcement of an internal review, an Imutran witness statement to the High Court had disputed Uncaged's argument that the Home Office could not be trusted to conduct an adequate inquiry. Instead, Imutran (2000) referred to the APC's likely role and stated:

Until the composition of the proposed inquiry is known, it clearly cannot be said that it will be lacking in independence. It can hardly be supposed that Ministers will appoint persons whose conduct is criticised by the Defendants to investigate their own conduct.

Yet, this is precisely what happened.

The Home Office attempted to reconcile their exclusion of the APC with the previous policy announcement by claiming that this stemmed from the fact that the Chief Inspector had been asked to conduct a 'routine review' of Imutran's compliance 'as part of the Inspectorate's normal statutory inspection and reporting function' (Home Office, 2003: 16). In other words, by not calling this exercise a 'special investigation', the Home Office were able to avoid the requirements of the earlier policy statement. However, there was no clearly explicable reason for this particular Home Office response, as the APC (2001a) later pointed out:

> Members of the Committee remembered the earlier allegations about Harlan-Hillcrest, where the Home Office response was an Inspectorate investigation carried out by Inspectors who had not been involved in dealing with Harlan-Hillcrest before. As the allegations about Imutran are arguably more serious than those about Harlan-Hillcrest, members were surprised that the Home Office chosen investigation was not given a more wide ranging remit.

The opacity of the Home Office's motivations was exacerbated by the department's inability to provide a coherent justification:

> The Chairman had twice asked for an explanation for the reasons in the case of the "Uncaged" allegations for deciding against commissioning an investigation by the Inspectorate, with a quality assurance panel provided by the Committee. There was general agreement that no satisfactory explanation had been given.... (APC, 2001b: paragraph 4.6)

The absence of a substantive explanation from the Home Office raises the suspicion that underlying political considerations affected the decision to exclude the APC or institute an independent inquiry. The Home Office's chosen response would tend to maximise the department's control over the investigation and restrain the emergence of information or conclusions that might embarrass the Home Office and/or Imutran, or lead to pressure for policy change deemed unwelcome by the department. An examination of the content of the Home Office's response will help to explore these possibilities.

The content of the Home Office's response

The Chief Inspector's Report (Home Office, 2001) on Imutran's regulatory compliance was published in June 2001 while the restrictive injunction remained in place and thus constrained any attempts by Uncaged to dispute the findings of the Home Office at that time. At that juncture, the legal proceedings were scheduled to progress to full trial. Therefore, the Home Office's statements would have the potential to affect those proceedings.

In relation to Imutran's compliance with endpoints and severity limits, the Home Office (2001: 18) claimed that 'where professional judgement was required with respect to the recognition and implementation of welfare-related endpoints it was generally properly exercised'. However, the report goes on to refer to 'perceived non-compliance' in respect of the implementation of endpoints in some unspecified renal xenotransplantation experiments where monkeys were suffering renal failure. In response, the Home Office issued letters of admonishment (Home Office, 2003: 2). Interestingly, the Chief Inspector went to some lengths to attenuate his conclusions, which, together with the arguably weak infringement action, indicates the Home Office's reluctance to challenge professional judgement and conduct:

> This finding is a matter of clinical judgement – and I offer it as my opinion rather than undisputed fact. The decisions that were taken by the surgical team were taken in good faith and based upon their clinical experience and judgement.

However, this appears to be a symbolic or token criticism as the Home Office's description of the circumstances surrounding such breaches does not correlate to the more severe adverse effects experienced by primates under this particular protocol, particularly where animals collapsed or were found dead. This is revealed by an analysis of the Home Office's relevant statements. As explained above, the Home Office (2003: 5) has stated that the reason for categorising the renal xenotransplantation experiments as 'moderate' was that:

> Untreated non-transient [i.e. irreversible] renal failure would result in gradual deterioration of the general health of the animal over several days – sufficient time for the problem to be identified by the routine blood tests and remedied or for the animal to be killed before the level of suffering merited a 'substantial' severity limit. (emphasis added)

The compliance review states that the infringements involved a delay of merely 'up to 24 hours' in 'humanely killing' the animals (Home Office, 2001: 18). But given that the deterioration of these animals due to renal failure was supposed to be a 'gradual' process lasting 'over several days', it is implausible that a delay of only 'up to 24 hours' in euthanasing the monkeys would cause them to be found dead, or collapsed and on the verge of death. However, the Home Office report did not mention any such outcomes for the animals, and neither did it discuss the incidents where immunosuppressive toxicity appears to have led to such adverse effects. Furthermore, the Home Office did not refer to any of the other protocols such as the cervical or abdominal heart transplants.

Therefore, it could be argued that the Home Office's response in this compliance review seeks to give the impression that the department conducted a rigorous review of Imutran's compliance while supporting the notion that the company was committed to compliance with the Act. Such an inaccurate impression would potentially prejudice the legal proceedings by supporting Imutran's promotion of the adequacy of the regulation of their research, and weaken Uncaged's defence that it was in the public interest to disclose the confidential documentation because of the evidence of maladministration and misconduct therein. Thus, the Home Office's response tends to reveal the closeness of its relationship with the researchers. Meanwhile, constrained by the injunction, it was impossible for Uncaged to publicly rebut the Home Office's claims at this juncture.

Similarly, in relation to the assessment of benefits, the Chief Inspector also claimed:

> In considering whether and on what terms to grant the project licence applications the Home Office judgment of 'potential benefit' was based upon new scientific insights that might be gained. Imutran did not advance, and the Home Office did not consider, *claims of imminent clinical trials* as a realistic short-term benefit. (Home Office, 2001: 7) (emphasis added)

Once again, this statement was pertinent to Uncaged's claims regarding the indulgent attitude of the Home Office towards Imutran when assessing the costs and benefits of their research. Yet, the evidence adduced above clearly contradicts this Home Office's assertion.

In trying to give the impression retrospectively of a competent and defensible benefit assessment, the Home Office contradicted its public cost-benefit justifications for the licensing of the research while it was being performed, which emphasised the potential benefits of the clinical application of Imutran's transgenic pig organs (Wilkes, 1998a, 1998b). But the discretion afforded the Home Office in its operation of the cost-benefit assessment means that, in practice, there is no formal impediment to the licensing of substantial suffering in non-human primates for the sole purpose of the advancement of knowledge. However, as shown above, the published guidance (Home Office, 1990: 9) and other public statements associate the most severe research such as Imutran's with high-value practical medical benefits such as life-saving therapies. Thus, in reality there is a mismatch between the implementation of the cost-benefit assessment and the public image given by the Home Office.

The evidence for a cooperative relationship between Imutran and the Home Office which made it difficult to secure accountability is further bolstered by examining the Home Office's actions in relation to the apparently use of additional baboons in 1997 'without the Home Office's prior knowledge or consent' (ND13.1). However, the Chief Inspector's Report, which was published before the Home Office documents were leaked, stated: 'All protected animals used by Imutran were...used with the knowledge and consent of the Home Office' (Home Office, 2001: 3).

In response to the accusation that the Chief Inspector's report was misleading, the Home Office claimed that the unexpected use of an additional sixteen baboons on one protocol by Imutran was not considered an infringement of their licences on the grounds of a technicality: '...Imutran explained that the precise distribution of animals between experiments was not specified on the licence and that no further Home Office consents had been required' (Home Office, 2003: 17). If this is true, then Imutran appear to have escaped punishment because the licence documentation itself, which forms the legal basis for the conduct of the experiments, was not drafted in such a way to reflect the detailed requirements set out in the APC recommendation. This indicates the failure of the regulatory system to exercise effective control over Imutran's research, thereby frustrating the intentions of the APC.

Presumably, a similar situation pertained to Imutran's subsequent use of two baboons after being asked to halt their research. Yet, no reference to this conduct appeared in the Chief Inspector's review. Even if, formally, this did not represent an actual breach of the law, the failure of the Home Office to discuss what the APC perceived as Imutran's 'cavalier attitude to the controls of the Act' could be seen as indicative of a lack of openness on the part of the Home Office and their affinity with Imutran's interests.

A further example of the Home Office's actions in defence of Imutran and policy stability can be discerned by comparing its statements that relate to events leading up to the cessation of Imutran's research. Firstly, in the Chief Inspector's compliance review, it is stated:

> Nevertheless in 1999, as the result of one study with an unexpectedly high technical failure rate, Imutran's operative surgery programme was halted whilst protocols and practice were reviewed and revised to ensure that the likelihood of problems had been minimised.

This moratorium was voluntarily proposed and implemented by Imutran management to address its own, and the Home Office's concerns. Work did not restart until Imutran and the Home Office were of the view that all reasonable steps had been taken to ensure that the likelihood of technical failures had been minimised. (Home Office, 2001: 15, emphasis added)

This gives the impression that Imutran were acting in accordance with their responsibilities as licence holders and voluntarily implemented a moratorium to deal with technical difficulties in their research. Moreover, the Home Office did not refer to the fundamental lack of progress achieved by this point in the research programme in terms of developing an effective immunosuppressive regime. From Imutran's perspective, these two factors were relevant to the question of whether there was a public interest in disclosure of their documents and provided a helpful impression for them to put before the High Court.

However, following the settlement of the proceedings, the Home Office provided a different version of events in response to enquiries from the House of Commons Home Affairs Select Committee (2003: 11):

By the middle of 1999 inspection findings, and supplementary enquiries made with respect to progress reports supplied by Imutran, indicated that Imutran was not making substantive progress...and the incidence of surgical failure was rising....

The Inspectorate advised Home Office licensing staff that the technical failure rate was a cause for concern, and that, from the findings to date, there appeared to be insufficient weight of evidence that Imutran's preferred strategy would ultimately yield success. *This was not resolved by negotiation with the project licence holder, and the Home Office implemented a moratorium on Imutran's main programme of work.* Some work was allowed to continue in pursuit of other secondary objectives. (emphasis added)

This divergent description of the same sequence of events portrays the Home Office as an independent, arm's length regulator, challenging Imutran and imposing a moratorium on Imutran's experiments. Furthermore, unlike the earlier version, it claims that the reasons for the moratorium included a fundamental lack of progress rather than merely technical failures.

The inconsistencies between various Home Office statements and the evidence of both regulatory breaches and Home Office misconduct raise further questions about the capacity of grievance procedures in this policy area to ensure democratic accountability.

An unresolved controversy: questions of accountability

In April 2003, Uncaged published over a thousand pages of confidential documents contained in two leaks from Imutran Ltd (Spring 2000) and the Home Office (October 2002), together with the report based on those documents. Uncaged had argued that the documents revealed breaches of legislation and inaccurate public statements on the part of Imutran, and official misconduct on the part of the Home Office in its implementation of animal research regulations. Therefore, Uncaged submitted, the public interest in revealing such wrongdoing outweighed the claims for commercial confidentiality. Legal aid had been awarded following a decision by the Legal Services Commission's Public Interest Advisory Panel (PIAP) that the case raised particularly significant matters of public interest, and that the Defendants' had a good chance of success in the case insofar as the documents appear to demonstrate Home Office misconduct. At this point in proceedings, Imutran and Novartis offered to settle out of court, and a new Court Order was agreed on 31 March 2003 authorising publication of over a thousand pages of documents listed by Uncaged as demonstrating the key public interest elements of the case.

While the outcome of the legal proceedings tends to support the accusations of Home Office misconduct, the Government continued to refuse to establish an independent inquiry. Following the submission of memoranda by Uncaged and the Home Office, in 2004 the Home Affairs Select Committee declined to launch their own inquiry, citing their existing workload, the complexity of the case and the time elapsed since the research programme. In December 2006, following a three-year investigation hampered by staff absences[28] and, as explained below, an apparent difficulty in understanding the regulatory system, the Parliamentary and Health Service Ombudsman (PHSO) laid before Parliament a final report that dismissed Uncaged's complaint of maladministration in respect of the Home Office's regulation of Imutran's research (PHSO, 2006).

Uncaged's complaint had argued that where procedures were allowed to continue until the point where the animals were found dead, instead of the animals being killed at or before the specified endpoint, then the endpoint and its corresponding moderate severity limit categorisation had been breached. In response, the Ombudsman stated:

> In considering the explanations given by the Home Office, the Ombudsman's staff have noted that death, in itself, does not appear

to constitute a breach of the moderate severity limit within 'The Guidance on the Operation of the Animals (Scientific Procedures) Act 1986'. This view is supported by the 'Report of the cost-benefit working group of the APC' which considered 'The weight assigned to "death of an animal" in itself (i.e. in absence of suffering)'. Within that heading they weighed various arguments put forward on whether a humane death (that is one without suffering) should be included in the cost/benefit assessment. A number of indirect 'harms' were put forward, which could be caused by the death of an animal, and that might be considered within the cost/benefit assessment. The report observed, however, that 'whilst these potential harms are important and should be considered within the cost-benefit assessment, they are not relevant to the question of whether death in itself is a harm'.

However, the Ombudsman's reasoning fundamentally misconstrued the complaint and the role of endpoints in limiting suffering. Uncaged's complaint was not based on the argument that *'death, in itself... constitute[s] a breach of the moderate severity limit'* (PHSO paragraph 14). Instead, the complaint was based on the fact that animals were allowed to suffer up until the point of death, i.e. that death was the *de facto* endpoint in some instances. Uncaged did not assert that 'death in itself' counted as harm in this regulatory context. On the contrary, they argued that the animals should have been killed earlier in the procedure in order to comply with the moderate severity limit. In reviewing their decision, the Ombudsman dismissed these submissions from Uncaged, claiming that they did not add to previous complaint submissions, despite the fact they were novel arguments specifically responding to the reasoning in the Ombudsman's final report.

Concerns about the adequacy of the PHSO investigation were intensified by subsequent comments by the then-Ombudsman, Ann Abraham, to the Public Administration Select Committee in February 2011:

We might [investigate], if there was a wider public interest. I am trying to think of an example that would help you, and most of the examples I can think of are where the wider public interest is in the possibility of the Ombudsman saying there was not maladministration. One of the cases we looked at some years ago was a complaint, again referred by many MPs and many complainants, about the regulation of the animal experimentation industry, and concerns that the regulation was not being carried out properly. There was a lot of outrage, a lot of concern, and a lot of distress about all this, and some very unhappy people. We did a very thorough investigation and we were satisfied that actually everything was being done reasonably, in accordance with the requirements of the regulatory regime. We produced a public report to say that. Now actually the wider public interest was in the Government Department concerned being able to say "The Ombudsman has looked at this, and this is being done properly." So there are lots of circumstances in which we would say that it is worth us doing this.[29]

However, given the basic error at the heart of the Ombudsman's exoneration of the Home Office and their apparent refusal to take Uncaged's submissions seriously, these comments raises questions regarding the point in the investigation when the Ombudsman decided there was a public interest in maintaining confidence in the Home Office. Relevant to this concern is the Ombudsman's position on the issue of severity limit breaches in their draft decision letter issued on 4 May 2005, which differs markedly from their final report:

7. When considering complaints concerning decisions that are reliant on professional judgements, the Ombudsman would not normally seek to substitute her judgment for that of relevant expert. You will appreciate that there are many such decisions within this case, and this confines our legitimate interest to the overall process governing this area.

13. Whilst there would appear to be an element of subjectivity involved in the assessment of severity, decisions about what severity limit to apply to the various procedures were a matter for the Inspectorate's professional judgement and expertise. I do not see any basis upon which this Office could seek to question their assessment of the position (paragraph 7). In the light of that, and as I can see no evidence of administrative fault by the Home Office in their handling of this matter, I can see no grounds for the Ombudsman's further intervention in the matter.

It is hard to reconcile the PHSO's original decision to exclude consideration of severity limits as they were beyond their competency, with the subsequent assertion that they 'did a very thorough investigation' and were able to legitimately claim 'The Ombudsman has looked at this and this is being done properly'. When considered in conjunction with the Ombudsman's conflation of the harm caused by 'death-in-itself' and that caused by 'death-as-an-endpoint' in the final report, it is hard to avoid the conclusion that whatever the Ombudsman's motivation for exonerating the Home Office was, it had little to do with the facts of the case. Rather, it appears to confirm in stark terms the structural power advantages enjoyed by animal research interests and Home Office Inspectors compared to animal advocates.

Uncaged also lacked the resources to launch Judicial Review proceedings against the Home Office and the Ombudsman, or build enough political pressure to persuade the Home Office to establish an independent inquiry. Crucially, the RSPCA did not appear to carry out any significant campaigning or lobbying on the basis of their own critical report (Jennings et al., 2002) that echoed Uncaged's conclusions. Therefore, although Uncaged has had some effect through publication of, and national media exposure for, information that indicates regulatory failure (Townsend, 2003), they have lacked sufficient economic and political resources to stimulate policy change – in terms of increasing consideration of animal welfare[30] – from this critical juncture. This indicates the significant constraints on the ability of grievance procedures to hold this policy network to account.

Conclusion

At the beginning of this chapter, it was noted that the Animals (Scientific Procedures) Act 1986 had introduced minor secondary policy changes and that an entrenched policy community dominated by pro-animal research economic and professional interests was likely to have remained in place. However, the discretionary nature of the policy framework meant that it was necessary to study the implementation of the 1986 Act to reach a firmer conclusion regarding policy outcomes and the character of the policy network. The Imutran xenotransplantation research case study presented in this chapter has provided a unique opportunity to circumvent the normal confidentiality constraints to uncover the animal research policy process.

In order to characterise the animal research policy network following the Animals (Scientific Procedures) Act 1986, lessons need to be drawn from this case study regarding three aspects of this policy process:

1. the operation of the cost-benefit assessment
2. the impact of the APC
3. the relative quality of access for interest groups.

Examining these processes allows an assessment of whether the post-1986 animal research network corresponds to either of the seven-fold criteria outlined at the beginning of this chapter that would indicate either an issue network or policy community model.

Firstly, the severity assessments, which were proposed by Imutran and accepted by the Home Office, did not take full account of the range and intensity of the adverse effects suffered by the monkeys and baboons. In 'moderate' severity procedures, many animals were found dead or in a collapsed state, and endured various systemic and debilitating consequences as a result of the surgery, organ rejection, infections and immunosuppressive toxicity. The severity of these effects clearly exceeded both the definition of the moderate categorisation in the Guidance and the Home Office's own stated derivation of the moderate severity limit in this case.

Furthermore, Imutran experiments were repeatedly approved by the Home Office on the basis of:

- claims regarding the 'success' of the earlier research in terms of understanding and overcoming the immune response to pig organs subsequent to the initial hyperacute rejection process, and
- the likelihood of further 'benefits' involving extending survival times of xenotransplanted organs sufficiently to warrant clinical trials of the organs.

Yet, in early 2000, Imutran were admitting that they had failed to overcome the same immunological obstacles that they had told the Home Office – four years previously – they had conquered. As in the case of cost assessments, this indicates that it was Imutran's benefit assessments that formed the basis of the regulation of this research programme.

The Home Office's acquiescence to Imutran's cost-benefit assessment, despite the existence of contrary evidence at the initial application stage and resulting from the first experiments, reveals an ideological consensus between these actors and a resource interdependency favouring researchers. Indeed, this would be consistent with observations in previous chapters concerning the lack of Inspectorate resources and the significant role of licence applicants in the assessment of their own projects. Furthermore, the lack of routine Home Office access to all research records and the evidence suggesting that Imutran may not have been entirely open with either the Home Office or the scientific community about their results both emphasise the Home Office's dependency on Imutran for information to enable regulation.

In this context, regulatory co-production took place that seems to have furthered Imutran's interests in extending the experiments as far as possible and attenuating potential scrutiny by the APC, instead of limiting animal suffering. In other words, the conflict between scientific and animal welfare goals was resolved in favour of scientific goals. Thus, when these conclusions are related to the first of the seven criteria for characterising the policy network, it can be seen that they correspond to the policy community model:

1. Researchers' assessments of costs and benefits were adopted as the basis for licensing decisions. Thus, economic and professional interests prevailed over the interests of animals and the goals of animal protection groups. Researchers and Inspectors cooperated, though researchers were the dominant party, but in a positive-sum relationship.

Likewise, the second criterion also implies a policy community type network:

2. Home Office assessments of the costs and benefits of project licences did not envisage refusal or revocation, even when the actual costs and benefits did not reflect those predicted in the original application.

This is important because it is relevant to the question of whether the Animals (Scientific Procedures) Act 1986 represented significant policy change. If refusal or revocation of project licences on cost-benefit grounds does not occur, then one of the essential mechanisms intended to reduce the number of animal experiments is not functioning. The fact that this contradicts the stated intentions of the formulators of this regulatory instrument (Hollands, 1995: 35, 37) brings up the third criteria, which is concerned with the consistency between public accounts of regulation and laboratory-level implementation. The Home Office, in line with the vague intimations in the Guidance concerning the balancing of costs and benefits, gave the impression that Imutran's research involved mainly 'moderate' severity experiments that were likely to achieve their aim of life-saving medical advance. In fact, the animals suffered adverse effects of substantial severity for the sake of the advancement of knowledge. Thus, the case study demonstrates that in this respect, a policy community is once again indicated:

3. Discrepancies existed between, on the one hand, public accounts of the operation of cost-benefit assessment and its components such as severity limits, and, on the other hand, actual implementation. Implementation appears to have favoured animal research interests.

This means that the animal research policy network is insulated from Parliament, the public and animal protection groups. The weaknesses in the implementation of formal policy requirements show that the 1986 Act did not, in reality, introduce significant refinements to the severity of animal experiments.

Another putative change said to have been introduced by the Animals (Scientific Procedures) Act 1986 involved the establishment of the APC and the increased accountability it is said to have introduced to the policy process. However, this study suggests that the Inspectorate and researchers may be able to either bypass the APC or filter information to influence its deliberations, potentially rendering it little more than a 'rubber stamping' body. Consequently, instead of the APC enhancing public accountability and thus indicating some degree of pluralistic policy change in the network, it would seem that its true impact is consonant with the policy community model:

4. The APC did not exercise effective oversight of the research or offer significantly improved accountability.

The maximum impact of the APC seems to have consisted of nuisance value to research in terms of affecting its pace rather than whether it actually took place. However, even in these instances, the Committee's oversight was constrained by the latitude granted to Imutran and the failure of the Home Office to fully implement the APC's recommendations in the form of detailed licence conditions. This reinforces the validity of the fifth postulated policy community indicator, which is also supported by the operation of the cost-benefit assessment as discussed above:

5. Conditions on research were broadly drawn. Regulatory infringements were difficult to detect and researchers were given the benefit of any doubt by the Home Office's discretionary interpretation of rules. There was a prevalent ideology of self-regulation.

It could be argued that the cooperative relationship between the Home Office and Imutran also extended to the form of the Home Office's response to the allegations of non-compliance and maladministration. Thus, the tight integration between researchers and the Home Office that characterises a policy community is signified by the validity of the sixth indicator of such a network:

6. Grievance procedures, which tended to bypass the APC, lacked independence and thoroughness. The Home Office appeared to defend researchers' interests and obstruct animal protection groups' goals. The Home Office and researchers were able to restrict the availability of information about controversial events. Infringement action appeared weak and tokenistic.

The scope of the Home Office response is related to the question of policy learning and policy change. The confidential Imutran documentation and the accompanying report raised questions about the adequacy of animal research policy implementation relative to formal policy requirements, and implied policy changes to rectify this discrepancy. However, in responding to this potentially anomalous information and external pressure for change, the Home Office and Imutran seemed to mitigate its impact through the limited scope of the Chief Inspector's review, the launch of legal proceedings and the promulgation of alternative interpretations. Subsequent to the publication of the confidential material, the Home Office has undertaken strategic action to mediate the impact of that information by continuing to advance contrary explanations of the documentation while refusing to release the information that might substantiate those contrary explanations (Home Office, 2004). Once again, these dynamics correspond to the policy community model of resistance to exogenous shocks:

7. Policy learning and change was significantly constrained.

This analysis enables the fourth research question to be addressed. As the Imutran research was subject to relatively close scrutiny compared to other projects, including particular attention from the APC, it represents a critical case most likely to contradict this study's hypothesis by indicating a pluralistic, issue network type policy area. It is therefore highly significant that this case demonstrates that the cost-benefit assessment has been implemented in a manner consistent with the pre-existing 'animal use' ideology that structured the animal research policy community prior to the assent of the 1986 Act. Thus, the new law has resulted in, at most, minor secondary change to the regulation of animal research. Therefore, it is possible to conclude that the implementation of the Animals (Scientific Procedures) Act 1986 reflects a policy community type of network.

The final chapter below relates these findings to this study's hypothesis: **The interests of animals are given scant consideration in**

an élitist policy process characterised by research interests' domination and the effective exclusion of animal protection groups. The concluding chapter also includes an update on the development of this policy network since the mid-2000s culminating in transposition of the updated EU Directive on animal experimentation regulation (2010/63/EU), summarises the findings of the study as a whole, assesses its limitations and contribution to political science and suggests further areas of research.

9
Conclusion: The Power Distribution in British Animal Research Politics

Introduction

This book has addressed the significant lacuna in public policy research in relation to animal experimentation. With Garner's 1998 study standing as the only previous work to address this policy issue, this work has re-evaluated his analysis and taken forward knowledge in this field.

The primary focus of this study has been the impact of the Animals (Scientific Procedures) Act 1986. Garner (1998) has proposed that this critical juncture introduced a core change in policy, which was associated with novel access for animal protection groups to state decision-makers during the formulation of the 1986 Act and its subsequent administration.

This putative policy change can be expressed in terms of a shift from an 'animal use' ideology to an 'animal welfare' ideology. The 'animal use' ideology encapsulates the core policy position of animal research interest groups and is defined by a belief in the routine ethical acceptability of causing pain to non-human animals in order to serve goals such as the advancement of knowledge and the development and testing of various products for human use, ranging from household cleaning products to pharmaceuticals. Self-regulation by animal researchers is another important facet of this position. In contrast, the 'animal welfare' position does not automatically assume that animals' interests may be sacrificed for human interests, and instead requires an *independent* utilitarian cost-benefit assessment of animal research proposals where the interests of animals are given significant weight in a balancing exercise against predicted benefits for humans.

However, the identification of various methodological and empirical constraints on earlier animal research policy research suggested the potential validity of an alternative hypothesis that has been addressed in this book: **The interests of animals are given scant consideration in an élitist policy process characterised by research interests' domination and the effective exclusion of animal protection groups.**

Understanding change and continuity in animal research policy

Policy network analysis

This work has adopted a policy network analytical framework on the basis that it:

- represents a realistic model that corresponds to the complex interactions that take place in diverse and disaggregated policy-making arenas (Parsons, 1995: 185)
- has been found to be a useful heuristic tool across a range of case studies that help to explain policy outcomes (Marsh, 1998b)
- has become a prominent framework in public policy research (Marsh and Smith, 2000: 4)
- was the approach adopted by Garner and thereby facilitates direct analysis of both empirical and theoretical aspects of the sole previous study of this policy area work.

The review of the policy network literature discovered that there had been insufficient attention paid to possible variability in the way different types of network – policy communities and issue networks – interact dialectically with exogenous factors and strategic agents to affect policy and network evolution. Initially, models were developed to distinguish where networks lay on the policy community-issue network continuum (Marsh and Rhodes, 1992b), but this tended to adopt a reified approach, and so theories of network change were underdeveloped (Marsh, 1998b: 192). More recently, attempts have been made to 'consider the mechanics and processes through which network formation, evolution, transformation and termination occur' (Hay and Richards, 2000: 25). However, this study has taken this agenda further to investigate how variability in these 'mechanisms and processes' of change might depend on where the network sits on the Marsh/Rhodes continuum. Beyond the broad observation that policy communities tend to promote continuity in policy-making, while issue networks are more unstable environments, what has been lacking is a combination of these two approaches, which has constrained understanding of policy network dynamics. Therefore, a heuristic framework was synthesised from the existing literature that postulated interactions between the various categories of exogenous and endogenous dynamics and the two ideal types of policy network (see Table 2.2). This potentially advances policy network analysis and, therefore, one major question to be addressed in this conclusion is: How useful has this framework been in terms of understanding policy making?

Developing an analytical framework to re-assess animal research policy-making

This study has self-consciously adopted a critical realist methodology that tries to reconcile structure and agency by:

- allowing for the possibility that institutions or structures that shape the nature of power relationships may be both directly and non-directly observable, and both formal and informal; and
- recognising that reflexive actors' interpretations of structures affect their behaviour and hence outcomes, and that those interpretations are influenced by social constructions of reality.

This method is also consistent with the modern 'dialectical' approach to policy network analysis. The application of this analytical framework to the previous animal research policy literature has generated four subsidiary research questions aiming to reconstruct the evolution of animal research policy and address this study's hypothesis:

1. Which group(s) interests were served by the assent of the Cruelty to Animals Act 1876?
2. Did the policy network that emerged during the passage of the 1876 Act evolve into a policy community in the subsequent years?
3. Did the passage of the Animals (Scientific Procedures) Act 1986 signify a core change in policy or an example dynamic conservatism?
4. Does the implementation of the Animals (Scientific Procedures) Act 1986 reflect an issue network or a policy community?

Answering the first and second research questions has involved the relatively novel application of policy network analysis to Victorian policy-making. This may be unusual because:

> Concepts and models in political science probably reflect the politics of the period in which they were first formulated.... An ambition to develop concepts more applicable to the realities of post-war British

politics was the foundation of the British origins of what is now termed the network approach. (Richardson, 2000: 1006)

However, applying policy network analysis throughout the life of the network has been useful because it facilitates comparisons across time. It also enables 'process tracing' that involves the building of an empirically-based narrative of the interactions that generate continuity and change in policy-making (Hay, 2002: 149). This type of 'diachronic' (Hay, 2002: 148–50) approach is essential to an optimal understanding of political processes that are recognised as constrained and enabled by their historical background that includes inherited institutions and structures that may privilege certain actions and outcomes over others.

Interestingly, Hay (2002: 149–50) remarks that the diachronic approach tends to eschew presumptions of clear temporal shifts in political processes. This stance therefore has affinities with scepticism (Judge, 2004: 697) about the perception that prior to either the second world war or the 1970s, the British polity was characterised by the 'Westminster model' of parliamentary and cabinet government, in contrast to the contemporary situation where interest groups in policy networks are the main causal factors (Richardson, 2000: 1006–7). In other words, there is no *a priori* reason why the policy network approach should not be used to study policy processes in the Victorian era. Indeed, by doing so, it generates one of the contributions of this study, which is to provide evidence relating to the validity of the Westminster model over time. An additional contribution arises from the fact that 'conducting a diachronic analysis in any rigorous fashion is a laborious and time-consuming exercise' (Hay, 2002: 150). The fruit of this labour is a thorough and comprehensive work which prioritises empirical detail over prejudgements of the processes of change, and applies this data to test, and thus develop, theoretical propositions. Indeed, this represents the optimum approach to understanding political processes over time (Hay, 2002: 150).

Serving the purpose of animal researchers: the Cruelty to Animals Act 1876

In addressing the first research question in Chapter 5, this study has undertaken a much more detailed examination of the inception of the animal research policy network than hitherto attempted.

This investigation found that although pro-animal research interests did not have pre-existing institutional relationships with the Home Office, they still enjoyed significant structural advantages as a consequence of their exogenous structural resources. These came to have an effect on the passage of the 1876 Act through the agency of leading members of the medical profession, who had already achieved institutionalised professional autonomy and had developed considerable organisational resources. Furthermore, a cultural context that included significant and growing lay deference towards technical expertise and the professions enhanced the structural resources of the animal research lobby in terms of influence over both state actors and the mass media.

The political power of this lobby manifested itself in the evolution of the statute. In its original form, the Bill had reflected the fact that it initially came about largely in response to anti-vivisection strategic action designed to introduce strict control and accountability on animal research. However, when animal researchers and their allies in the medical profession perceived the measure as potentially damaging to their interests, they skilfully deployed their resources to effect fundamental amendments. Although the anti-vivisection lobby had sufficient organisational and political resources to place the issue on to the policy agenda, they were found to be at a significant disadvantage relative to the experimenters and medical profession when it came to lobbying state actors and legislators during the passage of the Bill. The potency of the anti-vivisection lobby's resources may have been, to some extent, hindered by the tendency towards disunity, in addition to apparently weak strategic action on the part of the most powerful member of this lobby, the RSPCA. An unforeseen delay in the Bill's passage also gave the pro-research lobby an opportunity to intervene, a random event which emphasises the ineluctable potential for unpredictability and indeterminacy in political processes. The consequence was legislation which was generally interpreted as protecting experimenters' interests by potentially licensing unrestricted experiments on animals that may otherwise have breached other anti-cruelty legislation.

This is related to one of the key advances achieved by this book's examination of this period: the elucidation of the ideology and strategic actions of both the animal research lobby and the anti-vivisection movement. Given their relative dominance, the pro-animal research lobby is particularly interesting in terms of understanding policy outcomes. Thus, it emerges that one principle aim of this lobby was the provision of symbolic legitimacy for the practice by reassuring the public that vivisection was independently regulated, while in fact evading political supervision that may have hindered the pursuit of knowledge and was deemed to be an affront to their professional status. Relatedly, at the embryonic stage in the development of this policy network, despite a degree of fluidity in the goals of the relevant actors, two key differences between the two lobbies have been discerned.

The first of these involved the meta-policy issue of where the authority to make licensing decisions should reside. In respect of this fundamental question of power distribution, anti-vivisectionists sought public accountability, while animal researchers preferred self-regulation. The second conflict area concerned the relative priority of the goals of advancing knowledge and protecting animals from pain and suffering in experiments. In policy terms, these were to become the central arguments throughout the history of British animal research politics.

This finding also has analytical importance because the careful description of the lobbies' ideologies is essential to understanding power distribution and structures, as it allows more accurate determination of whose interests are served by policy-making institutions and policy outcomes. Thus, having found that both lobbies favoured some form of legislation, ascertaining whose interests were served by the Cruelty to Animals Act 1876 has necessitated a closer examination of the evolution and provisions of the law to establish the extent to which each lobby achieved their aims and, hence, the purpose of the Act. This study has found that the main effect of the 1876 Act was most likely to have been to enhance protection for researchers rather than animals.

Therefore, previous knowledge needs to be revised as a result of this study. First, instead of the incipient animal research policy network being structured by a legal framework designed to restrict animal experimentation, in fact, that structure was permissive of the practice. Second (and relatedly), this outcome came about because of structural resource advantages enjoyed by the animal research lobby that previously have not been fully acknowledged. This also alters the context of the formative stage of the network, with major repercussions for the understanding of this policy area.

However, despite the apparent pro-animal research purpose of the 1876 Act and relative dominance of animal researchers in this policy area, the formative nature of the network meant that, as Garner has observed, research interests had not yet institutionalised relations with the Home Office. Although the anti-vivisection lobby had largely lost the political battle over the provisions of the 1876 Act, they had achieved access to ministers and officials during its passage, though of a more sporadic nature than the pro-vivisectionists. It is also important to note the extensive discretion granted to the Home Secretary by the 1876 Act in relation to approving animal research licences. This meant that the future pattern of policy outcomes was not pre-determined and would depend on the attitude of the minister and lobbying from the groups. Therefore, this study has identified an issue network-type of policy network at this juncture.

But while this represents an aspect of agreement between this study and Garner's position, the aforementioned contextual differences fundamentally affect the respective understandings of the network's likely subsequent evolution. Furthermore, the account offered here suggests that the network was already beginning to evolve in the direction of a policy community, as the 1876 Act brought about a shift from consultative relations between the Home Office and researchers to one where the pro-animal research lobby was incorporated into the implementation of the licensing system. In addition, the resource and power distribution in the network favoured researchers over anti-vivisectionists.

A theoretical insight developed in Chapter 4, relating to variability in patterns of institutional change, assists with understanding the trajectory of the network from this point. Thus, it was proposed that the relatively weak degree of institutionalisation in issue networks potentiates their instability. Combining this proposal with the empirical findings leads to the conclusion that substantial uncertainty surrounds any analytical extrapolations into the future from this juncture. In contrast, any inference of a persistent issue network from this point onwards assumes institutional inertia and fails to recognise the dynamic potential of agency to change networks. Therefore, in tracing the evolution of this policy area, this study tests the dialectical conception of structure and agency and the postulation of broader, unobservable structural factors which are both pivotal aspects of its underlying critical realist methodology. These assessments are explored below.

The unstable issue network transforms into a policy community

Chapter 3 analysis raised the second research question which has considered whether the network changed into a policy community after the 1876 Act. In order to address this question, the evolution of the network until 1950 – the juncture marking the beginning of the modern period of animal research politicisation – was examined in Chapter 6. The first significant finding was that for the initial six years of the 1876 Act's administration, the Home Secretary, in likely contradiction of advice from the Inspectorate established by the Act, rejected 15 per cent of applications for licenses to conduct animal research on the grounds that the suffering of the animals was not warranted by the utility of the research. This demonstrates that although the pro-vivisection lobby was the more powerful grouping, it did not absolutely dominate the policy process: researchers were not entirely self-regulating, nor did the pursuit of knowledge, irrespective of potential practical utility, exclusively structure the network at the expense of animal protection. These outcomes were associated with an absence of organised lobbying by the medical profession. Thus, during this period, the network displayed significant issue network characteristics. The correspondence between the network interrelationships and the pattern of policy outcomes that has been established by this study provides an unusual indication of the utility of the policy network analytical framework in understanding policy processes during this era. It also supports the proposition that issue networks are associated with embryonic policy issues (Smith, 1993a: 10; Hay and Richards, 2000: 7).

However, the access for anti-vivisection values into the network was found to rest largely on the operation of the Home Secretary's discretion, thereby highlighting the tenuous status of anti-vivisection membership of the network and, at the same time, the unstable nature of the issue network. There were found to be no other institutional factors in or around the network that might have helped to stabilise a pluralistic policy-making environment. Consequently, animal researchers and the medical profession were able to apply their significant organisational and expertise resource advantages through strategic action in 1881–2 to effect network change. Thus, at a critical juncture, the Association for the Advancement of Medicine by Research (AAMR) – in reality a pro-animal research coalition – was formed by scientific and medical bodies. The AAMR succeeded in incorporating itself into the policy network and instituted itself as, in effect, a rubber-stamping agency for licence applications. Interestingly, it was found that while the AAMR propagandised for animal research on the basis of its general utility, it did not apply such criteria when approving applications. Moral considerations that reflected lay opinion regarding whether research should be permitted were removed from the policy process in favour of scientific self-regulation. Consequently, animal experimentation underwent rapid growth, compared to the 1876–82 period. Therefore, it has been established that, contrary to previous conceptions of this policy area, a fundamental network transformation took place in the early 1880s whereby animal research policy came to be made in an environment more akin to the policy community model.

These findings, in addition to further confirming the utility of policy network analysis to the study of this period, simultaneously support Moran's (2003) account of the formation of the regulatory state across many policy areas in the nineteenth century. This postulates that a powerful ideology of professional self-regulation combined with poorly resourced Inspectorates to create a situation where inspectors and regulatees practiced informal, cooperative policy implementation rather than formal, independent enforcement of the law. Hence, regulation became a symbolic phenomenon behind which the dominant values and interests of elite groups were advanced. These considerations also serve to undermine the key tenets of the Westminster model as they demonstrate the potentially peripheral role of Parliament in policy change, the hegemony of expert claims over public opinion and bureaucratic subordination to group interests.

The detailed analysis undertaken here has also sharpened the narrative of the impact of the 1876 Act and has provided novel insights for policy network analysis. Thus, closer inspection of the available data reveals that the significant impact of regulation on animal research was generally confined to the first six years of the 1876 Act's administration, rather than the entire life of the law as claimed in previous literature. In respect of policy network analysis, the consideration of the evolution of the embryonic animal research issue network highlighted the absence in the policy network literature of analysis of structural transformations in issue networks. However, relevant scattered observations in the literature concerning issue networks have been combined with the empirical findings presented here to generate the following theoretical proposition that invites further research:

Issue networks may transform towards a policy community network type *soon after their formation* if:

1. the interests of economic or professional groups are threatened and...
2. in the absence of institutions or structures that ensure broad access to the policy-making process and some degree of state neutrality.

Bearing in mind the concept of path dependency, testing this proposition may also advance understanding of the stability of mature policy communities that began life as issue networks.

Furthermore, it was observed that the Home Office perceived that it lacked the resources to resist the AAMR's claims of unique competence to judge animal research issues, and this resulted in the network transformation to a policy community. Moreover, animal research was considered to be a peripheral policy issue by the government. This therefore suggests two modifications to heuristic models engaged with in this study. First, in relation to the asymmetric power model of British politics, state actors are *not* always the dominant power in every policy network.

Second, in relation to the evolution of policy communities, state actors are *not* always the main drivers of the formation of government actively desires to intervene in.

This network transformation, not previously detected in the study of animal research policy, overturns the issue network thesis which was significantly premised upon the persistence of such a policy-making environment stemming from the institutional choices embodied in the formation of the network in 1876. However, this study has shown that the 1876 network was characterised by a relative absence of institutionalisation, in terms of routinized standard operating procedures and ideas (Peters, 1999: 64–6). In contrast, the policy community that emerged in 1882 was, by its very nature, a strongly institutional network. This was manifest in the routinized role of the AAMR in working with the Home Office to approve licence applications, and the entrenchment of an ideological structure that consistently reflected the beliefs and interests of the pro-vivisection lobby that prioritised the pursuit of knowledge and professional autonomy over animal welfare and public accountability. Therefore 1882 rather than 1876 is a more plausible starting point for the institutionalisation of animal research policy, which implies fundamental differences in its subsequent developmental path.

Nevertheless, this book has argued that a coherent concept of historical institutionalism and path dependency must reject a deterministic temporal extrapolation from initial institutionalisation, and instead acknowledges that all decisions emerge from within a context that includes inherited institutions and structures that may privilege certain actions and outcomes over others. Different conjunctions of institutions, exogenous structures and actors can lead to variations in path direction. The dialectical network model and the policy network dynamics table (2.2) developed in this study offer ways of conceptualising possible mechanisms of network change and stability. Subsequent events seem to confirm the usefulness of this model in terms of tracing the types of dynamic interaction that occur, although it should be noted that outcomes cannot be predicted with perfect confidence from such a model, such is the complex nature of social phenomena.

Thus, the advent of the second Royal Commission from 1906 to 1912 represented an instance of how the success of a policy community can stimulate potentially destabilising activity from excluded interest groups as they receive feedback about adverse policy outcomes. Anti-vivisectionists adopted the classic strategic activity of outsider groups, attempting to stimulate increased public opposition to vivisection as a form of external shock to the closed and stable policy community. This led to pressure on the Government that persuaded it to appoint the Commission in 1906.

However, the pattern of events surrounding the second Royal Commission mirrored the formulation of the 1876 Act. For while anti-vivisectionists had managed to politicise the issue to create a critical juncture, the research lobby prevailed. The anti-vivisection movement was vigorous, relatively popular among the general public and increasingly associated with progressive reform groups such as the suffrage cause. However, it was also divided, lacked support from scientific professionals, and was perceived as increasingly radical and hence marginalized from the values of policy-making élites such as the Royal Commission. On the other hand, the constraining power of the animal research lobby's knowledge, expertise, organisational and social status resources was such that the pro-vivisection ideology, such as the notion of scientific domination of the licensing system, had become hegemonic. The scientific lobby were thus able to protect network structures such as their *de facto* control over application approvals and prevent stricter controls on permissible pain that may have interfered with their experimental objectives. These dynamics also reflect the asymmetric power model where élites are able to advance their interests at the expense of competing values and goals. From this point until after the Second World War, the structural context offered even less encouragement for reformers, and the policy network remained relatively undisturbed. This analysis of the evolution of animal research policy throughout this period has identified new outcomes that have emerged from the dialectical interactions among agents, networks and the exogenous context, including unobservable structures, thereby appearing to confirm the utility of the dialectical network model and a critical realist method.

The gestation and assent of the Animals (Scientific Procedures) Act 1986

In order to address the third research question concerning the network and policy outcome implications of the Animals (Scientific Procedures) Act 1986, the modern path to the replacement of the 1876 Act has been examined in detail. The critical juncture which marked the re-emergence of animal research as a major policy issue was the Home Secretary's appointment of the Littlewood Enquiry in 1962.

Once again, the dialectical network model has helped to capture the complex interactions among outcomes, structures, networks and agency that gave rise to this critical juncture. The policy community instantiated by the AAMR and Home Office in 1882 and entrenched by the 1906–12 Royal Commission had allowed animal research to expand virtually unhindered. That expansion, from 95,731 experiments in 1910 to 1,779,215 in 1950, both reflected and promoted increased resources for the pro-animal research lobby, as the practice became more institutionalised within biomedical research and product testing. A further aspect of this expansion that increased the structural resources of animal researchers was the growth of a pharmaceutical industry that conducted animal experiments and whose commercial success was perceived by the government as vital to the health of the British economy. The massive expansion of animal research over this period exacerbated the resource shortfall of the Home Office Inspectorate, leading to increased autonomy of animal researchers and their associated groups in terms of assessing the merits of animal research proposals and controlling the performance of the experiments. Since at least the 1906–12, period the Home Office had deemed it unfeasible for inspectors to be expert in all areas of biological research. Consequently, the Home Office felt that it was not in a position to be able to make an informed scientific or ethical judgement about research applications. In other words, despite the Inspectorate's formal political resource of authority to licence, it was perceived to lack the necessary organisational, knowledge and information resources to scrutinise applications, and so its authority was significantly dependent upon the information supplied by researchers. Furthermore, the loosening of licensing controls and the Home Office's trusting attitude towards researchers had reinforced the latter's autonomy and entrenched a pattern of self-regulation.

However, the policy community's apparent success in furthering its interests combined with growing public concern for animal welfare to generate increased concern not only in anti-vivisection circles but also among the public and Parliament. Interestingly, prior to the Littlewood Enquiry, the policy community was discovered to have mediated these exogenous changes in attitudes by interpreting it through the lens of its ideological structure. In particular, while the 'scientific' approach to animal welfare claimed to reduce pain within experiments so long as that was consistent with scientific imperatives, the anti-vivisection lobby wished to highlight what they saw as the significant moral relevance of animal welfare in terms of its implications for whether experiments should be authorised at all. Thus, the ethical and policy lessons the research lobby drew were quite different to the animal protection and anti-vivisectionist lobby, and indicate a constrained instance of policy learning that involved secondary policy change at most. However, evidence of the failure of researchers to use analgesics to relieve serious pain, even when this would not interfere with their scientific objectives, emphasises that animal welfare concerns could not be entirely dissolved by animal researchers and may explain the ongoing politicisation of the issue.

Strategic action on the part of the most powerful animal protection group, the RSPCA, combined with a context of public concern to place the issue of an inquiry into the regulation of animal research onto the political agenda. Eventually, an apparent change in policy direction initially emerged when, in late 1962, the Advisory Committee (AC) was consulted by the Home Secretary. However, the constraining effects of policy communities on policy agendas were signified by the fact that the issues considered did not include the animal protection lobby's core policy demands relating to expanding the membership of the policy network and the introduction of utilitarian criteria to research applications.

The response of the AC corresponds to the postulated dialectical interaction between a policy community and exogenous public opinion. In particular, the policy community exhibited 'dynamic conservatism' in order to maintain network homeostasis. This involved downplaying the main concerns underlying calls for an enquiry, while noting the potential for it to enhance the network's legitimacy. Secondary changes were countenanced, in particular in relation to the size and composition of the inspectorate, which illustrates the policy community's mediation of exogenous pressure. Such changes were also compatible within the broader ideology of the network as they did not question the cooperative style of regulation.

Nevertheless, an enquiry was established. The evidence gathered here does not provide a full explanation for this, but given the rationale provided by the AC, it seems possible that the intention was to stabilise policy outcomes through enhancing the network's legitimacy rather than an open exploration of the issue. Indeed, the limited scope of the subsequent Littlewood Enquiry, which excluded core elements of the anti-vivisection position that raised fundamental questions about the necessity and justifiability of animal experiments, would seem to support the idea that the enquiry's main purpose was the symbolic reassurance of exogenous concern. Groups' strategic actions and resources ineluctably influence outcomes. It was therefore concluded that the potential for the Littlewood Enquiry to stimulate policy change was undermined by the fact that animal research interests were more numerous, more organised and, hence, better resourced than the animal protection lobby in their interactions with the enquiry.

Indeed, the consistent unanimity of researchers in support of animal experimentation and professional autonomy, and the corresponding lack of overt expert support for the animal protection lobby, appear to have placed a major constraint on the possibility of policy change. This virtual monopolisation of decisive expertise resources appears to have had further powerful network-to-context effects in terms of the Littlewood Enquiry's deliberations and recommendation. Thus, the Littlewood Report did not recommend any significant changes to this policy area, confining itself mostly to adjustments aimed at rectifying undeniable administrative anomalies that had developed because of policy stability, such as the obvious understaffing at the Inspectorate. In other words, the Report pointed to a continuation of a policy community and outcomes that favoured animal research interests.

However, the Home Office implemented very few of the Littlewood Report recommendations, (in)action which was associated with a suspicion of animal protection groups and strong cohesion with insider research groups who opposed reform. This failure to implement the minor recommendations of the Littlewood Report represents stark evidence of the extremely narrow scope of policy learning in the network and hence indicates the existence of a tight policy community rather than an issue network. However, this persistence of the policy community also provoked continuing exogenous action to reform the 1876 Act, which eventually resulted in the Animals (Scientific Procedures) Act 1986. Thus, the answer to the third research question, which has a major bearing on the validity of this study's hypothesis, relates to whether the policy community changed to allow effective participation by animal protection groups in the formulation and passage of the 1986 Act.

One of the key tasks essential to addressing this question is a detailed specification of the ideologies of the lobby groups and the network during this period. This leads to a rejection of the simplistic dichotomy between abolitionist groups and pro-animal research groups in favour of a more sophisticated three-fold typology that recognises a major distinction in the so-called 'pro-animal research' category. Interestingly, these positions correspond closely to the ideological terrain of the debate during the formation of the network in 1876. The distinction in the 'pro-animal research' category is between:

1. *Animal use*: Animal welfare secondary to research goals; animal experiments considered generally 'necessary' and hence permissible in the pursuit of knowledge without immediate or foreseeable human benefit; resistance to utilitarian scrutiny of experimentation

proposals; professional self-regulation and the avoidance of lay interference in animal experimentation; and
2. *Animal welfare*: Animal welfare may outweigh research goals; animal experimentation only considered – on a case-by-case basis – 'necessary' and hence permissible to satisfy urgent and pressing human needs; should be subject to independent utilitarian analysis; lay control required to ensure consideration of the wider public – and animals' – interests.

'Animal use' represents the position of the animal research interest groups, while 'animal welfare' represents the position of groups who seek policy reforms that impose some limits on animal experimentation, such as the RSPCA (Orlans, 1993: 22). Abolitionist animal rights anti-vivisection lobby groups have tended to argue on 'animal welfare' grounds as this strategy is perceived to be most likely to achieve change. This is relevant to an understanding of the breadth and inclusivity of the ideological consensus in the animal research policy network during the formulation of the Animals (Scientific Procedures) Act 1986. This study has established that in the mid-1970s, an entrenched animal research policy community persisted which was characterised by an 'animal use' ideological structure, and this has provided the starting point for understanding the policy change implications of the Animals (Scientific Procedures) Act 1986. In particular, this represents a narrower ideological consensus that conditioned access to policy-making than hitherto acknowledged.

One of the most important new findings is that the putative 'animal protection' actors who gained access to the Home Office in the formulation of the Act – the CRAE Alliance – were unrepresentative of the animal protection movement, in terms of both its composition and its goals for the 1986 Act. Despite their willingness to argue within an 'animal welfare' rather than 'animal rights' framework, the major anti-vivisection groups were denied access to formulation discussions. Indeed, while the CRAE Alliance, together with pro-animal research groups, generally supported the Bill during its passage, the marginalized animal protection movement opposed it because of its 'enabling' format, which was manifest in the absence of any definite bans on any type of experiment and its perceived failure to improve public accountability. In other words, they perceived that it failed to ensure any significant changes. However, the key test for policy change focuses on two formal innovations introduced by the 1986 Act which are related to the key areas of animal protection concern: the cost-benefit assessment and the Animal Procedures Committee (APC).

In relation to the cost-benefit assessment, the 'enabling' and discretionary nature of the legislation meant that, for at least the first decade of the Act's operation, all licensing decisions were still to be made on a project-by-project basis[1]. Furthermore, the statutory criteria, involving an assessment of costs and benefits and a subsequent 'weighing' of the two concepts, remained a subjective, expert-dependent process that required the discretionary application of vague statutory criteria to complex, multi-faceted cases. The Inspectorate continued to lack the resources to effectively question the expert judgments implicit in licence applications. Furthermore, inspectors and researchers shared professional backgrounds and an animal use ideology. This represented a policy community with a structure of resource interdependency that allowed animal researchers to dominate policy implementation through positive-sum power games. The review of the policy network literature in Chapter 2 proposed that these types of implementation structures promote major discrepancies between intended formal rules and actual policy outcomes that reflect the goals of regulatees. The secrecy surrounding the policy process reinforced the network's isolation from public accountability. Thus, it was proposed that the implementation structure for animal research policy envisaged in the Animals (Scientific Procedures) Act 1986 appears to have facilitated an élitist 'professional treatment' model of decision-making that serves the interests of 'regulated' professions and associated industries, rather than the pluralistic 'moral judgement' model envisaged by the CRAE Alliance that involves independent regulation that incorporates broader social values. The 'professional treatment' model corresponds to a policy community, while the 'moral judgement' model has an affinity with an issue network.

In respect of the APC, although, unlike its predecessor the AC, it was empowered to investigate and advice on its own initiative (instead of only acting at the behest of the Home Secretary) and lay an annual report before Parliament, scientific domination was enshrined by the new Act, and its initial membership did not reflect public opinion in favour of a ban on cosmetics testing, for example.

By formally incorporating these policy changes, the policy community appears to have maintained legitimacy resources and thus mitigated the potentially destabilising impact of exogenous public concern. However, it has been argued that the wide scope of discretion within an unchanged implementation framework structured by an animal use ideology meant that core policy change was unlikely to ensue from the Act's operation.

Assessing the implementation of the Animals (Scientific Procedures) Act 1986

Indeed, the final and most salient stage in addressing this study's hypothesis has involved an analysis of the implementation of the Animals (Scientific Procedures) Act 1986. Previous research on animal experimentation policy was hampered by a lack of primary data relating to policy outcomes, due to the legally-enshrined confidentiality requirements surrounding such sources. However, this study has overcome this fundamental obstacle by presenting unique confidential primary data relating to the results of Imutran's primate xenotransplantation experiments and related regulatory interactions.

The case study focussed on the implementation of the Animals (Scientific Procedures) Act 1986 to examine whether the network and policy outcomes had changed to resemble the issue network model. It was found that, on the contrary, Imutran researchers and Home Office inspectors cooperated to implement a skewed cost-benefit assessment which, throughout the lifetime of the project, underestimated animal suffering while exaggerating the potential benefits of the research. When scientific goals conflicted with the goals of reducing animal suffering, scientific goals tended to prevail. Consequently, the policy outcomes involved suffering that exceeded regulatory limits, with minimal benefits that failed to achieve the objectives that had formed the justification – both administratively and in public statements – for the Home Office's approval of the research. There is also evidence that, on occasions, these actors worked together to minimise APC scrutiny.

The evidence of maladministration and regulatory breaches relative to published policy laid before Parliament indicates the insulation of the network from Parliament, the public and animal protection groups. This is a key sign of a policy community. Furthermore, the failure to implement the main formal regulatory innovations introduced by the 1986 Act suggests that the new law did not represent policy change. Other features of policy communities to emerge from the case study are the weakness of grievance procedures and the subsequent actions of researchers and regulators that aimed to frustrate accountability.

One way of explaining this continuity is through assessing the ideological structure of the policy community and relating this to the way groups' resources are perceived and valued. It has been established that the animal research lobby's 'animal use' ideology continued to structure the network. The goals of policy remained the facilitation of research and the advancement of knowledge through the instruments of self-regulation and cooperative relations between inspectors and researchers, while assuaging public concern about animal welfare and unnecessary experiments by positing a cost-benefit assessment with severity limit requirements. In reality, animal welfare remained a decidedly secondary consideration to these fundamental goals. Thus, the 'ethical' resources of animal protection groups, which seek to represent animals' interests in order that when they are weighed against research goals at least some research may not be permitted on cost-benefit grounds, are not valued, because they are not perceived to be instrumental to the network's goals. Thus, apart from some legitimacy, animal protection groups have no resources to exchange in the network in its current form, as they are not perceived to be relevant to the resolution of policy goals and the implementation of policy as defined by the animal use ideology. The lack of policy change achieved by reform groups such as CRAE and FRAME during the formulation of the Animals (Scientific Procedures) Act 1986 reveals that they were merely peripheral insiders, experiencing illusory network membership. They possessed some political and knowledge resources that increased the network's legitimacy resources and assisted with animal welfare in the strictly secondary sense, compatible with the 'animal use' ideology. Thus, the 'animal welfare' ideology and its associated groups remained excluded. This does seem to explain the evolution of the network and thus indicates the utility of the Marsh/Rhodes model that asserts that resource distribution is, indeed, a highly significant influence on network interactions and outcomes.

In drawing conclusions about the claims that can be made on the basis of this research, it is necessary to recognise the inherent limitations posited by the critical realist epistemology that underpins this study. Those limitations relate to the subjectivity of the researcher that ineluctably affects the interpretation of data, especially in relation to assessing the implications of unobservable structures. Nevertheless, it seems to reasonable to claim that the findings of this study have demonstrated the plausibility of the hypothesis that, except for the period of 1876 to 1882: the interests of animals are given scant consideration in an élitist policy process characterised by research interests' domination and the effective exclusion of animal protection groups. This conclusion should now serve as a starting point for further research in this policy area that may either enhance or challenge the validity of this hypothesis.

Animal research policy since 2000

The Imutran research programme ended in 2000, however there have been no significant, core-level changes in the character of the animal research policy network in the last thirteen years. As noted towards the end of the previous chapter, in 2006 the Home Office continued to deny accusations of regulatory failure in the Imutran case, and their position remains constant at the time of writing. The false exoneration provided by the Ombudsman has helped to stifle consequent policy change. In response to queries from the APC regarding their handling of the case, the Home Office confirmed that it saw no need to change its operation of the cost-benefit assessment (APC, 2002: 120).

Since 2000, the annual statistics for animal experiments rose significantly from 2.7 million to 3.7 million in 2011,[2] while project severity bands indicate a slight increase in severity over that period.[3] This trend reflects the continuing dominance of animal research interests and the persistent ability of the policy network to minimise external constraints on animal experimentation. APC recommendations for significant policy changes have not been implemented by the Home Office. For example, in 2003, the APC's chairperson wrote to the Home Secretary to suggest the exploration of targets for the reduction of animal experimentation though stakeholder dialogue. This was rejected by the-then Home Office Minister Caroline Flint MP (APC, 2006: 44–5) on the grounds that, firstly, it would be unnecessary as the thrust of the legislation and the commitment of the animal research community to the replacement and reduction of animal experiments would achieve the results intended by targets. But it is hard to accept this reasoning, given the significant increases in animal experimentation since 2000. Secondly, the Minister indicated that policy should be led by scientific developments, thereby implicitly rejecting the role of animal welfare or public accountability in the structure of this policy sector. This overarching approach has been confirmed by subsequent Home Office references to the 'demand-led' nature of the policy area and, contradicting the above Home Office reason for rejecting targets, assertions that the 1986 Act contains no mechanisms to reduce animal experimentation, which would appear to discount the wide discretion the Act affords the Home Secretary.

The announcement in the Coalition Government's Programme for Government that they would 'work to reduce the use of animals in scientific research' (HM Government, 2010: 18) appeared to introduce a historically significant policy change involving external policy constraints that reflect democratic and ethical considerations. However, over two years later, the announcement regarding how this would be implemented revealed an essentially 'business-as-usual' approach that relied on the development of alternative methods, with no targets or strategy to provide drive for such a reduction in animal experimentation.

Therefore, in reality, no noticeable policy change has occurred and, given previous trends under this approach, this policy pledge is unlikely to be realised.

Finally, new EU Directive 2010/63/EU on the 'Protection of Animals Used for Scientific Purposes'[4] came into force in the UK on 1 January 2013,[5] signalling a potentially critical juncture in the evolution of this network. At the time of writing, it is impossible to ascertain the true impact of this legislative change. However, it is possible to identify key indicators from the formulation process and text of the new law.

During the transposition process, the 2010 Coalition Administration stated its preparedness to lower existing stricter UK animal welfare measures to those in the Directive, despite Article 2's provision for the retention of such measures (Home Office, 2011). This announcement contributed to considerable concern in animal welfare circles that the Government perceives the Directive as an opportunity to weaken regulation:

> Taken together, the RSPCA believes these factors mount a serious attack on the standards of regulation of animal experiments in the UK. Indeed, we could see the longstanding, tried and tested system being systematically dismantled. (RSPCA, 2011: 2)

While such warnings might, at first sight, appear alarmist, it should be noted that the aim of deregulation would be consistent with this policy community's 'animal use' ideology.

In broad terms, the scope for policy change appears limited because the new Directive incorporates two of the cornerstones of the 1986 UK legislation: a 'harm-benefit' assessment (Article 38(2)(d)) and a national advisory committee (Article 49). The new regulations maintain requirements for project licence applications to include severity classifications of intended experiments, categorised as 'mild', 'moderate' or 'substantial/severe'. The anticipated level of animal suffering is then supposed to form the 'harm' part of the 'harm-benefit' analysis of animal research applications.

The drafting of the Directive is imprecise with regard to the subsequent enforcement of severity classifications as severity limits, although Article 36(1) requires that projects be carried out 'in accordance with the authorisation'. Indeed, it would seem highly unlikely for the Directive to require severity classifications if it did not intend them to be enforced in practice. It would mean that the initial harm-benefit analysis and any favourable project evaluation would be based on unreliable, unenforceable criteria, as researchers would be free to exceed the level of suffering that they had originally submitted in their application. However, the new UK regulations appear to abolish the specific legal obligation upon the Secretary of State to ensure that severity classifications submitted in authorised licence applications under the new regime are enforced as limits in practice.

Interestingly, in response to concerns raised by the author and animal welfare groups, the Home Office has claimed that severity limits will be applicable.[6] However, close examination of the regulations shows that these are contained under 'Transitional Provisions' that are only applicable to pre-existing licenses issued before the new regulations came into force on 1 January 2013, and which expire five years after being granted. At the time of writing, the Home Office has not managed to identify the clause(s) in the new regulations that maintain the explicit requirement for the Secretary of State to enforce severity limits. In particular, there are no provisions in Schedule 2C of the 2012 Regulations[7] – which legislates for the standard licence conditions that must be applied by the Secretary of State – that are equivalent to Standard Project Licence Conditions 6 and 8, and Personal Licence Condition 13 in the previous 1986 legislation:

> 6. For any procedure, the degree of severity imposed shall be the minimum consistent with the attainment of the objectives of the procedure, and this shall not exceed the severity limit attached to the procedure. The minimum number of animals of the lowest physiological sensitivity shall be used in procedures causing the least pain, suffering, distress or lasting harm.
>
> 8. It is the responsibility of the project licence holder to ensure adherence to the severity limits as shown in the listing of procedures/ protocols (Section 19a [of the project licence application form]) and observance of any other controls described in the procedure/protocol sheets (Section 19b). If these constraints appear to have been, or are likely to be breached, the project licence holder shall ensure that the Secretary of State is notified as soon as possible.
>
> 13. It is the responsibility of the personal licensee to notify the project licence holder as soon as possible if it appears either that the severity limit of any procedure listed in the project licence (Section 19a) or that the constraints upon adverse effects described in the protocol sheets (Section 19b) have been are likely to be significantly exceeded. (Home Office, 2000: 78–9, 82)

Draft Guidance on the operation of the 2012 regulations, issued by the Home Office in December 2012, does include standard licence conditions concerning severity limits. However, the text indicates that a breach of the licence may only happen if the personal and project licence holder simply fails to subsequently notify the Secretary of State of suffering that breached the severity limit: 'The conditions of your licence will be breached if you do not notify us promptly when an animal suffers, or is likely to suffer, more than is authorised.' (Home Office, 2012: 47). In other words, a breach of the severity limit/classification does not, in itself, appear to be unlawful or a breach of licence. There is also provision for the 'temporary authorisation' of a higher severity limit (Home Office, 2012: 47). There appears to be a danger that licence holders will, in practice, be able to move the severity 'goalposts' at will, regardless of the initial classification and harm-benefit assessment.

It may be significant that the Home Office has stated that it does not believe that enforcement of severity classifications is legally required by the Directive. Not only does this interpretation of the Directive seem implausible, but, taken alongside the decision to omit such an explicit requirement from the transposed regulations and the Draft Guidance's text, it raises the possibility that the Home Office may be quietly sabotaging its ability to enforce severity limits on animal researchers?

Once again, while this interpretation of the Home Office's intentions may seem overly cynical to some readers, it would be consistent with both the historic Home Office policy as exemplified in the Imutran case study and also this study's policy network analytical framework. The Home Office's manoeuvrings appear to be a case of constrained policy learning by a policy community dominated by an 'animal use' value system, at the expense of animal welfare values. Specifically, the lesson the Home Office seems to have learnt from the Imutran case is that it needs to weaken the impact of severity limits in order to protect the freedom of researchers and avoid accusations of maladministration in future. Yet, sensitive to public perceptions of weakening animal welfare standards, it has tried to conceal such actions: an example of the politics of symbolic reassurance. Parliament's ability to scrutinise the critical details of the new regulations has been undermined by the Home Office's decision to amend the 1986 Act through secondary rather than primary legislation.

The validity of this interpretation will only be established through the passage of time, and with the assistance of reliable data about policy outcomes – though it should be noted that the Home Office has retained Section 24 of the 1986 legislation, which prohibits officials from releasing any information without the consent of researchers, despite this being in potential conflict with the Directive's requirements for the publication of project licence summaries. This author's prediction is that the apparent weakening of severity limits will become clear in future political and legal battles surrounding emerging evidence of severity limit breaches.

In summary, this study's identification of an animal research policy community heavily dominated by animal research interests remains valid, and there appears no likely change of trajectory in the foreseeable future. However, the gaping chasm between the spin of tightly-regulated animal research and the *laissez-faire* reality generates potential instability in this regime. Therefore, animal protection could be improved if the major national animal advocacy groups were to coordinate to execute a sustained, strategic effort to highlight this gap. Indeed, barring some external shock to the system, this type of strategy is the only one that offers tangible hope of significant advances for animal protection in this field.

Additional contributions, limitations and further research paths

Pluralistic or asymmetric power relations?

What further insights for animal research policy and policy network analysis can be gleaned from this study? It has been found that, as in the case of North Sea petroleum health and safety policy (Cavanagh, 1998) (and, indeed, earlier critical junctures in animal research policy), pressure from non-economic/professional interests created a reaction from state actors in the form of the Animals (Scientific Procedures) Act 1986. However, firstly, the formulation process and the eventual content of the Act substantially reflected the pre-existing constraints imposed by the policy community. Furthermore, the potential impact of the new measures on the pattern of policy outcomes was constrained by discretionary exemptions and implementation structures that effectively excluded non-economic/professional actors who sought policy change. Consequently, exclusive relationships between researchers and regulators continued to affect policy outcomes in such a way that research imperatives were and are given influence that is disproportionate, compared to the stated intentions of formulators and the formal policy requirements, thereby indicating the network's insulation from democratic accountability.

This insulation from democratic accountability is further emphasised when the impact of Labour's 1997 General Election victory is considered. Despite significantly greater support for progressive animal welfare measures amongst Labour MPs, compared to Conservative members, the case study demonstrates no evidence that a core change in policy occurred after 1997. This suggests that shifts in parliamentary support do not affect the animal research policy network, which is another indicator of a policy community in this policy area. One specific animal research policy change was made in 1998, with the announcement that animal testing of cosmetics and their ingredients would no longer be licensed (Nuffield, 2005: 231). Interestingly, this cessation came about following a voluntary agreement between the Home Office and the relevant licence holders (APC, 1999: 2), and this category of testing represented a tiny proportion – less than 0.1 per cent – of animal research, with 2,200 procedures in 1992, and 600 in 1998 (Home Office, 2006: 65).[8]

The implication that the pace of policy change has continued to be dominated by the relationship between the Government and animal researchers can be seen by the reaction of the major scientific research institutions such as the Wellcome Trust and pharmaceutical industry to the 1997 Labour Government's early intimations of a desire to reduce animal experimentation (Appleyard, 2006). These intentions were perceived as a threat to the hegemony of the animal research lobby, who wished to increase animal experimentation (Hinde, 1999). Consequently, the pharmaceutical industry threatened to move their operations out of the UK unless the regulation of animal research were relaxed (Hughes, 2000). This strategic action was spearheaded by the Association of the British Pharmaceutical Industry and came through the Prime Minister, the Department of Trade and Industry – particularly the Science Minister Lord Sainsbury – and the Department of Health rather than the Home Office (Hughes, 2000; Pharmaceutical Industry Competitiveness Task Force, 2001). Direct lobbying by the three largest pharmaceutical companies led to an audience with the Prime Minister in 1999 and the subsequent establishment of a *Pharmaceutical Industry Competitiveness Task Force* (PICTF) that issued action points across a number of policy areas, bypassing the statutory Animal Procedures Committee. The task force, which was set up through the Department of Health and excluded any animal welfare representation, issued directions related to animal research, including a 'streamlining of the licensing process' and additional investment in animal-based university courses (PICTF, 2001: 17, 12). These measures would have the effect of increasing the autonomy of researchers at the expense of Home Office scrutiny, and are likely to have been a major contributory factor in the increases in animal experimentation throughout the 2000s. The powerful constraints on Government policy imposed by the structural economic resources of the pharmaceutical industry are clearly expressed by the Prime Minister in his Foreword to the PICTF's report (2001: 1):

> A successful pharmaceutical industry is a prime example of what is needed in a successful knowledge economy.... It has provided tens of thousands of high quality jobs, substantial investment in research and development, and a massive contribution to the UK's balance of trade.... I am committed to ensuring that the UK retains the features that have made it an attractive location for investment.

The change of government in 1997 had the potential to be a critical juncture in animal research policy, with the Labour Party representing 'the blade for prizing apart the mollusc's shell of Whitehall and the policy networks' (Marsh and Rhodes, 1992b: 257). However, contrary to pluralistic assumptions regarding the responsiveness of policy networks to public opinion and Parliament, it is clear that the animal research lobby deployed its resources to prevent policy change once again.

Furthermore, the case study data tend to validate the dialectical conception of network evolution and relationship between structure and agency rather than a deterministic extrapolation from the 1876 network and context. Moreover, it supports the critical realist method and the accompanying acknowledgement of the role of structural constraints that are not directly observable.

In relation to testing for pluralism in this network, it is perhaps significant that the case study's micro-level analysis of particular, observable policy outcomes implicitly involves a narrow pluralist version of power (Lukes' first dimension of power) that brackets off questions of the 'mobilisation of bias' that keeps issues off the agenda. However, this observation highlights a limitation in the study and the possible utility of conducting further research on the second and third dimensions of power (Hay, 2002: 178–87) in this policy area that appear to contribute to the exclusion from the policy agenda of more radical demands such as abolition, as seen in the limited scope of the Littlewood Enquiry.

Thus, the findings of this study also have implications for the understanding of power in British politics. The existence of an animal research policy community contradicts the Westminster model because it has shown that Parliament is generally excluded (due to implementation gaps) and that civil servants are not neutral insofar as the Inspectorate has been found to be susceptible to group pressure. Far from the government promoting equality of arms by positively assisting the weak and disorganised, the state has been seen to cooperate with the more powerful lobby in order to disempower the animal protection movement and facilitate the painful and lethal exploitation of animals. However, the promulgation of the myth of substantive state neutrality is key to preventing its realisation, as the discursive legitimisation of the network constrains pressure for change. The secrecy surrounding animal research policy-making resonates with the asymmetric power model, though to some extent it is a double-edged sword in terms of its contribution to undermining legitimacy. For, on the one hand, it helps to conceal the divergence between the public discourse of animal research policy – strict regulation and a cost-benefit assessment – and the reality of policy-making, while on the other hand, secrecy in itself breeds suspicion and erodes legitimacy.

Because the pluralistic presumption of state neutrality is a fundamental tenet of the issue network thesis, a key contribution of this study has come from testing this assumption. Thus, the findings presented here correspond closely with the postulations of the asymmetric power model where weaker groups who challenge insider groups' interests are excluded from policy networks, while '[those] powerful economic and professional groups that have the greatest resources to exchange with government...are evident in policy communities' (Marsh et al., 2003: 318).

This book has also afforded insights into the scope for changes in ideological context that might stimulate network and outcome change. At the time of the Animals (Scientific Procedures) Act 1986, animal welfare was not a new ideology as such, though it became more popular from 1960s onwards. But this study reveals that, in fact, the animal research policy network has been a relatively hostile environment for the animal welfare ideological 'virus' (Richardson, 2000: 1018). The potential tension between broader support for the notion (as evidenced by the intentions behind the cost-benefit assessment in the passage of the Animals (Scientific Procedures) Act 1986) and the policy community's appreciative system has been mitigated by a combination of secrecy and symbolic politics in terms of formal regulatory requirements and peripheral insiders in the network. This seems to confirm Smith's (1997) postulation of policy communities' dynamic conservatism, where the limited accommodation of new types of actor can largely maintain existing power structures by appearing to internalise legitimacy resources of new actors.

Limitations and further research

One of the basic theoretical insights of policy network analysis is that in order to explain network characteristics and policy outcomes, 'the meso-level policy network model needs to be integrated with both micro-level and macro-level analysis' (Daugbjerg and Marsh, 1998: 54). The longitudinal structure of the empirical section of this study engages with these three levels of analysis, though the emphasis alters as the diachronic analysis proceeds: the first three chapters tend to emphasise meso- and macro-level analysis, while the final case study chapter has a stronger emphasis on the micro-level, due to its focus on individual decisions and their specific outcomes. Thus, although this study combines the three levels of analysis, that integration is not always synchronised chronologically. Given the longitudinal approach, this is an inevitable and necessary compromise within a book-length study. Nevertheless, that integration still allows a useful degree of understanding. For example, the micro-level elements of the case study have tended to confirm the policy community implications that were deduced from examination of the meso-level network patterns associated with the assent of the Animals (Scientific Procedures) Act 1986. However, data collection issues permitting, this research could be usefully complemented by more 'static' case studies throughout the evolution of the animal research policy network.

Furthermore, the case study's micro-level analysis adopted a descriptive starting point and did not assume a theory of individual behaviour such as rational choice (Daugbjerg and Marsh, 1998: 67). However, in the context of a critical realist method that appreciates the interaction between structure and agency, the empirical findings of the case study permit inductive reasoning towards the conclusion that the 'animal use' ideology has acted as a 'shorthand guide for processing information and making strategic decisions' (Daugbjerg and Marsh, 1998: 69). This observation implies that this study could usefully be augmented by research involving interviewing in order to elucidate individuals' perceptions about their context, the likely outcomes of their actions, and how their subjective valuations of those consequences affect their choices (Daugbjerg and Marsh, 1998: 69). Such research has the potential to increase understanding of how network and exogenous structures dialectically interact with agents' behaviour. Sadly, though, such research would be subject to potentially severe obstacles due to the political sensitivity and associated confidentiality surrounding this policy process.

Another useful future research agenda would involve studies that compare the British animal research policy network (particularly in relation to this case study) with animal research policy networks in other countries. For example, the author's receipt of an invitation from the Norwegian Government's research ethics advisory committee to give a presentation on the Imutran case study (Ekern, 2004: 18–20) – in marked contrast to the antagonistic and distant stance of British policy makers – suggests a hypothesis regarding the possible impact of different types of macro-level political culture or state power structures on the breadth of access in animal research policy networks.

The recent formulation, passage and transposition of EU Directive 2010/63/EU offers important opportunities for comparative studies of UK and other EU countries' animal research policy networks, as well as an examination of policy-making at the EU level itself. The considerable similarities between the UK's 1986 legislation (relatively unique in the EU) and the 2010 Directive suggest an example of the UK 'uploading' its policy approach to the EU level. Furthermore, while the implications of the 2012 amendments to UK legislation have been outlined, clearly much more in-depth analysis of this critical juncture is required.

Another potentially fruitful research path would be the application of a different theoretical perspective to the analysis of change and continuity in animal research policy. One such approach, exemplified by Hajer (1995), sees politics as institutionalised discursive struggle. Hajer applies discourse analysis in an attempt to explain environmental policy since the early 1970s, focussing particularly on acid rain policy in the UK and the Netherlands over that period. Hajer (1995: 6) asserts that because acid rain is not directly visible, its effects are not immediately obvious and are slow to materialise, it is: 'typical of a new generation of environmental issues that depend on their discursive creation'. From this social constructivist perspective, politics is conceived as a struggle for discursive hegemony between discourse coalitions whose members share a view of reality with significant commonalities. These coalitions attempt to define policy problems and their potential solutions, and persuade other actors of the legitimacy of their definition. Hence, Hajer's approach shares a research focus with policy network analysis insofar as the interaction between actors against a background of structural parameters is conceived as the primary determinant of policy evolution.

Hajer develops the concept of 'story-lines' which are said to represent a 'subtle mechanism of creating and maintaining discursive order' (1995: 56). They achieve this by bringing a perceived unity and relatedness to disparate composite discourses and 'providing actors with a set of symbolic references that suggest a common understanding' (Hajer, 1995: 62). Hajer argues that it would be practically impossible for even the most gifted and experienced actor to hold a detailed understanding of all the discourses that make up the acid rain issue. Story-lines help to reduce arcane scientific positions, with their inherent uncertainties and conditionalities, to metaphorical representations such as graphs, diagrams or aphorisms. They may incorporate historical references, appeals to collective fears and implicit worldviews or value commitments.[9]

In respect of environmental policy and the specific issue of acid rain, Hajer identifies two main discourse-coalitions, traditional-pragmatist and ecological modernization, each with their own story-lines and involving participants from politics, science, regulatory bodies, NGOs, journalism and academia. He then proceeds to investigate the regulation of acid rain in both the UK and the Netherlands in order to elucidate how and to what extent a new discourse of ecological modernization achieved discursive structuration and institutionalisation. He finds that the entrenched traditional-pragmatic discourse coalition, led on the acid rain issue by the Central Electricity Generating Board (CEGB), the Royal Society and the Cabinet, held sway over both discourse structuration in the core policy domain and the actual decision-making processes, thanks to well-institutionalised practices and their ability to stretch the discourse in such a way as to claim legitimacy for their construction of the new issue (1995: 161).

There are a number of considerations arising from the application of discourse analysis to acid rain and ecological modernization that suggest fruitful parallels to UK animal research policy. Both policy sectors have arguments about science, economics and ethics at their heart. Animal welfare, like ecological modernization, has emerged prominently in the last thirty years to challenge an existing paradigm characterised by empirical research science and economic considerations, and advanced by corresponding professional and industrial discourse-coalitions. One common feature of both policy debates is a commitment by 'traditional' coalitions to science as exemplified by established institutions, accompanied by a moral legitimisation. In animal research, this occurs by reference to 'vital benefits' for humans in terms of the treatment of diseases and the provision of environmental and consumer protection through animal-based risk assessments of chemicals. In acid rain, moral legitimisation was expressed in terms of the detrimental impact on the standard of living that would be caused by imposing extra costs on the generation of electrical power (Hajer, 1995: 161). In fact, the three 'essential' issues focussed upon by Hajer in his examination of the discursive framing on acid rain are all highly salient to the controversy surrounding animal research, and in particular, xenotransplantation experimentation: 'the image of damage, the role of science, and the issue of regulation' (Hajer, 1995: 125).

In both cases, the practical effectiveness of the new paradigm appears to have been disappointing in relation to the wider social support they seem to enjoy. Thus, the same discursive mechanisms that have frustrated the hegemony of ecological modernization may also shed light on the skewed character of UK animal research policy.

Notes

1 Introduction

1. By the Animals (Scientific Procedures) Act 1986 Amendment Regulations 2012.
2. This primary material comprises confidential material that came into the writer's possession through two unauthorised disclosures: from within Imutran Ltd in spring 2000, and then the Home Office in October 2002. The documents had originally included thirty-nine final draft study reports that detailed the design, materials, methods and results of various xenotransplantation procedures on forty-nine baboons (*Papio anubis*) and 424 cynomolgus monkeys (*Macaca fascicularis*, also commonly known as 'crab-eating macaques'). Other documents included correspondence with HLS, suppliers and Home Office Inspectors, meeting minutes, feasibility studies and internal reports concerning many aspects of the conduct of and plans for xenotransplantation research. The Home Office documents comprise correspondence between Imutran and both the Home Office Animals (Scientific Procedures) Inspectorate (ASPI) and the APC, reports submitted by Imutran to both those bodies in support of their licence applications, and actual project licence authorities. The disclosed documents can be viewed at www.xenodiaries.org.

2 Towards a Dynamic Model of British Policy Networks

1. Hereby acknowledging that some network theorists 'treat the concept merely as a heuristic device, while others see it as having explanatory utility' (Marsh, 1998a: 3).
2. This raises questions regarding the underlying ontological position of this study, which is explored explicitly in Chapter 4.
3. Marsh and Rhodes' approach, which posits informal and non-observable structures as constraining policy processes and individual decision-making, indicates an underlying 'realist' social ontology (Marsh and Furlong, 2002: 38), which is discussed in more detail in the Chapter 4.
4. BPM stands for 'Best Practicable Means'. Rather than prescribing specific limits, BPM gave flexibility to site-level Inspectors and industry operators to exercise discretion relevant to local circumstances and available pollution-control technologies, when operationalising BPM's principles (Smith, 1997: 54–5).
5. Here, by virtue of his discussion of 'informal institutions', Smith is using the term 'institutional' in a broad sense which, according to Bulkeley (2000: 731), reveals 'the influence of ... "new institutionalist" concepts'. New institutionalism is discussed in Chapter 4.
6. The qualification 'effective' is used here because, according to Cavanagh (1998: 93–5), labour organizations did sporadically apply pressure to the

government, but the responses, such as the Mineral Workings (Offshore Installations) Act 1977 were attenuated by discretionary exemptions and other weaknesses. In addition, although the TUC was represented on the Health and Safety Executive (HSE) after 1977, and their remit was simultaneously extended to cover offshore matters, the government stymied any potential threat to existing policy by restricting the HSE's influence to broad policy while delegating detailed implementation to the Petroleum Engineering Division (PED) within the Department of Energy. The PED's relationship with oil companies is described as 'cosy' and 'mutually supportive' through, for example, a predominance of former oil company employees in the Division.

7. These issues of the relationship between structure and agency raise important underlying ontological, epistemological and methodological questions that are examined in Chapter 4.
8. Putnam, R. (1976) *The Comparative Study of Political Elites*. Englewood Cliffs, N.J.: Prentice Hall. Cited by Sabatier (1993: 292).
9. Sabatier's discussion is in the context of his advocacy coalition framework, which has some broad similarities with the policy network approach. For example, he focuses on 'policy subsystems' which, like policy networks, are a middle-range concept, though they are wider in the sense of including actors not directly involved in the policy process.
10. MAFF was merged with the environment and countryside sections of the then Department for Environment, Transport and the Regions (DETR) to form the Department for Environment, Food and Rural Affairs (DEFRA) in June 2001.
11. Most prominently advanced by Ulrich Beck (1992). However, both Moran (2003: 30–1) and Smith (2004: 328–9) qualify the proposition of a 'risk society' as a universal narrative of modernity, arguing that perceptions of risk are contingent, and vary across time, location and policy area.
12. Richardson's discussion appears to be ambiguous on this matter, because he goes on to say that one possible international, cross-venue structure may involve expert groups or 'epistemic communities', but these are not policy communities, or, as he puts it, 'a system of transnational governance'.
13. Indeed, Smith cites the absence of a monopoly over a policy area (1993b: 83) as a feature of issue networks.
14. The 'at least partly' qualification acknowledges the debate over models of individual behaviour, and that rational actor models, while containing some valid points, are probably over-simplified models of human motivation.
15. This also relates to the three faces of power argument. Thus, on this account, those actors with macro-level structural advantages enjoy the benefits of (and possibly contribute towards) hegemonic modes of agenda-setting and preference-shaping.

3 The 'Animal Research Issue Network' Thesis: A Critique

1. This section is retained in the amended 2012 legislation.
2. According to official publications cited by Garner, the 23 personal licences issued in 1877 had increased to over 17,000 in 1991.

3. Departmental 'sponsorship' of economically-significant industries was widespread in British politics, particularly between 1945 and 1979. Symptomatic of a political system known as 'corporatism', this was partly due to successive governments' desire for an interventionist industrial and economic policy that sought to achieve growth and full employment, often through the additional incorporation of trade unions into a tripartite relationship (Grant, 2008). Other commentators have emphasised the role of corporate power in the creation of such 'sponsorship', with industries' capturing state policy processes relevant to their interests (Wainwright, 2004: 144).
4. Animal rights and animal liberation are, strictly speaking, quite different ethical frameworks, but the important point from a policy perspective is that they both demand a greatly enhanced degree of moral consideration for animals relative to the status quo.
5. An additional factor in discouraging the RSPCA from participation in a coalition is cited by Garner from interview data with a senior RSPCA employee that reveals the RSPCA's desire to maintain a separate identity as the leading 'brand' of animal protection.
6. For example Garner (1998: 180) describes the reaction of the then-recently retired Home Office Chief Inspector to a CRAE reform proposal in 1977 as 'an astonishing public attack'. Initial CRAE submissions are said to have been largely ignored by the Home Office (Garner, 1993: 206).
7. This quote is taken from an annual review of one of the main organisations behind CRAE, the Scottish Society for the Prevention of Vivisection, whose leader, Clive Hollands, is said to have been a key figure behind CRAE.
8. Though clearly there would be some room for discretionary judgment regarding what counts as 'exceptional circumstances', for example.
9. This discussion is in the context of Garner's criticism of attacks by some animal rights advocates on the concept of 'animal welfare', which he says is mischaracterized and does, in fact, offer the opportunity for achievable policy change that favours animals' interests.
10. The definitions of 'substantial' and other categories of severity are discussed in more detail below in the case study chapter.
11. Discussed in more detail below.
12. Directive 86/609/EEC.
13. Interestingly, Garner notes that the one area of animal welfare policy-making where Parliament does play a role, and where the executive is not dominant, is hunting, and that this policy area: 'would seem to resemble a classic case of an open and pluralistic issue network' (1998: 111). The implication is that policy areas where Parliament does not play a role are more likely to fit the policy community model. This is consistent with most network analysts' view of the relationship between Parliament and policy communities.
14. It might be argued that counting MPs' signatures on EDMs might give a more comprehensive assessment of animal welfare commitment.
15. 48 per cent of the MPs interested in animal welfare were Conservatives, although they comprised 58 per cent of MPs. 41 per cent of animal welfare MPs were Labour, who filled 35 per cent of the seats, and 7 per cent of the concerned MPs were from the Liberal Democrats, who comprised 3 per cent of all MPs.

16. For example, the relative weight of the welfare of human and non-humans, the ability of technology to solve problems, and the relative value of knowledge.

4 Theory and Method in the Study of Animal Research Policy

1. In particular, Garner describes these institutions from a comparative perspective examining both the US and UK, as well as outlining the structures for farm animal welfare policy.
2. Peters (1999: 6) remarks on 'old' institutionalism: 'Despite their being characterized or even stereotyped, as being atheoretical and descriptive, it is still important to note that there were theories lurking in this research'.
3. This period of the policy process is examined in Chapter 6.
4. Hay (2002: 113).
5. Hill (1997: 24) also suggests that the existence of secrecy in a policy process may mask the exercise of power, implying that power may be exercised in ways that may attract public concern. It is, therefore, unfortunate that Garner does not explore this hypothesis.
6. The present application of policy network analysis within a critical realist approach iteratively combines deductive and inductive reasoning, constantly relating theoretical notions to empirical observations (and vice versa) in order to optimise explanation. As Read and Marsh (2002: 234) observe, {AQ: 'I' in source?}'it is evident that all researchers use both inductive and deductive approaches in constructing explanations or developing understanding. In all research we move from ideas to data and from data to ideas'.
7. This conception implies a 'process' rather than 'arena' definition of politics (Hay, 2002: 3).
8. The only accessible public records at the National Archives relate to Home Office out-letters concerning the granting of licences between 1876 and 1921. (See page http://www.nationalarchives.gov.uk/catalogue/displaycatalogue details.asp?CATID=7711&CATLN=3&Highlight=&FullDetails=True#index, accessed 19 November 2005.)
9. In any case, the Home Office has operationalised the spirit of the FoI Act by publishing brief 'abstracts' of project licences that are written by the project licence holders themselves (Home Office, 2005: 4). Many such abstracts do not refer to the adverse effects experienced by animals, no primary documentation has been released, and no retrospective information has been published to allow an assessment of the actual costs and benefits of projects and a comparison with the information in the abstracts.
10. The use of the term 'tertiary' to categorise documentary sources largely follows Burnham et al.'s (2004: 165) usage: '"tertiary sources" [consist] of material written afterward to reconstruct the event … in the public domain'. The other two categories are 'primary' and 'secondary'. Primary sources refer to data which are part of, or produced by, the political event in question, while secondary sources relate to documents that were written and put into the public domain shortly after the event.
11. Uncaged Campaigns Ltd ('Uncaged').

12. Injunction Order of Master Price in the High Court of Justice, Chancery Division, dated 31 March 2003.
13. The term 'xenotransplantation' refers to procedures where *live* cells, tissues or organs are transplanted from one species of animal into another (McLean and Williamson, 2005: 41–4).
14. It was agreed that the following details, deemed irrelevant to the public interest issues, should be redacted from the documents: names of individuals and companies linked with the research, details of drugs and dosage details.
15. These documents are listed at Schedule 2 to the Injunction Order of Master Price in the High Court of Justice, Chancery Division, 31 March 2003. The agreed bundle comprises those documents published by Uncaged Campaigns (2003).
16. See beginning of clinical signs appendix for study ITN3: p. 16 at http://www.xenodiaries.org/studies.pdf. An abbreviated version of the same explanatory note also appears in the other study reports.

5 The 1876 Cruelty to Animals Act: Protection for Animals or Animal Researchers?

1. These resources comprise legal/authority (both formal and discretionary), economic/financial, political legitimacy (access to policy-makers, public opinion), information (especially control over its generation and distribution) and organisational (resources that enable a group to engage in direct policy-related action).
2. By far, the most detailed examination of the events surrounding the 1876 Act is provided by French's *Antivivisection and Medical Science in Victorian Society* (1975). In fact, French is referenced by almost all the other commentators discussed in this chapter, and another historian of vivisection, Rupke, describes French's book as an: 'exemplary study' (1987a: 3).
3. For example, the first principle suggested that alternatives to animal experiments should be employed if the information required can be obtained without using animals, and the fourth principle recommended the minimisation of suffering: '... compatible with the success of the experiment' (Hall, 1831, cited in Smith and Boyd, 1991: 249). The possible ongoing relevance of this belief system is highlighted by Paton (1993: 1–2), a Professor of Pharmacology and proponent of animal experimentation, when he asserts that Hall's principles 'remain the objectives today'.
4. The anaesthetic *ether* had been discovered in 1847 (Monamy, 2000: 21).
5. Interestingly, at the same time, Rupke postulates the persistence of this structure of beliefs in the scientific lobby (up until the publication of his study in 1987).
6. For example, the scientists' bill was supported by the philanthropic Lord Shaftesbury, who later played a major role in the anti-vivisection movement (French, 1975: 73).
7. See Table 2.1.
8. From the anti-vivisection side, although the RSPCA had not been directly involved in Henniker's Bill, the Society had played an essential role in putting the issue on to the political agenda, particularly through the Norwich

prosecution. Furthermore, it should be noted that Cobbe had beseeched the RSPCA to initiate legislation and it was the Society's cautious response that had persuaded Cobbe to bypass it. Moreover, the scientific response in the shape of the Playfair Bill had been anticipatory of RSPCA legislative activity.

9. There is a disagreement between the commentators over this issue. Contrary to French (whose discussion is much more meticulous), both Monamy (2000: 23) and Ryder (1989: 114) erroneously state that it banned painful procedures. The error might be explained by the fact that Carnarvon maintained that such certificates would be very rare, combined with Cobbe's incorrect account of the bill as containing such a prohibition (French, 1975: 115–17).
10. For example, the cabinet minister and future Prime Minister Lord Salisbury, supported the scientists and opposed the bill (Ryder, 1989: 116; French, 1975: 123–4).
11. One member of the GMC, George Rolleston, supported Carnarvon's bill.
12. The remaining clause that frustrated the vivisection lobby brought frogs under the purview of the Act by changing the scope of the proposed legislation from 'warm-blooded' animals to 'vertebrates'.
13. The following extract from an 1876 article in Nature that is quoted by French (1975: 119) gives a flavour of this position:And yet our leading politicians, in introducing the above quoted Bill, are bold enough to advance, as a motive for the legal machinery they are endeavouring to enforce, the idea that there is any real substantiality in the notion that the lengthening of human life can form any direct stimulation to physiological work. In so doing they show how little they are capable of appreciating the spirit of the higher philosopher, whose thoughts and temptations to investigate, however much they may be disguised by secondary motives, are but the involuntary secretion, as it may be termed, of his individual brain.
14. Derived from French (1975: 80–91) and comprising the London Anti-Vivisection Society, the Society for the Abolition of Vivisection (French (1975: 89) notes that this group was founded by George Jesse: 'a retired civil engineer of prodigious idiosyncracy. His Society ... never became more than a vehicle for one man's quixotic tilting at various vivisectionist windmills') and the International Association for the Total Suppression of Vivisection.
15. See Table 2.2 in Chapter 2.

6 The Evolution of the Animal Research Policy Network: 1876–1950

1. See next section for reference to their approach to this role.
2. Interestingly, Ryder (1983: 135) has opined that this is 'a comment highly pertinent to the modern situation'.
3. Unfortunately, there is no discussion in this account of the scientific validity of the experiments or the level of suffering experienced by the animals, but given the highly invasive and potentially damaging nature of the procedures, it is possible that the resultant suffering would have been severe, which appears to underline the leniency of the Home Office's response.

4. Richards himself is a physiologist and is clearly sympathetic to their professional interests and animal experimentation, so he cannot be accused of portraying an unfair picture of pro-vivisection ideology.
 5. Given Richards' background and perspective, his analysis could be taken as a further indication of professional opposition to the idea that animal experimentation should be subjected to political scrutiny.
 6. The Physiological Society had been founded the year before during the passage of the Act by leading vivisectionists.
 7. He could be prosecuted because he was not a licensee; the 1876 Act stipulated that the Home Secretary's permission was required to prosecute licensees.
 8. French appears to make this claim regarding decreased public concern on the basis of changes in the volume and tone of newspaper coverage of the issue.
 9. See 'Endogenous network factors' section of Chapter 2.
10. As related in previous chapter.
11. Most of literature focussing on policy change (e.g. Jordan and Greenaway, 1998) examines how it occurs in the context of apparently stable policy communities, through, for example, the accumulation of anomalous information that undermines hegemonic paradigms or ideologies and thus causes elite dissensus.
12. Kean (2003: 363) reports that due to ongoing attacks by medical students from University College London, Battersea Council were forced to allocate £700 per year to guard the brown dog statue. However, after the Conservatives took control of the Council in November 1909 the statue was quietly removed overnight in March 1910 on cost and political grounds.
13. Rogers was Secretary of the pro-vivisection lobby group that succeeded the AAMR, the Research Defence Society (RDS). In trying to portray the AC as balanced, Rogers emphasises the fact that AC members could not be licenceholders 'at the same time', but does not deny they may previously have been licensed vivisectors. If this were not the case it is unlikely that Rogers would have omitted to mention it. Unfortunately, no studies provide a specific analysis of the background of named advisors for this period, but given Rogers' wording in a self-proclaimed pro-vivisection propaganda vehicle, this seems a highly plausible contention.
14. Kean provides these references for this contemporary quotation: J. Howard Moore, *The Universal Kinship* (London, 1906): 329; Henry S. Salt, *Seventy Years among Savages*, (London, 1921): 133.)
15. At this time, the term referred to drugs used to treat infection by microorganisms rather than the modern usage which refers to cancer treatment.
16. Vaccines and sera differ from pharmaceuticals in that they contain some biological rather than purely chemical material.

7 The Animals (Scientific Procedures) Act 1986: Emergence of an Issue Network or Policy Community Dynamic Conservatism?

1. The RCVS and the British Veterinary Association (BVA).
2. The Home Office issued licences to unqualified persons to perform relatively routine procedures if an inspector considered that the applicant 'both under-

stands the significance of the experiments described in his application and can carry out the required manipulations' (Littlewood, 1965: 32).
3. See Chapter 6.
4. '... that painful experiments should not be authorised save for recognisably worthwhile purposes'.
5. 'Many may think that [the AC] is a body that frequently meets, whereas being advisory it only meets when its advice is sought.... Thus apart from meeting to discuss the matters now being considered the Committee has only been summoned twice in about the last four years.' (Littlewood, 1965: 211)
6. The Littlewood Report (1965: 79) expressed these three questions in the following terms:(a) Who can say whether, if certain biological tests were forbidden, satisfactory chemical or other methods of testing would not be developed?(b) Who is responsible for establishing whether modern techniques, with their emphasis on immunology and drug therapy, both of which are inseparable from animal experimentation, are developing medical practice in the right direction?(c) Who is to take responsibility for moral or ethical judgment in the use of animals for experimental purposes as such?
7. The animal protection lobby also called for tighter regulation of the supply and breeding of animals for research, but consideration of this area is omitted due to space constraints and, in any case, the debate over research practices is sufficient to represent the essential aspects of the policy process.
8. Specifically the SSPV.
9. 'We think that it has been a most valuable part of Home Office supervision that inspectors have used their opportunities to question and examine such proposals for research as have seemed to them to be not self-evidently justifiable' (1965: 121).
10. On the other hand, experiments where pain was not expected to ensue required a briefer application with only one sponsor rather than two, and could be presumed to be authorised unless prohibited by an inspector within fourteen days because he or she believed that the research was likely to cause pain. In which case, the researcher could re-apply under the route for painful experiments.
11. For the main non-experimental research usage of animals – the production and testing of biological products – the Littlewood Report recommended no further restrictions beyond the requirement that the bodies formulating such tests should consult the Home Office and the AC in the formulation of mandatory tests.
12. Since merged with the Cancer Research Campaign to form Cancer Research UK.
13. The RDS submission incorporated many other research interest groups, as discussed earlier in the subsection 'Group Actors'.
14. Though one could reasonably speculate that scientific witnesses argued that groups with a critical stance towards animal experimentation would disrupt the network, which in itself would indicate the institutionalised nature of the pro-vivisection ideology in the policy community.
15. Garner (1993: 126) states: 'No fundamental changes were envisaged'.
16. 'It is clear to us that there has been an appearance of secrecy about the practice of animal experimentation in the past. The public has little information to go on except the Home Office Annual Return ... [which was] described by

many lay witnesses as incomprehensible to the average reader' (Littlewood, 1965: 164). Groups complained about a lack of information concerning the broad categories of experiments, the numbers of animals used in each type of test and the actual procedures and their effects.
17. Expert knowledge is particularly favoured in highly technical policy areas (Smith, 1993b: 81).
18. As demonstrated by the Home Office presumption of compliance in the absence of contrary evidence.
19. See policy network dynamics Table 2.2.
20. See Chapter 2.
21. These participants were the Committee for the Reform of Animal Experimentation (CRAE), which composed a small number of leading animal welfare figures, and subsequently FRAME and the BVA.
22. Godlovitch, S., Godlovitch, R. and Harris, J. (eds) (1971) *Animals, Men and Morals: An Enquiry into the Maltreatment of Non-humans*. London: Gollancz.
23. This is an edited version of Singer's original review in the *New York Review of Books* in 1973.
24. These ethical challenges in turn stimulated a counter critique that sought to defend a privileged moral status for humans (e.g. Leahy, 1994; Carruthers, 1992) and/or the practices deemed beneficial to humans that involve harm to animals (Frey, 1980).
25. See Table 2.2 on policy network dynamics
26. The pro-animal research position, articulated by Paton (1984: 154–5), while defending such experiments, acknowledged that such research was particularly controversial because of the animal protection point. This book by Sir William Paton, at the time Professor of Pharmacology at Oxford University, is widely considered to be the most significant statement of the pro-animal research argument (Smith and Boyd, 1991: 26, 28; Monamy, 2000: 71; Rupke, 1987a: 2).
27. One might speculate that the participation of FRAME and the BVA in the CRAE Alliance introduced a trusting attitude towards scientists.
28. Professor Patrick Wall, University College London.

8 Imutran Xenotransplantation Research Case Study

1. Imutran was a Cambridge-based biotechnology company founded in 1984. In 1996, it was acquired by the Swiss pharmaceutical company Sandoz, who in 1997 merged with Ciba-Geigy to form Novartis Pharma. Imutran was closed by Novartis in autumn 2000.
2. David Mellor MP.
3. Robin Corbett MP.
4. Straughan (1995: 47) observes that by 1992, the size of the Inspectorate (twenty-one) relative to the scale of animal research not only remained below the minimum recommendation made by the Littlewood Report in 1965, but this perceived under-staffing was exacerbated by the fact that the 1986 Act formally required more intensive regulatory activity. By 1997, staffing levels had dropped to 18 inspectors.
5. One study did not involve xenograft surgery, but investigated the effectiveness and toxicity of a combination of immunosuppressive drugs intended for use in conjunction with xenotransplantation procedures.

6. i.e. the original 2000 edition of Lyons (2003).
7. It is unclear whether the personal licence condition adequately transposed the 1986 EU Directive, as it allowed animals to experience severe pain, albeit for a minimum period of time, whereas the 1986 Directive prohibited it altogether.
8. A 'heterotopic' transplant refers to a transplant that has been placed in an abnormal anatomical position.
9. An 'orthotopic' transplant refers to a transplant that has been placed in the normal anatomical position.
10. The distant evolutionary relationship between pigs and humans leads to an extremely rapid and virulent immune response to pig-to-human transplants called *hyperacute* rejection. HAR cannot be controlled with immunosuppressant drugs (Kennedy et al., 1997: 21).
11. The application also sought permission for concordant xenotransplant experiments – cynomolgus monkeys to baboons – as a model for possible baboon-to-human organ transplants (ND1.7, 1.9). This research was not pursued beyond an initial study (see ND2), probably due to the practical difficulties in supplying sufficient numbers of primate organs for transplants (ND5.3) and concerns about potentially lethal viruses in baboons (Concar, 1994: 28).
12. Allotransplantation refers to human-to-human transplantation.
13. Although this study was carried out after the commencement of Imutran's research, it was a relatively simple task that could have been undertaken beforehand.
14. This is consistent with the Home Office's (2003: 5) retrospective account of its moderate severity assessment derivations that overlooked the adverse effects of the immunosuppressive regimes due to the same false assumption.
15. The Imutran witness statement claimed, 'The materials in question are misappropriated and include a range of documents which would not in the ordinary way be required by or provided to official bodies'.
16. It should be noted here that detailed records exist for all animals that shed further light on their condition, but remain confidential under the terms of the High Court Order. These records include important data such as: dosing regimes; bodyweight; food consumption; haematology; biochemistry; comparative donor organ weights between transplant and necropsy. No independent body has yet had the opportunity to examine these records.
17. Indeed, a lobbyist acquaintance of the author was present at a conference where Imutran researchers' highlighted this conflict of interest.
18. Due to size considerations, Imutran believed that orthotopic procedures should be conducted in baboons rather than cynomolgus monkeys. However, Imutran now acknowledged that immunological differences between these two species meant that before the orthotopic procedures could take place, 'a small number' of heterotopic experiments in baboons were necessary to ensure that their immune response to hDAF organs was substantially similar to cynomolgus monkeys (ND5.7, ND1.52).
19. This has been suggested to the writer informally.
20. These are the first five procedures under ITN9.
21. W211m is 'Animal 5' in this report.
22. These procedures were licensed as Protocol 19b10 of Project Licence PPL 80/848: 'heterotopic abdominal heart xenografting'

23. However, the licence was not formally suspended under section 13 of the 1986 Act. This will be discussed near the end of the chapter in the context of the Home Office's response to allegations of regulatory breaches and maladministration.
24. These possibilities are envisaged by Standard Condition 18 on personal licenses, paragraph 3.2(iv) of record-keeping requirements for project licence holders and Standard Condition 15(iii) on project licenses (Home Office, 1990: 55, 52, 50).
25. A simultaneous depletion of red and white blood cells and platelets.
26. These procedures were performed by Imutran under study numbers ITN4, ITN12, ITN13, ITN16, ITN18, ITN21, ITN26, IAN001, IAN001, IAN002, IAN004, IAN005, IAN008, IAN010, IAN013, IAN017, IAN018, IAN020 and procedures at the beginning of the last study report, IAN022 (these studies are those for which reports and clinical signs are available) between July 1995 and when this licence elapsed in April 1999. The final few procedures were performed from April 1999 under project licence number PPL 80/1366 (ND24), and formed the latter procedures in study IAN022. In addition to the nineteen reported studies, at least four more took place for which final draft reports have not become available. Leaked internal communication between Imutran and Novartis suggests that such studies continued until approximately February 2000 (Lyons, 2003: 84).
27. These are supposed to specify the worst-case scenario, seen in exceptional circumstances (Home Office, 1990: 10–11).
28. The PHSO's draft decision letter dated 4 May 2005 stated: '[T]he investigation of the complaint has been interrupted on more than one occasion by illness, which has meant that the case has had to be considered by several different officers, which has led to significant delays' (paragraph 2).
29. http://www.publications.parliament.uk/pa/cm201011/cmselect/cmpubadm/781/11020902.htm accessed on 29 November 2011.
30. See following chapter though for indications that some policy learning has taken place that appears to reinforce the existing policy community.

9 Conclusion: The Power Distribution in British Animal Research Politics

1. In 1998, a small number of definitive policy 'bans' were introduced relating to categories of experiment that no longer occurred or represented a tiny fraction of the practice: for cosmetics products or the development of military weapons and on great apes.
2. http://www.homeoffice.gov.uk/publications/science-research-statistics/research-statistics/other-science-research/spanimals11/spanimals11?view=Binary (accessed 2 January 2013)
3. 2000 breakdown: 39 per cent mild, 55 per cent moderate, 2 per cent substantial, 4 per cent unclassified. 2011 breakdown: 36 per ent mild, 61 per cent moderate, 2 per cent substantial, 2 per cent unclassified.
4. Available from http://eur-lex.europa.eu/LexUriServ/LexUriServ.do?uri=OJ:L:2010:276:0033:0079:EN:PDF (accessed 16 January 2012)

5. UK legislation has been updated by Statutory Instrument: SI 2012/3039 The Animals (Scientific Procedures) Act Amendment Regulations 2012. A consolidated version of the new regulations is available at http://www.homeoffice.gov.uk/publications/science-research-statistics/animals/transposition_of_Eudirective/consolidated_aspa?view=Binary (accessed 3 January 2013)
6. E.g. Parliamentary Under-Secretary of State, Home Office, Lord Taylor of Holbeach, Hansard, 13 December 2012, Column GC380. http://www.publications.parliament.uk/pa/ld201213/ldhansrd/text/121213-gc0001.htm#12121339000271 (accessed 5 January 2013)
7. A consolidated version of the new regulations is at http://www.homeoffice.gov.uk/publications/science-research-statistics/animals/transposition_of_Eudirective/consolidated_aspa?view=Binary (accessed 5 January 2013)
8. The Home Office also announced on 6 November 1997 that experiments on great apes would not be licensed as a matter of policy, although no such procedures had been licensed under the 1986 Act (House of Lords Select Committee on Animals in Scientific Procedures, 2002: 9).
9. The role of story-lines in providing a simplified version of reality to act as a guide to action has interesting parallels with the concept of 'bounded rationality' (though story-lines have a less rationalistic basis) which, according to Daugbjerg and Marsh (1998: 68), has been introduced into rational choice theory and potentially helps to explain individual action in policy networks.

Bibliography

Abraham, J. (1995) *Science, Politics and the Pharmaceutical Industry*. London: Routledge.
Anon. (1998) 'Affairs of the heart'. *The Economist*, 4 April 1998: 106.
Anon. (2002) 'Waiting for a miracle – Time is running out for organ transplants from animals'. *New Scientist*, 12 January 2002: 3.
APC (1996) *Report of the Animal Procedures Committee for 1995*. London: The Stationery Office.
APC (1997) *Report of the Animal Procedures Committee for 1996 (Cm 3777)*. London: The Stationery Office.
APC (1998) *Report of the Animal Procedures Committee for 1997*. London: The Stationery Office.
APC (1999) *Report of the Animal Procedures Committee for 1998*. London: The Stationery Office.
APC (2000a) *Report of the Animal Procedures Committee for 1999*. London: The Stationery Office.
APC (2000b) *Minutes for April 2000 Meeting*. London: APC. http://www.apc.gov.uk/reference/apr00.htm (accessed 26 June 2006).
APC (2001a) *Letter from Chairman of APC to Mike O'Brien MP, Parliamentary Under Secretary of State, Home Office*, 20 March 2001. Annex to http://www.apc.gov.uk/reference/feb01.htm (accessed 28 June 2006).
APC (2001b) *Minutes for October 2001 Meeting*. London: APC. http://www.apc.gov.uk/reference/oct01.htm (accessed 26 June 2006).
APC (2002) *Report of the Animal Procedures Committee for 2001*. London: The Stationery Office.
APC (2003) *Review of Cost-Benefit Assessment in the Use of Animals in Research*. London: Home Office.
APC (2005) *Report of the Animal Procedures Committee for 2004*. London: The Stationery Office.
APC (2006) *Report of the Animal Procedures Committee for 2005*. London: The Stationery Office.
Appleyard, B. (2006) 'The animals, the madmen, and me'. *The Sunday Times*, 5 March 2006.
Balls, E.M. (1989) 'The Moral Status of Animals and the Animals (Scientific Procedures) Act 1986'. *Alternatives to Laboratory Animals*, 16: 353–7.
Balls, M. (1986) 'Animals (Scientific Procedures) Act 1986: The Animal Procedures Committee'. *Alternatives to Laboratory Animals*, 14: 6–13.
Bean, D. and Afeeva, M. (2000) *Defence*, Imutran Ltd v. Uncaged Campaigns Ltd & Daniel Louis Lyons, No. HC0004406 (High Court of Justice, Chancery Division).
Beck, U. (1992) *Risk Society*. London: Sage.
Bevir, M. and Rhodes, R. (1999) 'Studying British government: reconstructing the research agenda'. *British Journal of Politics and International Relations*, 1 (2): 215–39.

Bevir, M and Richards, D. (2009) 'Decentring policy networks: a theoretical agenda'. *Public Administration*, 87 (1): 3–14.
Bharati, S. et al. (1991) 'The conduction system of the swine heart', *Chest*, 100(1): 207–12.
Bhatti, F.N.K. et al. (1999) 'Three-month survival of HDAFF transgenic pig hearts transplanted into primates'. *Transplantation Proceedings*, 31: 958.
Birch, A. (1964) *Representative and Responsible Government*. London: Allen & Unwin.
Blanco, I., Lowndes, V. and Pratchett, L. (2011) 'Policy networks and governance networks: towards greater conceptual clarity'. *Political Studies Review*, 9: 297–308.
Bomberg, E. (1998) 'Issue networks and the environment: explaining European Union environmental policy'. In: Marsh, D. (ed.) *Comparing Policy Networks*, 167–84. Buckingham: Open University Press.
Brooman, S. and Legge, D. (1997) *Law Relating to Animals*. London: Cavendish.
Bulkeley, H. (2000) 'Discourse coalitions and the Australian climate change policy network'. *Environment and Planning C: Government and Policy*, 18, 727–48.
Burnham, A. (2006) Written Answer to PQ 55454, 6 March 2006, Columns 1131W–1132W, *Hansard* (HOC). London: HMSO.
Burnham, P. et al. (2004) *Research Methods in Politics*. Basingstoke: Palgrave Macmillan.
Butler, D. (1998) 'Last chance to stop and think on risks of xenotransplants'. *Nature*, 391: 320–4.
Carruthers, P. (1992) *The Animals Issue: Moral Theory in Practice*. Cambridge: Cambridge University Press.
Cavanagh, M. (1998) 'Offshore health and safety policy in the North Sea: policy networks and policy outcomes in Britain and Norway'. In: Marsh, D. (ed.) *Comparing Policy Networks*, 90–109. Buckingham: Open University Press.
Compston, H. (2009) *Policy Networks and Policy Change*. Basingstoke: Palgrave Macmillan.
Concar, D. (1994) 'The organ factory of the future?' *New Scientist*, 18 June 1994: 24–9.
Concar, D. (2002) 'Gagged and bound'. *New Scientist*, 16 March 2002.
Crick, S.Y. et al. (1998) 'Anatomy of the pig heart: Comparisons with normal human cardiac structure'. *Journal of Anatomy*, 193 (1): 105–19.
Dark, J. (2001) *Presentation to UKXIRA Open Meeting, Wed 7 Feb 2001* (copy of transcript at Appendix I, taken from recording supplied by UKXIRA which is available from author).
Daugbjerg, C. (1998) 'Similar problems, different policies: policy network and environmental policy in Danish and Swedish agriculture'. In: Marsh, D. (ed.) *Comparing Policy Networks*, 75–89. Buckingham: Open University Press.
Daugbjerg, C. and Marsh, D. (1998) 'Explaining policy outcomes: integrating the policy network approach with macro-level and micro-level analysis'. In: Marsh, D. (ed.) *Comparing Policy Networks*, 52–71. Buckingham: Open University Press.
Department of Health (1999) *Good Laboratory Practice Regulations 1999 (S.I. 3106)*. London: The Stationery Office.
Devine, F. (2002) 'Qualitative Methods'. In: Marsh, D. and Stoker, G. (eds) *Theory and Methods in Political Science*, 2nd ed. Basingstoke: Palgrave Macmillan.
Dickson, D. (1995) 'Pig heart transplant 'breakthrough' stirs debate over timing of trials'. *Nature*, 377, 21 September 1995: 185.

Dolowitz, D, and Marsh, D. (1996) 'Who learns what from whom: a review of the policy transfer literature'. *Political Studies*, 44: 343–57.
Dunleavy, P. and O'Leary, B (1987) *Theories of the State: The Politics of Liberal Democracy*. Basingstoke: Macmillan Education.
Edelman, M. (2001) *The Politics of Misinformation*. Cambridge: Cambridge University Press.
Ekern, L. (2004) 'Ulovlige Dyreforsok Oppdaget' ['Illegal Animal Research Discovered']. *Forskningsetikk [Research Ethic]*, 18–20. Oslo: Norwegian National Research Ethics Committee ['Nasjonale Forskningsetiske Komiteer']
Elston M.A. (1987) 'Women and anti-vivisection in Victorian England'. In: Rupke, N. A. (ed.) *Vivisection in Historical Perspective*, pp. 259–94. London: Croom Helm
Featherstone, L. (2011) Written Answer to PQ 78306, 8 November 2011, Column 167W–168W, *Hansard* (HOC Deb). London: HMSO.
Food and Drug Administration (1999) *Transcript of Xenotransplantation Subcommittee 4 June 1999*. Washington D.C.: Department of Health and Human Services, http://www.fda.gov/ohrms/dockets/ac/99/transcpt/3517t2a.pdf; http://www.fda.gov/ohrms/dockets/ac/99/transcpt/3517t2b.pdf; http://www.fda.gov/ohrms/dockets/ac/99/transcpt/3517t2c.pdf (accessed on 26 June 2006).
FRAME Trustees (1996) 'The first ten years of the Animals (Scientific Procedures) Act 1986'. *ATLA*, 24: 639–47.
French, R.D. (1975) *Antivivisection and Medical Science in Victorian Society*. Princeton: Princeton University Press.
Frey, R. (1980) *Interests and Rights*. Oxford: Oxford University Press.
Garner, R. (1993) *Animals, Politics and Morality*. Manchester: Manchester University Press.
Garner, R. (ed.) (1996) *Animal Rights: The Changing Debate*. Basingstoke: Macmillan.
Garner, R. (1998) *Political Animals: Animal Protection Politics in Britain and the United States*. Basingstoke: Macmillan.
Garner, R. (2002) 'Political science and animal studies'. *Society and Animals*, 10 (4): 395–401.
Grant, W. (1995) *Pressure Groups, Politics and Democracy in Britain*. Hemel Hempstead: Harvester Wheatsheaf.
Grant, W. (2000) *Pressure Groups and British Politics*. Basingstoke: Macmillan.
Grant, W. (2008) 'The changing patterns of group politics in Britain'. *British Politics*, 3: 204–22.
Greer, A. (2002) 'Policy networks and policy change in organic agriculture: a comparative analysis of the UK and Ireland'. *Public Administration*, 80 (3): 453–3.
Hall, P. (1993) 'Policy Paradigms, Social Learning and the State'. *Comparative Politics*, 25 (3): 275–96.
Hajer, M.A. (1995) *The Politics of Environmental Discourse: Ecological Modernization and the Policy Process*. Oxford: OUP.
Hammer, C. (1991) 'Evolutionary, physiological, and immunological considerations in defining a suitable donor for man'. In: Cooper, D. et al., *Xenotransplantation – The Transplantation of Organs and Tissues Between Species*, 429–38. Springer-Verlag: London and Heidelberg. Cited in Langley and D'Silva (1998).
Hammer, C. (1994) 'Xenotransplantation and its future'. *Forensic Sci Int*, 69 (3): 259–68.

Hammer, C. (2003) 'Xenotransplantation: the good, the bad, and the ugly or how far are we to clinical application'. *Transplant Proceedings*, 35 (3): 1256–7.

Hampson, J. (1987) 'Legislation: a practical solution to the vivisection dilemma'. In: Rupke, N.A. (ed.) *Vivisection in Historical Perspective*, 314–39. London: Croom Helm.

Hampson, J. (1989) 'Legislation and the Changing Consensus'. In: Langley, G. (ed.) *Animal Experimentation: The Consensus Changes*, 219–51. Basingstoke: Macmillan.

Hay, C. (1995) 'Structure and agency'. In: Marsh, D. and Stoker, G. (eds) *Theory and Methods in Political Science*, 1st ed. Basingstoke: Macmillan.

Hay, C. (1998) 'The tangled webs we weave: the discourse, strategy and practice of networking'. In: Marsh, D. (ed.) *Comparing Policy Networks*, 33–51. Buckingham: Open University Press.

Hay, C. (2002) *Political Analysis*. Basingstoke: Palgrave.

Hay, C. and Richards, D. (2000) 'The tangled webs of Westminster and Whitehall: the discourse, strategy and practice of networking within the British core executive'. *Public Administration*, 78: 1–28.

Hay, C. and Wincott, D. (1998) 'Structure, agency and historical institutionalism'. *Political Studies*, 46 (4): 951–7.

Hill, M. (1997) *The Policy Process in the Modern State*. Hemel Hempstead: Prentice Hall/Harvester Wheatsheaf.

Hinde, J. (1999) 'Science faces troubled future'. *The Times Higher Education Supplement*, 30 April 1999: 4.

Hindmoor, A. (2009) 'Explaining networks through mechanisms: vaccination, priming and the 2001 foot and mouth disease crisis'. *Political Studies*, 57: 75–94.

HM Government (2010) *The Coalition: Our Programme for Government*. London: Cabinet Office. http://www.direct.gov.uk/prod_consum_dg/groups/dg_digitalassets/@dg/@en/documents/digitalasset/dg_187876.pdf (accessed 16 January 2012).

Hollands, C. (1989) 'Trivial and questionable research on animals'. In: Langley, G. (ed.) *Animal Experimentation: The Consensus Changes*, 118–43. Basingstoke: Macmillan.

Hollands, C. (1995) 'Achieving the achievable: a review of animals in politics'. *Alternatives to Laboratory Animals*, 23: 33–38.

Home Office (1990) *Guidance on the Operation of the Animals (Scientific Procedures) Act 1986*. London: HMSO.

Home Office (1998a) 'Cost-benefit assessment: a note by the chief inspector'. In: APC (1998) *Report of the Animal Procedures Committee for 1997*, 50–59. London: The Stationery Office.

Home Office (1998b) 'Assessment of benefit and severity: a 1993 note by the then chief inspector'. In: APC (1998) *Report of the Animal Procedures Committee for 1997*, 60–62. London: The Stationery Office.

Home Office (2000) *Guidance on the Operation of the Animals (Scientific Procedures) Act 1986*. London: The Stationery Office. http://www.official-documents.gov.uk/document/hc9900/hc03/0321/0321.pdf (accessed 5 January 2013).

Home Office (2001) *Imutran Ltd: Compliance With Authorities Issued Under the Animals (Scientific Procedures) Act 1986*. London: Home Office. http://scienceandresearch.

homeoffice.gov.uk/animal-research/publications/publications/reports-and-reviews/imutranreport.pdf?view=Binary (accessed 21 June 2006).
Home Office (2002) *Aspects of Non-human Primate Research at Cambridge University, A Review by the Chief Inspector (October 2002)*. London: Home Office. http://scienceandresearch.homeoffice.gov.uk/animal-research/documents/chief_insp_animals_review.pdf?view=Binary (accessed 26 November 2005).
Home Office (2003) *Animals (Scientific Procedures): Imutran Xenotransplantation Research*. London: Home Office. http://scienceandresearch.homeoffice.gov.uk/animal-research/publications/publications/reports-and-reviews/horesponseimutranjun2003.pdf?view=Binary (accessed 21 June 2006).
Home Office (2004) *Home Office Response to Twelve Questions Raised by Dan Lyons, Uncaged Campaigns, About Imutran Xenotransplantation Research Licensed Under the Animals (Scientific Procedures) Act 1986*. London: Home Office. (Copy of this document with covering letter from Home Office at Appendix X to this thesis).
Home Office (2005) *Animals (Scientific Procedures) Inspectorate Annual Report 2004*. London: Home Office. http://scienceandresearch.homeoffice.gov.uk/animal-research/publications/publications/reports-and-reviews/annual-report?view=Binary (accessed 16 January 2006).
Home Office (2006) *Statistics of Scientific Procedures on Living Animals: Great Britain 2005*. London: HMSO.
Home Office (2011) *Animals Scientific Procedures Division and Inspectorate – Annual Report 2010*. London: Home Office. http://www.homeoffice.gov.uk/publications/science-research-statistics/769901/annual-reports/animals-annual-report-2010?view=Binary (accessed 24 November 2011).
Home Office (2012) *ASPA Draft Guidance*. Published at http://www.homeoffice.gov.uk/publications/science-research-statistics/animals/transposition_of_Eudirective/quick_start_guide?view=Binary (accessed 5 January 2013).
Hopley, E. (1998) *Campaigning Against Cruelty: The Hundred Year History of the British Union for the Abolition of Vivisection*. London: British Union for the Abolition of Vivisection.
House of Lords Select Committee on Animals in Scientific Procedures (2002) *Volume I – Report*. London: HMSO.
Hu, K-C. (1995) *Policy Networks in Democratic and Authoritarian Regimes* (University of Sheffield, Unpublished PhD).
Hughes, C. (2000) 'Drug firms confident of faster approval for tests on animals'. *The Independent*, 4 October 2000, 16.
Imutran (undated) *Imutran*. Cambridge: Imutran.
Imutran (2000) *Second Witness Statement in Imutran v Uncaged Campaigns and Anr* [Case No. HC 0004406], 11 October 2000. London, High Court of Justice, Chancery Division.
Jenkins-Smith, H. C. and Sabatier, P.A. (1994) 'Evaluating the advocacy coalition framework', *Journal of Public Policy*, 14 (2): 175–203.
Jennings, M. et al. (2002) *RSPCA Report: Non-Human Primates in Xenotransplantation Research in the UK*. Horsham: RSPCA.
John, P. (1998) *Analysing Public Policy*. London: Pinter.
Johnston, L. and Calvert, J. (2000a) 'Terrible despair of animals cut up in name of research'. *Daily Express*, 21 September 2000.

Johnston, L. and Calvert, J. (2000b) 'Animal tests probe after we expose suffering'. *Daily Express*, 22 September 2000.

Johnston, L. and Calvert, J. (2000c) 'Animal lab shuts down after we reveal horrors'. *Daily Express*, 28 September 2000.

Jordan, A. and Greenaway, J. (1998) 'Shifting agendas, changing regulatory structures and the "New" politics of environmental pollution: British coastal water policy, 1955–1995'. *Public Administration*, 76: 669–94.

Jordan, A. (2002) *The Europeanization of British Environmental Policy: A Departmental Perspective*. Basingstoke: Palgrave Macmillan.

Jordan, G. and Richardson, J. J. (1987) *Government and Pressure Groups in Britain*. Oxford: Clarendon Press.

Judge, D. (2004) 'Whatever happened to parliamentary democracy in the United Kingdom?' *Parliamentary Affairs*, 57 (3): 682–701.

Kean, H. (1998) *Animal Rights: Political and Social Change in Britain Since 1800*. London: Reaktion Books.

Kean, H. (2003) 'An Exploration of the Sculptures of Greyfriars Bobby, Edinburgh, Scotland, and the Brown Dog, Battersea, South London, England'. *Society & Animals* 1(4): 353–73.

Kennedy, I. [the Advisory Group on the Ethics of Xenotransplantation] (1997) *Animal Tissue into Humans*. London: Department of Health & HMSO.

Langley, G. (1989) 'Plea for a Sensitive Science'. In: Langley, G. (ed.) *Animal Experimentation: The Consensus Changes*, 193–218. Basingstoke: Macmillan.

Langley, G. and D'Silva, J. (1998) *Animal Organs in Humans*. London and Petersfield: British Union for the Abolition of Vivisection and Compassion in World Farming.

Leahy, M. (1994) *Against Liberation: Putting Animals into Perspective*. London: Routledge.

Lindblom, C. and Woodhouse E. (1993) *The Policy-making Process*. Upper Saddle River, N.J.: Prentice Hall.

Littlewood, S. (1965) *Report of the Departmental Committee on Experiments on Animals [Cmnd. 2641, 'The Littlewood Report']*. London: HMSO.

Lowndes, V. (2002) 'Institutionalism'. In: Marsh, D. and Stoker, G. (eds) *Theories and Methods in Political Science*, 90–108. 2nd ed. Basingstoke: Palgrave Macmillan.

Lyons, D. (2003) *Diaries of Despair* [redacted 2nd ed.]. Sheffield: Uncaged Campaigns Ltd. http://www.xenodiaries.org/report.pdf.

McAnulla, S. (2002) 'Structure and Agency'. In: Marsh, D. and Stoker, G. (eds) *Theories and Methods in Political Science*, 271–91. 2nd ed. Basingstoke: Palgrave Macmillan.

MacDonald, M. (1994) *Caught in the Act: The Feldberg Investigation*. Oxford: John Carpenter.

Maehle, A. and Trohler, U. (1997). In: Rupke, N. A. (ed.), *Vivisection in Historical Perspective*, pp. 14–47. London: Croom Helm.

McLean, S.A.M. and Williamson, L. (2005) *Xenotransplantation: Law and Ethics*. Aldershot: Ashgate.

Marsh, D. (1998a) 'The development of the policy network approach'. In: Marsh, D. (ed.) *Comparing Policy Networks*, 3–17. Buckingham: Open University Press.

Marsh, D. (1998b) 'The utility and future of policy network analysis'. In: Marsh, D. (ed.) *Comparing Policy Networks*, 185–97. Buckingham: Open University Press.

Marsh, D., Richards, D. and Smith, M. (2001) *Changing Patterns of Governance in the United Kingdom: Reinventing Whitehall?* London: Macmillan.
Marsh, D. (2002) 'Pluralism and the study of British politics: it is always the happy hour for men with money, knowledge and power'. In: Hay, C. (ed.) *British Politics Today*, 14–37. Cambridge: Polity.
Marsh, D., Richards, D. and Smith, M. (2003) 'Unequal plurality: towards an asymmetric power model of British politics'. *Government and Opposition*, 38 (3): 306–32.
Marsh, D. (2008) 'Understanding British government: analysing competing models'. *British Journal of Politics and International Relations*, 10: 251–68.
Marsh, D. and Furlong, P. (2002) 'A skin, not a sweater: ontology and epistemology in political science'. *In*: Marsh, D. and Stoker, G. (eds) *Theory and Methods in Political Science*, 2nd ed. Basingstoke: Palgrave Macmillan.
Marsh, D. and Rhodes, R.A.W. (1992a) 'The implementation gap: explaining policy change and continuity'. In: Marsh, D. and Rhodes, R.A.W. (eds) *Implementing Thatcherite Policies: Audit of an Era*, 170–87. Buckingham: Open University Press.
Marsh, D. and Rhodes, R.A.W. (1992b) 'Policy communities and issue networks: beyond typology'. In: Marsh, D. and Rhodes, R.A.W. (eds) *Policy Networks in British Government*, 249–68. Oxford: Oxford University Press.
Marsh, D. and Smith, M. (2000) 'Understanding policy networks: towards a dialectical approach'. *Political Studies*, 48: 4–21.
Marsh, D. and Smith, M. (2001) 'There is more than one way to do political science: on different ways to study policy networks'. *Political Studies*, 49: 528–41.
Marsh, D. and Stoker, G. (1995) 'Conclusions'. In: Marsh, D. and Stoker, G. (eds) *Theory and Methods in Political Science*, 1st ed. Basingstoke: Macmillan.
Marsh, D. and Stoker, G. (2002) 'Conclusion'. In: Marsh, D. and Stoker, G. (eds) *Theory and Methods in Political Science*, 2nd ed. Basingstoke: Palgrave Macmillan.
Martin, E.A. (ed.) (1996) *Concise Medical Dictionary*, 4th ed. Oxford: Oxford University Press.
Matfield, M. (1992) 'Animal research within an ethical framework'. *Alternatives to Laboratory Animals*, 20: 334–37.
Monamy, V. (2000) *Animal Experimentation: A Guide to the Issues*. Cambridge: Cambridge University Press.
Moore, S.D. (1997) 'Novartis picks up pace in xenotransplant race'. *Wall Street Journal (Europe)*, 10 October 1997: 4.
Moran, M. (2003) *The British Regulatory State: From Stagnation to Hyperinnovation*. Oxford: Oxford University Press.
Moran, N. (1995) 'Pig-to-human heart transplant slated to begin in 1996'. *Nature Medicine*, 1 (10): 987.
Nicholson, R. (1997) 'If pigs could fly'. *Nursing Times*, 3 February 1997: 20.
Nuffield Council on Bioethics (1996) *Animal-to-Human Transplants: The Ethics of Xenotransplantation*. London: Nuffield Council on Bioethics.
Nuffield Council on Bioethics (2005) *The Ethics of Research Involving Animals*. Nuffield Council on Bioethics: London.
O'Brien, M. (2000a) Written Answer to PQ 125262, 28 June 2000, Columns 520W–523W, *Hansard* (HOC). London: HMSO.

O'Brien, M. (2000b) *Letter from Mike O'Brien MP, Parliamentary Under Secretary of State, Home Office, to Dan Lyons, Uncaged*, 29 September 2000.

O'Brien, M. (2000c) Written Answer to PQ 136225, 1 November 2000, Column 517W, *Hansard* (HOC). London: HMSO.

O'Riordan, T. and Jordan, A. (1996) 'Social institutions and climate change'. In: O'Riordan, T. and Jager, J. (eds) *Politics of Climate Change: A European Perspective*, 65–105. London: Routledge.

Orlans, F. (1993) *In the Name of Science: Issues in Responsible Animal Experimentation*. New York: Oxford University Press.

Parliamentary and Health Service Ombudsman (PHSO) (2006) *Investigation into the Home Office's Regulation of Animal Experimentation*. London: The Stationery Office [accessed from http://www.ombudsman.org.uk/__data/assets/pdf_file/0007/3877/Home-Office-regulation-of-animal-experimentation.pdf on 29 November 2011]

Parsons, W. (1995) *Public Policy*. London: Edward Elgar.

Paton, W. (1993) *Man and Mouse: Animals in Medical Research*. 2nd ed. Oxford: Oxford University Press.

Peters, B.G. (1999) *Institutional Theory in Political Science: The 'New Institutionalism'*. London: Pinter.

Pharmaceutical Industry Competitiveness Task Force (2001) *Final Report 2001*. London: Department of Health and the Association of the British Pharmaceutical Industry (http://www.advisorybodies.doh.gov.uk/pictf/pictf.pdf, accessed 26th July 2006).

Pierre, J. and Peters, B.G. (2000) *Governance, Politics and the State*. Basingstoke: Macmillan.

Platt, J. (1998) 'New Directions for organ transplantation'. *Nature*, 392 (Supp): 11–17.

Poulsen. K., Haber, E. and Burton, J. (1976) 'On the specificity of human renin. Studies with peptide inhibitors'. *Biochim. Biophys. Acta*, 452: 533–37. Cited in Langley and D'Silva (1998).

Radford, M. (2001) *Animal Welfare Law in Britain: Regulation and Responsibility*. Oxford: Oxford University Press.

Read, M. and Marsh, D. (2002) 'Combining quantitative and qualitative methods'. In: Marsh, D. and Stoker, G. (eds) *Theory and Methods in Political Science*, 2nd ed. Basingstoke: Palgrave Macmillan.

Regan, T. and Singer, P. (eds) (1976) *Animal Rights and Human Obligations*. Englewood Cliffs, NJ: Prentice Hall.

Rhodes, R.A.W. (1990) 'Policy networks: a British perspective'. *Journal of Theoretical Politics*, 2 (3): 293–317.

Rhodes, R.A.W. (1997) *Understanding Governance: Policy Networks, Governance and Accountability*. Buckingham: Open University Press.

Rhodes, R.A.W. and Marsh, D. (1992) 'Policy networks in Britain'. In: Marsh, D. and Rhodes, R.A.W. (eds) *Policy Networks in British Government*, 1–26. Oxford: Oxford University Press.

Richards, D. and Smith, M. (2002) *Governance and Public Policy in the UK*. Oxford: Oxford University Press.

Richards, D. and Smith, M. (2004) 'Interpreting the world of political élites'. *Public Administration*, 82 (4): 777–800.

Richards, S. (1987) 'Vicarious suffering, necessary pain: physiological method in late nineteenth-century Britain'. In: Rupke, N.A. (ed.) *Vivisection in Historical Perspective*, 125–48. London: Croom Helm.
Richardson, J. (2000) 'Government, interest groups and policy change'. *Political Studies*, 48: 1006–25.
Richardson, J. and Jordan, G. (1979) *Governing Under Pressure*. Oxford: Martin Robertson.
Ritvo, H. (1984) 'Plus ça change: anti-vivisection then and now'. *Science, Technology, & Human Values*, 9 (2): 57–66.
Ritvo, H. (1987) *The Animal Estate*. Cambridge (US-MA): Harvard University Press.
Rogers, L. (1937) *The Truth About Vivisection*. London: Churchill.
Rose, R. (1993) *Lesson-drawing in Public Policy*. Chatham (NJ): Chatham House.
Rowan, A.N. (1996) 'The use of animals in experimentation: an examination of the technical arguments used to criticize the practice'. In: Garner, R. (ed.) *Animal Rights: The Changing Debate*, 104–11. Basingstoke: Macmillan.
RSPCA (2011) *Amending the UK Animal Experimentation Law – A Threat to UK Standards*. Horsham: RSPCA. http://content.www.rspca.org.uk/cmsprd/Satellite?blobcol=urldata&blobheader=application%2Fpdf&blobkey=id&blobnocache=false&blobtable=MungoBlobs&blobwhere=1232999239968&ssbinary=true (accessed 16 January 2012)
Rupke, N. (1987a) 'Introduction'. In: Rupke, N.A. (ed.) *Vivisection in Historical Perspective*, 1–13. London: Croom Helm.
Rupke, N. (1987b) 'Pro-vivisection in England in the early 1880s: arguments and motives'. In: Rupke, N.A. (ed.) *Vivisection in Historical Perspective*, 188–213. London: Croom Helm.
Rupke, N.A. (ed.) (1987c) *Vivisection in Historical Perspective*. London: Croom Helm.
Ryder, R.D. (1983) *Victims of Science: The Use of Animals in Research*. London: National Anti-Vivisection Society Limited.
Ryder, R. (1989) *Animal Revolution*. Oxford: Basil Blackwell.
Ryder, R.D. (1996) 'Putting animals into politics'. In: Garner, R. (ed.) *Animal Rights: The Changing Debate*, 166–93. Basingstoke: Macmillan.
Sabatier, P.A. (1993) 'Top-down and bottom-up approaches to implementation research'. In: Hill, M. (ed.) *The Policy Process: A Reader*. Hemel Hempstead: Harvester Wheatsheaf.
Sabatier, P.A. (1998) 'The advocacy coalition framework: revisions and relevance for Europe'. *Journal of European Public Policy*, 5 (1): 98–130.
Sachs, D.H. (1994) 'The pig as a potential xenograft donor'. *Vet Immunol Immunopathol*, 43 (1–3):185–91.
Sharpe, R. (1989) 'Animal experiments – a failed technology'. In: Langley, G. (ed.) *Animal Experimentation: the consensus changes*, 88–117. Basingstoke: Macmillan.
Schmoekel, M., Bhatti, F.N.K., Zaidi, A., Cozzi, E., Waterworth, P.D., Tolan, M.J., Chavez, G., Warner, R., Langford, G., Dunning, J.J., Wallwork, J. and White, D.J.G. (1998) 'Orthotopic heart transplantation in a transgenic pig-to-primate model'. *Transplantation*, 65 (12): 1570–77.
Singer, P. (1975) *Animal Liberation*. London: Jonathan Cape.

Singer, P. (1996) 'Animal liberation'. In: Garner, R. (ed.) *Animal Rights: The Changing Debate*, 7–18. Basingstoke: Macmillan.
Smith, J.A. and Boyd, K.M. (1991) *Lives in the Balance: The Ethics of Using Animals in Biomedical Research*. Oxford: Oxford University Press.
Smith, A. (1997) *Integrated Pollution Control: Change and Continuity in the Industrial Pollution Policy Network*. Aldershot: Ashgate.
Smith, M. (1992) 'The agricultural policy community: maintaining a closed relationship'. In: Marsh, D. and Rhodes, R.A.W. (eds) *Policy Networks in British Government*, 27–50. Oxford: Oxford University Press.
Smith, M. (1993a) *Pressure, Power and Policy: State Autonomy and Policy Networks in Britain and the United States*. Hemel Hempstead: Harvester.
Smith, M. (1993b) 'Policy networks'. In: Hill, M. (ed.) *The Policy Process: A Reader*, 76–86. Hemel Hempstead: Harvester Wheatsheaf.
Smith, M. (2004) 'Mad cows and mad money: problems of risk in the making and understanding of policy'. *British Journal of Politics and International Relations*, 6: 312–32.
Sperling, S. (1988) *Animal Liberators*. Berkeley, Calif..: University of California Press.
Stephens, M. (1989) 'Replacing animal experiments'. In: Langley, G. (ed.) *Animal Experimentation: The Consensus Changes*, 144–68. Basingstoke: Macmillan.
Straughan, D. W. (1995) 'The role of the Home Office Inspector'. *Alternatives to Laboratory Animals*, 23: 39–49.
Straw, J. (2000) Written Answer to PQ 140726, 29 November 2000, Columns 684W–685W, *Hansard* (HOC). London: HMSO.
Toke D. & Marsh D. (2003) 'Policy networks and the GM crops issue: assessing the utility of a dialectical model of policy networks'. *Public Administration*, 81 (2): 229–51.
Townsend, M. (2003) 'Exposed: secrets of the animal organ lab'. *The Observer*, 20 April 2003. http://observer.guardian.co.uk/uk_news/story/0,6903,940033,00.html
Turner, J. (1980) *Reckoning with the Beast: Animals, Pain, and Humanity in the Victorian Mind*. Baltimore: John Hopkins University Press.
UKXIRA (2001) *Third Annual Report, September 1999–November 2000*. London: Department of Health.
Uncaged Campaigns (2003) *The Documentary Evidence*. http://www.xenodiaries.org/evidence.htm.
Van den Bogaerde, J. and White, D.J.G. (1997) 'Xenogeneic transplantation'. *British Medical Bulletin*, 53 (4): 915.
Vial, C., Ostlie, D.J., Bhatti, F.N.K., Cozzi, E., Goddard, M., Chavez, G., Wallwork, J., White, D.J.G. and Dunning, J.J. (2000) 'Life supporting function for over one month of a transgenic porcine heart in a baboon'. *Journal of Heart and Lung Transplantation*, 19 (12): 224–9.
Wainwright, H. (2004) 'Reclaiming "The Public" through the People'. In: Gamble, A. and Wright, T. (eds) *Restating the State?* 141–56. Oxford: Blackwell.
Ward, H. and Samways, D. (1992) 'Environmental policy'. In: Marsh, D. and Rhodes, R.A.W. (eds) *Implementing Thatcherite Policies: Audit of an Era*, 117–36. Buckingham: Open University Press.
Wilkes, S. and Wright, M. (1987) *Comparative Government-Industry Relations*. Oxford: Clarendon Press.

Wilkes, S. (1998a) *Home Office Letter*, 3 March 1998.
Wilkes, S. (1998b) *Home Office Letter*, 28 April 1998.
Zaidi, A., Schmoekel, M., Bhatti, F.N.K. Waterworth, P.D., Tolan, M., Cozzi, E., Chavez, G., Langford, G., Thiru, S., Wallwork, J., White, D.J.G. and Friend, K. (1998) 'Life-supporting pig-to-primate renal xenotransplantation using genetically modified donors'. *Transplantation*, 65 (12): 1584–90.

Index

AAMR, *see* Association for the Advancement of Medicine by Research
Aberdeen Group, 22
ABPI, *see* Association of the British Pharmaceutical Industry
Abraham, Ann, 295–6
 see also Parliamentary and Health Ombudsman
Abraham, John, 61, 97–9, 103, 191, 225–6
Advisory Committee (AC), 59, 174–5, 196–200, 213–14, 223, 315
advocacy coalition framework, 26–7
Advocates for Animals, 82
agency, 6, 46
 relationship with structure, 6–7, 48–51, 93–7, 221–2
agenda-setting, 143, 174, 187, 195, 201, 216, 223–4
agricultural policy, 17, 38–9
Animal Aid, 63, 65, 82, 228
Animal Procedures Committee, 71–2, 75, 77–9, 81–2, 84, 110, 231, 236, 254, 268, 271–5, 279–81, 287–9, 292, 300, 318, 321, 326
animal protection lobby, 82, 203–5, 212–14, 223, 227–9, 231, 325
animal research groups, 67, 81–2, 121–2, 125, 129, 132–6, 141–2, 144–5, 154–5, 161, 173, 175–6, 178, 181, 189, 202–5, 307–9
 see also animal use ideology
animal rights
 groups, 65, 67, 222
 philosophy of, 63, 178, 221–2
animal use ideology, 73, 121–2, 126, 155, 165–6, 175, 184, 197–8, 200, 203, 220, 230, 237, 243, 303, 312, 316–17, 320
 see also self-regulation
animal welfare, 156, 193, 255–6, 264–6, 270–3, 275–6, **281–4**

ideology, 67, 71, 194, 203, 220, 242, 314–15, 317, 320
Animals (Scientific Procedures) Act 1986, xiii, 2, 7, 62, 65–81, 110, 230–9, 241–9, 317, 319–20
 Section 24 ('confidentiality clause'), 3, 56–7, 75, 95, 101, 231
Animals (Scientific Procedures) Act 1986 Amendment Regulations 2012, 8, 323–4
anti-racism, 178
anti-vivisection movement, 124–5, 128–9, 141, 143–5, 155–6, 165, 168, 178–81, 192–3, 201, 203–4, 218, 307–9
 see also animal protection lobby
APC *see* Animal Procedures Committee
Association for the Advancement of Medicine by Research, 59, 61, 91, 149, 150, 158, 161–6, 167, 168, 171–5, 177, 310–12, 314, 339, *see* Research Defence Society
Association of the British Pharmaceutical Industry, 61, 66, 81–2, 202, 326
asymmetric power model, 13–14, 74, 169, 179, 238, 312, 328
attentive publics, 22, 46, 74
autonomy, restrictions on, 49–50

BAAS, *see* British Association for the Advancement of Science (BAAS)
Balls, Ed, 342
Balls, Michael, 132, 136, 164
Bean QC, David, 104
behaviouralism, 94
benefit assessment, 257–61, 269–71, 284–6
Bentham, Jeremy, 115
Bernard, Claude, 116
Best Practical Means (BPM), 18
Bevir, Mark, 11, 48, 345–6

biotechnology industry, 29, 49, 83
Blanco, Ismael, 16
bloodsports, 179
Bomberg, Elizabeth, 16, 20, 41–3, 73, 170
Boyd, Kenneth, 117–21, 132
BPM, *see* Best Practical Means (BPM)
British Association for the Advancement of Science (BAAS), 119–20
British Medical Association (BMA), 82, 120, 133–5, **144**, 202
British Medical Journal, 121, **144**
British Pharmacological Society, 72
British Union for the Abolition of Vivisection, 63, 65, 80, 82, 180, 182, 203, 228
British Veterinary Association, 67, 75, 202, 205, 211, 227, 232
Brooman, Simon, 115, 136, 154, 164, 171
Brown Dog statue, 173, 178
BUAV, *see* British Union for the Abolition of Vivisection
Bulkeley, Harriet, 22, 45, 333
Burnham, Andy, 265
Burnham, Peter, 90, 97–9, 102
BVA, *see* British Veterinary Association

Carnarvon, Earl of (aka Lord), 128, 131–6, 156
Cavanagh, Michael, 18, 325, 333–4
chemical industry, 18, 39–40, 61–2
class, 124, 156, 160
club government, 12, 74, 139
Coalition Government, 321–2
Cobbe, Frances Power, 122–3, 125–6, 129, 131, 180
Coleridge, Stephen, 172–3, 180
Committee for the Reform of Animal Experimentation (CRAE), 64–71, 72, 75, 225–7, 229–30, 232, 237–8, 317, 320
Compston, Hugh, 15
Conservative Party, 83, 226
contract research organisations, 62
corporatism, 9
Cosmetic, Toiletry and Fragrance Association, 82

cosmetics testing on animals, 210, 229, 326
cost-benefit assessment, 67, 70–1, 81, 104–6, 151, 189, 195, 209–10, 227, 229, 232–5, 241–3, **245–9**, 281–6, 298–300, 318
see also harm-benefit assessment; Animals (Scientific Procedures) Act 1986
Countryside Alliance, 51
CRAE, *see* Committee for the Reform of Animal Experimentation (CRAE)
critical realism, 7, 95–8, 102, 159, 183–4, 305, 327, 329
CROs, *see* contract research organisations
Cross, Richard, 128–9, 134, 150
Cruelty to Animals Act (1876), 2, 7, 56–62, 91, 131–48, 150, 201, 306–11
certificates, 188–90
prosecutions under, 152–3, 160, 191

Daily Mail, 30
Darwin, Charles, 125, **144**
Daugbjerg, Carsten, 11, 14, 34, 47, 329
democratic accountability, 13, 35, 151, 161, 234–6, 254, 287–97, 299–300, 325–6
Department of Health and Social Security, 225
sponsorship of pharmaceutical industry, 61–2
Department of Transport, 51
Descartes, Rene, 114–16
Devine, Fiona, 97–8, 102
DHSS, *see* Department of Health and Social Security
'Diaries of Despair' report, 103–4, 250
differentiated polity model, 12–13
disaggregation, 14, 34
discourse, 166, 181, 196, 328
analysis, 330–2
Dolowitz, David, 45

Early Day Motions (EDMs), 83

economic factors, 28, 47, 61–2, 191, 225, 226–7, 257, 326–7
Eckstein, Harry, 99
élitism, 14, 169, 234, 243, 313, 318
Elston, Mary Ann, 178
English Nature, 29
environmental policy, 20, 28–9, 330–2
epistemic communities, 45
see also power, of experts
ethics, 233, 236, 320, 330
European Union, 20, 34–6, 39, 41–2, 82–3, 230, 238
  Directive 2010/63/EU, 2, 302, 322–5, 330
  Directive 86/609/EU, 229–30, 251
  Directorate General XI, 20
  monetary policy, 36
  uploading of policy to, 36, 330
Europeanisation, 36
expertise, *see* power, of experts

farming interests, 29, 58
see also agricultural policy
feminism, *see* suffragism
Feldberg, Wilhelm, 80
Ferrier, David, 160
Flint, Caroline, 321
FRAME, *see* Fund for the Replacement of Animals in Medical Experiments (FRAME)
Freedom of Information Act 2000, 101
French, Richard D., 56, 101, 102, 117–26, 128–46, 150–2, 153–6, 159–66, 171, 175, 180–1
Friends of the Earth, 29
Fund for the Replacement of Animals in Medical Experiments (FRAME), 3, 67, 75, 79, 82, 227, 232, 320
Furlong, Paul, 96

Garner, Robert, 1–2, 5–7, 9, 15, 19, 22, 34, 55–91, 96, 98, 99, 106, 114, 116, 138–9, 169, 186, 191, 195, 206, 217–19, 222, 226, 303, 304, 308–9, 334–6, 340, 347
General Medical Council (GMC), 134, **144**, 155

genetically-modified (GM) crop policy, 29–31, 33, 42, 49–50
globalisation, 28
Good Laboratory Practice, 104
Grant, Wyn, xiii–xv, 22–3, 51, 80
Greenaway, John, 17, 26–7, 30, 32, 43–4, 193–4
grievance procedures, 47, 76, 85, 107, 152–3, 234, 242, 276, 294, 297, 301, 319

Hajer, Maarten, 330–2
Hall, Marshall, 117
Hall, Peter, 16–17, 26
Hampson, Judith, 1, 125, 134–6, 187, 230–1, 233
Handbook for the Physiological Laboratory, 120, 129
Harcourt, Sir William, 151–5, 158
harm-benefit assessment, 322
Hay, Colin, 6, 24–7, 32–4, 49, 73, 91–8, 127, 304, 306
health policy, 17
health and safety policy, 18, 325
Heclo, Hugh, 19
Henniker, Lord, 125–6
Her Majesty's Inspectorate of Pollution (HMIP), 39–40
Hill, Michael, 26, 46–7, 95, 151–2
Hindmoor, Andrew, 37
historical institutionalism, 6–7, 90–5, 147, 312
HM Treasury, 39, 191
Hollands, Clive, 224, 227
Home Affairs Select Committee, 107, 255, 294
Home Office, 4, 86, 103–7, 109
  Chief Inspector, 106, 173, 289–94
  infringement action, 80–1, 152–4, 190–1
  Inspectorate 80, 103, 176, 188, 212–13, 233, 266–70, 279–81, 287
  leak from, 249
  pro-animal research bias, 153–4, 173–4, 213, 281–94, 319, 322–4, 328–9
  refusal of licence applications, 69, 151, 154–5, 157

relationships with groups, 58–9, 65–9, 80, 104, 106–7, 132, 135, 139–40, 150–5, 157, 162, 173, 190, 212, 216, 218, 224–5, 255–6, 267–8, 273–5, 279–81, 286–294, 308–9
  secrecy, 56, 276
  Secretary of State (Home Secretary), 70, 71–2, 78, 85, 105, 106, 128, 138, 151
Hopley, Emma, 223, 228
Houghton, Lord Douglas, 64, 69, 224, 230
House of Lords, 5
Huntingdon Life Sciences, 6, 103, 249–50, 277
Hutton, Richard, 130
Huxley, Thomas Henry, 125, **144**

ideas, 31, 43–6, 91
  'virus' metaphor, 31, 328
ideology, 30, 43–4, 70
implementation gap, 46–8, 233, 268, 281–4, 325
Imutran Ltd, 4, 103–5, 109, 249, 254–7, 259–61, 266–70, 273–6, 284–7, 292–3, 319
inputism, 94–5
insider groups, 31, 33, 45, 50, 218, 328
  core, 22–3
  peripheral, 22–3, 41, 50, 74, 238, 320
  status, 19, 229
Inspectorates, 12, 139, 311
intentionalism, 48, 94, 237
International Medical Congress (1881), 155, 159–61
interpretation, 27–8, 32–3, 37–8, 44–6, 48–9, 53, 96–8, 101, 166, 179, 193, 221, 249, 300, 305, 314
issue networks, 5–6, 18–21, 41–3, 140–8, 149, 156–9, 170–1, 183–4, 206, 241–2, 310, 312

Jenkins-Smith, Hank, 26–7
Jennings, Maggy, 250, 267, 278–9
Jordan, Andrew, 17, 26–7, 30, 32, 43–4, 193–4

Jordan, Grant (A.G.), 10
Judge, David, 11, 306

Kean, Hilda, 172, 178–80
keynesianism, 191, 225
Kingdon, John, 31
Klein, Emanuel, 129, 143

Labour Party, 83, 181, 224, 230, 326
  New Labour, 32
Lancet, The, 121–2, 126, **144**
Langley, Gill, 224
Legge, Debbie, 115, 136, 155, 164, 171
lesson-drawing, 45–6, 53, 75, 194, 234–5, 314, 324
Liberal Democrat Party, 83
Lindblom, Charles, 27, 44
Linnean Society, 134, **144**
Littlewood Report, 59, 61–2, 64, 75, 164, 171, 185–219, 236
Lowndes, Vivien, 16, 57, 90, 92–3, 107

MacDonald, Melody, 80
macro-level, 25, 156
  see also policy networks, exogenous factors (interaction with), structural context
Magendie, François, 116–17
Magnan, Eugene (Norwich trial 1874), 120–1
Maloney, William, 22
Marquand, David, 12
Marsh, David, 5–6, 9–19, 21–39, 44–6, 48–50, 62–3, 74, 95–8, 238, 320, 327–9
Matfield, Mark, 2
McAnulla, Stuart, 22, 48, 50
media, impact of, 51, 223
Medical Research Council, 62, 80, 82, 182
Mellor, David, 68–9
meso-level, 10, 14, 19, 24–5, 34, 40, 46, 95, 184, 240, 329
meta-policy, 166, 308
micro-level, 10, 19, 24, 25, 46, 95, 327, 329
  see also agency

Ministry for Agriculture, Fisheries
 and Food (MAFF), 29
Ministry of Health, 61–2, 191
'model of justice' typology, 47–8, 76,
 107, 152, 234
Monamy, Vaughan, 114–15, 117, 233
monetarism, 225, 227
Moran, Michael, 12, 74, 139–40, 153

narratives, 31
 see also ideas
National Anti-Vivisection Society, 63,
 65, 82, 172, 203, 228
National Farmers Union (NFU), 38
Nature, **144**
NAVS, *see* National Anti-Vivisection
 Society
new institutionalism, 89–90, 95–6
new social movements, 30
North Sea oil and gas industry, 18
Novartis Pharma, 285

O'Brien, Mike, 105, 260, 287–8
O'Riordan, Tim, 27
Orlans, F. Barbara, 220
outsider groups, 22, 31, 50–1, 172

pain regulations, 68, 70, 130,
 176–7, 194, 206–7, 211,
 227, 251–7, 322–4
Parliamentary and Health Service
 Ombudsman (PHSO), 107,
 294–7, 321
 see also Abraham, Ann
Parsons, Wayne, 10
path dependency, 6–7, 57–8, 90–4,
 107–8, 147, 184, 194, 311–12
Paton, William, 2, 117
patronage, 15
Peters, B. Guy, 90–2, 94–5
pharmaceutical industry, 61–2, 67,
 83, 182–3, 188, 191, 225, 326–7
 see also Association of the British
 Pharmaceutical Industry
PHSO, *see* Parliamentary and Health
 Service Ombudsman (PHSO)
Physiological Society, **144**, 155, 159
pluralism, 9, 13, 45, 50, 159, 238,
 242, 326

pluralisation, 12–13
policy change
 measuring degrees of, 25–7
policy communities, 6, 16–18, 192,
 215, 243, 297–302
 dynamic conservatism, 41, 53, 186,
  193–4, 198–9, 236, 315, 328
 formation by groups, 171
 resistance to change, 38, 40–1,
  43–4, 175, 184–5, 218–19, 222–4,
  235, 318, 321, 327
 success causing destabilisation, 172,
  183, 314
policy learning, 43–6, 193–4,
 206, 301
policy networks, 5–8, 9–54
 applicability to Victorian era, 306
 appreciative system, 31, 70, 163,
  194, 216, 232, 320
 boundaries, 22, 157
 core-periphery distinction, 22
 dialectical model/interactions, 6,
  23–5, 49–51, 159, 166, 187, 218,
  313–16
 dynamics, 23–5, **53–4**, 84–5, 92–3,
  312–13
 exogenous factors (interaction
  with), 27–36, 177–81, 195
 group dominance over state, 171
 ideological structure, 32, 158,
  189, 193, 199–200, 215–16,
  232, 234, 301
 impact of other policy networks,
  34, 73
 Marsh/Rhodes typology, 5–8, 10,
  15–21, **21**, 87, 127, 140, 167–8,
  241, 304, 320
 mediation of information, 32,
  197–8, 276, 301
 mutual exclusivity of issue
  networks and policy
  communities, 22
 structural context, 24, 184,
  240, 312
 structures, 36–43, 214–15, 233,
  236–8, 248–9, 253, 208
 see also policy networks, ideological
  structure; policy paradigms
policy paradigms, 17, 26–7, 44

political parties
  impact on policy networks, 6, 29–30, 85–6, 326
pollution policy, 18, 39, 46–7
post-war consensus, 28
power
  distribution of, 15–17, 21, 48–9, 71, 153, 159, 167, 169, 233, 241
  of experts, 17, 35, 44–5, 47–8, 150, 162, 174, 194, 212, 215, 224, 296, 313
  state vis-à-vis economic/professional interests, 13, 28–9, 40, 46–7, 139–40, 153, 162–3, 166, 215, 225, 227, 326–7
  see also asymmetric power model
Pratchett, Lawrence, 16
precautionary principle, 31
Prime Minister, 32, 326–7
project licences, 103, 105, 229–30, 249, 254–5, 259–64, 268–70, 272–3, 276–81, 292, 323–4
public interest, 104, 249, 291, 293–6, 337
public opinion, 153, 156, 161–2, 177, 180, 193
  impact on policy networks, 6, 10, 30–1, 33, 40–1, 162, 172, 221, 223–4

Queen Victoria, 117, 123, 156

Radford, Mike, 115, 122–4, 129–30, 134, 174
RDS, *see* animal research groups; Research Defence Society
Read, Melvyn, 98
Regan, Tom, 63
Research Defence Society, 61, 66, 69, 82, 188, 202, 212, 217, 226
  see also Association for the Advancement of Medicine by Research
Research for Health Charities Group, 82
research outcomes, 178, 182
  see also cost-benefit assessment; utility (of animal experiments)

resources, 14–15, 17, 19–21, 37, 39–40, 44–51, **53–4**, 62–3, 65–7, 73, 75, 77, 81–2, 86–7, 96, 127, 132–3, 140, 142–3, **144–5**, 146–8, 158, 161–2, 166–7, 169, 175–6, 181, 184, 191–2, 215–17, 297
  legitimacy, 30, 67, 123–4, 158, 171, 194
re-use (of animals in experiments), 68–9
Rhodes, Rod, 5, 9–13, 17, 19, 25, 90, 94, 99–100
  see also policy networks, Marsh/Rhodes typology
Richards, David, 6, 11, 13–14, 19, 24–7, 30, 32–4, 49, 222, 304
Richards, Stewart, 120, 152–3, 160
Richardson, Jeremy, 9, 15, 28, 31, 35, 39, 44, 51, 172, 305–6, 328, 334
risk society, 30
Ritvo, Harriet, 136, 156, 177–8
roads policy, 51
Roberts, Angela, xvi
Rogers, Leonard, 174, 176, 339
Rose, Richard, 45
Royal College of Physicians, 134, **144**, 161
Royal College of Surgeons, 119, **144**, 161
Royal College of Veterinary Surgeons, 188
Royal Commission (1875–6), 127–31
Royal Commission (1906–12), 59, 172–81, 312–13
Royal Society, 134, 144
Royal Society for the Prevention of Cruelty to Animals, 60–1, 63, 65, 117–18, 121, 123–5, 128, 130, 132–3, 135–6, 140–1, 143, **144–5**, 146, 152, 155–6, 176, 179, 195, 208–9, 212, 228, 231, 250, 267, 278–9, 297, 322
Royal Society for the Protection of Birds (RSPB), 29, 49
Rupke, Nicolaas, 118, 124, 143, 150, 155, 159–61
Ryder, Richard, 64, 117–18, 123–4, 164–5, 171, 188, 221, 224, 237–8

Sabatier, Paul, 26–8
Salt, Henry, 180–1
Sanderson, John Scott Burdon, 120, 125
Sandoz, 267
secrecy, 14, 39, 99, 101, 156, 171, 195, 222–3, 301
  see also Animals (Scientific Procedures) Act 1986, Section 24 ('confidentiality clause')
self-regulation, 12, 47, 73, 80, 117–20, 126–7, 139–40, 142, 152, 162–4, 175–6, 182–3, 190, 198, 211, 243, 303, 310–11, 313
severity assessments, 10, 106, 110, 252–7, 265–8, 277–84, 322–4
  see also animal welfare; cost-benefit assessment; pain regulations
Schafer, E.A., 152
Shambles of Science, The, 172
Sharpe, Robert, 1
Shaw, George Bernard, 181
Singer, Peter, 63, 221–2
Smith, Adrian, 15, 17–18, 22, 30, 38–41, 46, 198, 215–16, 328, 333
Smith, Jane, 117–21, 132
Smith, Martin, 10, 16–20, 24–5, 27–8, 30, 32–4, 37–9, 43–4, 97, 192, 222
social construction, 37, 50, 96–7, 166
socialism, 178, 180–1
Soil Association, 29
Sperling, Susan, 168–9
Stoker, Gerry, 95
strategic-relational approach, 24
Straw, Jack, 288
structural inequality, 13, 179, 238, 296–7
structure, 6, 184, 306
  and agency, 7, 24–5, 31, 36–8, 221–2, 237, 305, 309
  of belief systems, 26–7
  see also agency
suffragism, 178–80

Thatcher Government, 28, 226
Three Rs, 78–9, 321
Times, The, 135, 144
Toke, David, 27
Townsend, Mark, 3, 103, 297
Turner, James, 120–1, 129, 162, 164

Uncaged, 249, 286–91, 294–7
United Kingdom Xenotransplantation Interim Regulatory Authority (UKXIRA), 250, 286
Universities Federation for Animal Welfare (UFAW), 182, 193, 203, 207, 212
utility (of animal experiments), 1–2, 129, 141, 144, 151, 154, 160, 182, 197–8, 201, 209

Victoria Street Society (VSS), 131, 135–6, 140, 143, **144–5**, 146, 155–7, 160, 172

Westminster model, 9, 11–12, 14, 74, 169, 306, 311, 327–8
wildlife protection, 29
Wincott, Daniel, 91–4
Woodhouse, Edward, 44

xenotransplantation, 3–4, 103, 259–61
  adverse effects on primate recipients, 255–6, 264–6, 270–3, 275–6, 278–9, **281–4**, 290–1, 294
  clinical trials predictions, 259–60, 266–70, 274
  heart, 255–6, 262–73, 275–7
  immunology, 259–64, 269, 271, 285–6
  kidney, 255–6, 267–8, 277–9
  predicted benefits, 259–61, 266, 269, 272, 274, 284–6, 291–2
  physiology, 260–1
  technical failures, 262–3, 271–2, 275, 284, 292–3

Yeo, G.F., 155, 160

Printed and bound in Great Britain by
CPI Group (UK) Ltd, Croydon, CR0 4YY